PRODUCTION SYSTEMS
FOR
ARCHITECTS
AND
DESIGNERS

A HANDBOOK BY

FRED A. STITT, ARCHITECT
Editor/Publisher, Guidelines

VNR VAN NOSTRAND REINHOLD
New York

Library of Congress Catalog Card Number 93-36615
ISBN 0-442-00920-8

I(T)P   Van Nostrand Reinhold is an International Thomson Publishing company.
        ITP logo is a trademark under license.

Printed in the United States of America

Van Nostrand Reinhold   ITP Germany
115 Fifth Avenue   Königswinterer Str. 418
New York, NY 10003   53227 Bonn
                                     Germany

International Thomson Publishing   International Thomson Publishing Asia
Berkshire House,168-173   38 Kim Tian Rd., #0105
High Holborn, London WC1V 7AA   Kim Tian Plaza
England   Singapore 0316

Thomas Nelson Australia   International Thomson Publishing Japan
102 Dodds Street   Kyowa Building, 3F
South Melbourne 3205   2-2-1 Hirakawacho
Victoria, Australia   Chiyada-Ku, Tokyo 102
                                     Japan

Nelson Canada
1120 Birchmount Road
Scarborough, Ontario
M1K 5G4, Canada

16 15 14 13 12 11 10 9 8 7 6 5 4 3 2 1

Library of Congress Cataloging in Publication Data
Stitt, Fred A.
    Production systems for architects and designers / Fred A. Stitt.
        p.   cm.
    .Includes index.
    ISBN 0-442-00920-8
    1. Architectural design—Data processing. 2. Architectural
drawing—Data processing. 3. Computer-aided design. I. Title.
NA2728.S75   1993
720'.28'40285—dc20   93-36615
                                               CIP

DEDICATION

To my brother,
Steve Stitt,
an inspiration.

ACKNOWLEDGMENTS

Thanks to those who make it happen:
Ernest Burden, Kathy Thompson, Ginnie Meyers, Marti Barrett, Patrick Nelson, Monika Haaf, Monique Viggiano, and ever-supportive VNR editor Wendy Lochner and editorial assistant Kelly Francis.

CONTENTS

CHAPTER FOUR
PERSONNEL MANAGEMENT 50

CHAPTER FIVE
THE IDEAL DRAFTING STUDIO 59

CHAPTER SIX
USER-FRIENDLY WORKING DRAWINGS 75

PREFACE -- READ ME FIRST

Here's some simple arithmetic to illustrate the primary benefits of using this handbook, namely cutting production time and cost by 30% to 40%. The savings are real, but they don't come in one big package; it's an accumulation of little things. For example:

--Notation takes about 10% of total drafting time in working drawings. If you save 50% of notation time with standardized and systematic notation, you can cut 5% of the total time and cost on every working drawing set.

--Overdone graphics also add about 10% to the cost of a typical set of drawings. That includes oversized drawings, excessive linework, and redundancies. Eliminate half the waste of overdrawing and redundancy, and you save another 5%. (In sets of the most poorly managed drawings, your savings would be more like 20%.)

--Details take anywhere from 10% to 40% of a working drawing budget; 20% is a reasonable average. Cut that in half with a standard detail system, and you have a net 10% savings.

--Redrawing design drawings from scratch to convert them to working drawings can take up to 15% of the working drawing schedule. Eliminate two-thirds of the redrawing by reusing design drawings, and you have another 10% gain.

If you go further:

--Use efficient layering systems to avoid unnecessary redrawing of plans and other base sheet information. That can be worth 5% to 10% of total working drawing time.

--Use pasteup drafting or CADD equivalents to avoid redrawing repetitive elements. This can be your biggest single time saver on some jobs -- up to 30% on plans for multi-housing unit projects, for example. The average overall savings for design drawings and working drawings is about 5% to 10% of "normal" time.

These numbers are from our working drawing time and cost surveys, plus numerous individual office reports that have come to us over the years. The same numbers come up time and again from every type of office, and they have proven to be quite reliable.

The bottom line, as systems enthusiasts have reported for years: 30% to 40% time and cost savings. If you have a project that would normally require 1,000 hours of drafting time (for example, a large residential or small commercial project with 20 architectural sheets at a budgeted 50 hours of manual drafting each), you can conservatively estimate a potential savings of 300 to 400 hours. If half that time could be allotted to improvements on design and problem solving, and the other half to profit margin, you could have an outstanding project.

It doesn't end there, but the savings of other systems are harder to quantify.

You might use a design database of office standards to cut back on much of the repetitive searches for data through catalogs, Sweets, Graphic Standards, and old job files that eat up so much design and production time. How much time is involved has been hard to measure, but a conservative estimate is 3% of total production time.

You might use instruction checklists, minimock-ups, and maximock-ups to reduce errors, backtracking, and general slowdowns while drafters await instructions and clarifications. What would be a reasonable average savings to expect from this? Perhaps another 3 to 6%.

Photodrafting, specifications as checklists, freehand drafting, photocopier reuse of graphics and text from previous drawings and outside sources . . . add all these into the pot, and what could you expect? A total of an additional 10% to 15% savings, perhaps, depending on the type of project. Once it's all added up, your savings can be as much as a 50% reduction of standard production cost. (We've known many firms to do even better, of course, and reported on them countless times over the years.)

A 50% reduction in the working drawing example noted on the previous page means 500 hours salvaged. If the production drawings are 40% to 50% of the job budget, then we have potentially saved a total of 20% to 25% of the total fee. A few projects like that each year could put any office on a very solid financial footing.

Unfortunately, in most offices it doesn't seem to work that way.

The reason is that there is minimal attention to an overall strategy of production. Principals are distracted trying to bring in the work and deal with the clients. Designers are occupied designing. Project managers are focused on the day-to-day problem solving, not the totality of office project management procedures.

On top of that, the office may be charging fees or billing work in a way that penalizes the office when it saves time and money on a project. Or there may be other management impediments -- aspects of office tradition that discourage or prevent an office from being efficient.

This book deals with all the details and with the overriding organizing principles required to make it work. Once you make it all work, whether you only use CADD or a mixture of production procedures, you'll be able to save your 30% to 40% and provide extra service and value to the clients in the process. Besides all that, you'll make substantially more money.

It's doable; it's being done all the time by design offices of every kind and size. There's no reason why you shouldn't benefit from the hard-won lessons of all those innovative architects, engineers, drafters, and designers who made it work so well in their offices. The one, single, key however, is **organization**. Just applying time savers at random won't do it. But if you plan, strategize, and add Production Management to the job roles in your office, it'll all work fine. Read on for details.

Fred A. Stitt, Architect.
Editor/Publisher, Guidelines.
Director, San Francisco Institute of Architecture.

CHAPTER ONE

SIMPLE REFORMS SAVE HUNDREDS OF WORK HOURS EVERY YEAR

IF CADD ISN'T THE SOLUTION, WHAT IS?

A WORKING DRAWING DISASTER, STEP-BY-STEP

THE BENEFITS OF HAVING A PRODUCTION MANAGER

WHILE AWAITING THE CADD MILLENNIUM

WORKING DRAWING EFFICIENCY REFORMS IN
A LARGE OFFICE

FULL-SIZE MOCKUPS ARE THE CURE FOR A MAJOR
PRODUCTION HEADACHE

IF CADD ISN'T THE SOLUTION, WHAT IS?

Computers and CADD are in near universal use now but offices still have the same old problems with their drawings:

--There are errors and omissions on virtually every sheet in every set of CADD drawings. Traditional drawings used to average four to six errors per drawing and the average hasn't changed. This problem precedes CADD but most users still don't know how to apply the computer effectively to help reduce the errors.

--Every set of drawings is loaded with unnecessary data and redundancies -- the same data is repeated in different places in the drawings and often in contradictory ways. (Redundancies and non-essential data on drawings are now considered neon-bright signals of flawed drawings by contractors who make their profits on change orders and extras.)

--CADD drawings are often unreadable with odd combinations of oversized lettering on some sheets and flyspeck size text and faded spiderweb linework on others. They're clean enough and should be easy to read, but while many are plotted on oversized sheets, parts of them can't be read without a magnifying glass.

--Construction information is not organized in ways that best facilitate bidding and construction.

--Too much information in the drawings is unnecessary to bidders or builders. Data they need either isn't there or it's so hard to find, they stop trying.

--Specifications still aren't particularly well coordinated with their drawings. This problem continues after years of promises of integrated documents as one of the primary benefits of CADD.

That's a lot of problems. They do a lot of harm and they're the reason why this book was created.

The problems are many but the causes are few.

They boil down to four primary principles of form and content of production documents.

First principle:

Design and working drawings are the record of a huge set of design decisions. The decisions number in the hundreds, sometimes thousands, and are strung out through many sheets of a set of working drawings. The decisions aren't recorded in any easy to search central checklist or database. They're scattered throughout the documents and the same decision may be shown in dozens of places. The management problem is compounded when there are changes and it's necessary to track down not only the decisions that are changed, but other decisions that may be affected by that change.

Second principle:

The decisions need to be validated by comparison with office standards. There has to be a reliable database used office-wide which is updated and upgraded by the experience gained from each project.

Third principle:

Design decisions must be cross coordinated with all related decisions. For example, decisions on ceiling heights and structural systems need to be coordinated with clearances for duct work and ceiling lighting fixtures.

Fourth principle:

The design decisions must be communicated with a minimum of extraneous data that distracts from the message -- just the facts and figures, clean and clear. That means no overdrawing, no boiler plate, no redundancies that fill up a set of drawings but tell the contractor nothing.

Some argue that all you have to do is hire the right people. If you do, they'll solve the problems. They'll produce, no matter what.

Their point is that efficiency systems in themselves won't do much for you if the expertise and motivation of your staff members are below par.

Granted.

But, in design firms all across the country, the time and ability of top employees is terribly wasted. The office's income and employee pay suffer accordingly.

Pay is directly linked to productivity. Most design professionals aren't paid much because, in terms of what they produce hour by hour, they aren't worth much.

It's not the people that are the problem, it's the systems they use, or lack of systems, that are the problem.

You'll see what this means in a multitude of examples throughout the pages that follow.

In terms of reforms, we'll start with the simplest and proceed to the most sophisticated. Reforms that pertain equally to every kind of production method: manual drafting, reprographics, or CADD.

A WORKING DRAWING DISASTER, STEP-BY-STEP

We all know the over-optimistic self-deception that dominates most project scheduling. The same deception works to persuade people not to bother with efficient design and production methods.

Projects look to be considerably less work than they will be.

So the methods of making the work easier and faster appear to be less valuable than they really are. That's the origin of such famous last words of wisdom as: *"Why go to all that trouble just to save a few minutes."*

For example:

A project architect was advised of a number of time and cost savers to apply to a university student housing project. Strongly recommended: Draw overall floor plans at 1/8" scale on relatively small sheets and do a few of the repetitive student rooms at 1/4" scale to show detailed room dimensions, symbols, detail references, etc.

Instead the decision was made to do plans at 1/4" on larger sheets because "It's the same basic work anyway and it will look better." The large scale floor plans started looking "empty" as soon as they were drawn, so the word came down to add the "gypsum board lines" on each side of the partition lines to "make the drawings look more finished." At one week into production, a crew of four were nearly done with the partitioning and overall dimensions on the floor plans and all appeared to be on schedule, about 15% done.

The added finish lines on the partitions did make the drawings look "more finished." They also doubled the linework that had to be drafted. The "four-line" partitions necessitated the addition of little jamb symbols at doors and openings. Single line door swings looked "weak" in that context, so every door had to be drawn with two long and two short lines as literal plan views of each door. That looked "much better."

The large scale floor plan rooms "looked empty," so every room was individually dimensioned, noted, and symboled instead of being referenced back to typical room plan drawings. Two more weeks passed and the job now appeared 20% finished.

It was inevitable that the exterior elevations would also be done at the overly large scale on the big sheets. They especially "looked empty," which led to double and triple lining the outlines to show frames around doors and windows on the elevations. Detail keys were added on the doors and windows too, instead of on a separate door and window frame schedule, and so were repeated, correctly and incorrectly, on all views of the building.

Overall cross sections followed suit: Too large, "empty," then filled with extraneous detail.

At six weeks into the project, the scheduled half-way point, it was looking 30% completed. A couple of additional drafters were put on the job, at which point the project architect fell behind in red-lining the check prints -- causing everyone else to slow down.

Then came some "minor" plan changes. That is, they would have been minor if all the repetitive rooms hadn't been hardlined throughout the building floor plans. As the project neared a scheduled 60% submittal, it was approaching 40% actual completion and bogging down completely. The project architect who had lost control left the office, but assured management that the project only required some "finish up work." An office principal and another project architect, already preoccupied with projects, took over the job to do the finish work and then themselves became bogged down.

Final net result: The project went to bid five weeks late, even with overtime. Since no-

4

body had time to do detailed cost estimates along the way, the lowest bid on the project was 15% over budget. The design had to be simplified, and the working drawings cost the office $85,000 in uncompensated salary and overhead.

Key lessons from this episode:

1) The potential value of the originally proposed timesavers was not visible to the inexperienced project architect or the design/drafting team. None of them had been exposed to such values in school or previous work experience and so they couldn't visualize all the implications and repercussions of their early decisions.

2) There was no detailed preplanning of the working drawings which could have led to visualization and solution of problems before they were built into the drawings.

3) The project was done "hardline" from the start (to make the job "look finished") so that all changes necessitated lots of heavy duty erasing and redrawing. That intensified the demoralization and grinding slow-down towards the end of the work as people realized virtually everything they drafted had to be redrawn at some point.

We're learning from the episode only because of a special circumstance that allowed me to observe the project. Plus, we have the knowledge to understand and explain most of what went wrong. **Most of those involved, including the original project architect, had no idea what was going wrong or why. They still don't.** They just blame various participants and say that the project was "jinxed," "mismanaged," and "had irrational clients." These opinions obviously close off any chance of learning from the experience.

As in most offices, there was never any review of the steps that lead to the problems and why they accumulated so severely. Most of the original project team are dispersed in other offices and perfectly capable of repeating the same errors all over again. Which leads to the next topic:

THE BENEFITS OF HAVING A PRODUCTION MANAGER

The production nightmare described on the previous page happened in that office on four jobs in a three-year period.

There's every reason to expect it will happen again in the years ahead. This is not a particularly inept office; in fact it does good buildings. Its chief principal is nationally famous and justifiably so.

Its problem, as in many firms, is that it has no collective way of learning from experience. That's because there is no one in charge of making sure that happens.

Offices that have the best production and quality control records invariably have such a person. He or she may be a partner, associate, or what used to be called a "chief draftsman." That person analyzes the pluses and minuses of every project, preplans new projects on the basis of analysis, holds informal seminars and crits to train designers and drafters in that knowledge, manages a standard detail library, checks all production drawings, and manages a feedback system from jobsites and completed buildings.

The alternative to having such a person is to have inexperienced people starting production drawings from scratch time and again. That's what happens in most firms. Project team members may have a lot of accumulated experience, but most haven't learned anything about efficient production. So they proceed with no outside source of direction.

Imagine if a design team had only themselves to rely on -- no design program, no client, no feedback from office principals. Imagine that situation, and imagine how most production teams work: no production plan or mockups, no direction in how to do the drawings most efficiently, and no monitoring or critiques of the production methods. That's the way it is, and that's why CADD or no CADD, projects are increasingly not getting done on time or on budget.

5

WHILE AWAITING THE CADD MILLENNIUM

A funny thing happened in the years just prior to 1,000 A.D. People in Europe wouldn't do anything on any large scale to improve their lot.

They wouldn't do anything because it was assumed that the world was about to end and the Second Coming was nigh. What would be the point of doing anything? After New Years passed without incident and after a suitable waiting period "just in case," the European Christians pulled themselves together and started climbing out of the dark ages.

A lot of design firms have behaved the same way towards their practices -- their design procedures, working drawings, and project management. They had faith that their savior, in the form of an advanced and economical CADD systems, was always about to come along any day and make everything all right.

Some have been waiting for ten years. Meanwhile they don't bother with even the easiest-to-use timesavers: *"Why go to the trouble? The computer will take care of all that."*

So they continue to lose hundreds of dollars on small jobs, thousands of dollars on large ones. In the years when work expanded and they should have done well, they were often lucky to break even.

Meanwhile the most productive firms keep ahead of budgets and schedules by using every drafting trick and shortcut in the they can muster (such as those which fill this book).

When they use CADD, they use it well by applying the same kinds of shortcuts and, as a result, are among the few who come out ahead with computers.

Here is a comprehensive survey of the most successful working drawing time- and cost-savers in use today, listed by drawing types.

SITE PLANS -- Avoiding Overdrawing:

-- Don't trace or re-draw the site survey. Have the original copied as a screened background sheet. If the original print is damaged or at the wrong scale, your reprographics shop can repair it and modify the size to match your preferred Site Plan scale.

-- Don't overdraw the foliage. Planting tends to be overly elaborate in linework and texture. Just provide outlines. Keynote any landscaping identification to avoid duplicate notes.

-- Avoid repetitive dimensions. Repetitive sitework dimensions can be handled with instructions like: "All curbs at 6' radius unless noted otherwise."

-- Avoid repetitive notes. Site plans are especially suitable for keynotes, or guide notes like: "Typical."

SITE PLANS -- Coordination:

-- Add a reduced-size copy of the floor plan to the site plan rather than drawing an outline or shadow "footprint." When you add up-to-date reduced ground plan images to the site plan it helps consultants keep their sitework up-to-date to match. Ground plans that show kitchens, baths, mechanical rooms, etc., are useful for accurately routing utilities, drains, and sewers to their best hookup points at the building.

-- In final printing, consider including separate site plans for each major trade. When you show each sitework contract as a solid line image combined with a multi-layer screened shadow print of all other contracts, each subcontractor can avoid conflicts and duplication of other work. For example, show trenching for water supply as solid on one sheet and show all other work as screened background. Show drainage and storm sewers in solid line on a Drainage Plan sheet with water supply and other work screened, etc. This can prevent a lot of confusion in complex sitework such as large, multi-building projects.

-- Don't forget the old office copier helpmates. Use photocopies of ready-made data such as road maps for your location and vicinity maps; planning and building code regulations for General Information sheets; property legal descriptions; abbreviations and symbols legend; etc.

-- Site photos, such as those often taken during the predesign and site analysis phase, are useful to show contractors what existing elements are to be kept or removed, trees or other items requiring special protection, etc.

FLOOR PLANS -- Avoiding Overdrawing:

-- Floor plans for smaller buildings tend to be over-scaled and over-sized. If the design firm can avoid making drawings "look big" to impress the client, there's a noticeable savings in doing plans at 1/8" scale instead of 1/4". That helps discourage drafters from drawing double-line door symbols and double lines on walls to indicate wall finishes -- a real time killer!

-- As with Site Plans, avoid repetitive dimensions. Duplicate dimensions can be handled with **notes** like: "All ceilings are 8' high unless noted otherwise," "Typical," etc. Don't include dimensions of structure, such as column spacings, that will be completed by the time the contractor is working on that particular floor. Such dimensions either duplicate the structural work already in place, or contradict it. Either way is wasteful.

-- Handle repetitive notes with keynotes or guide notes. A strip of standard or general notes can sometimes cover over half the notes that will normally be required on a floor plan.

-- Minimize textures and crosshatching. A little extra texturing doesn't look like much of a clutter problem when it's first added to a drawing. Early, when the drawing is relatively empty, some patterning may improve the appearance. But when other plan data is added in, the

excess poche and patterning just becomes visual noise. So apply it in small doses at the end of the job.

-- When doing repetitive rooms, don't repeat the notes and dimensions. Drafters find it very easy to repeat notes and re-do dimensions, room after room. After all, when you get one right, why not keep going and make the drawings look full of data? It's one of the most common time wasters in floor plans, especially in housing projects, hotels, medical buildings, and others with many repetitive or similar rooms.

FLOOR PLANS -- Coordination:

-- For the best coordination, the magic word is OVERLAYS. There is still no better way to keep all the disciplines coordinated than to keep basic architectural building information on a base sheet, and all particularized engineering data and architectural data (notes, dimensions, symbols) on overlays.

-- Avoid duplicate drawing during the latter stages of design development by doing original plans, elevations and sections as base sheets. Do design variations and rendering on overlays. Later, the base sheets will be clear and uncluttered, ready for direct reuse in working drawings. Besides saving time, using the base sheet-overlay system early in design development does wonders for overall project cross-checking and coordination.

-- If doing specialized work as different contracts, make separate plans for different subcontractors as suggested for Site Plans.

For example, if doing extensive custom cabinet work, draw it on a separate overlay and provide a screened background image of all other work with the cabinet work in solid line. HVAC design and construction is clarified by showing all duct work in solid line and **ALL** framing and plumbing in the same region in screened shadow print. Similarly, it helps the plumber to see **ALL** the HVAC, structural work, and electrical chases that might impinge on the piping.

-- Have the foundation and subfloor

7

framing printed with a screened image of the ground floor plan, and the roof framing with the upper floor plan. That shows consultants and specialty contractors what's expected to happen right below and right above their work.

REFLECTED CEILING PLANS --
Efficient Drafting and Coordination:

-- **The worst and least excusable time-waster is drawing and redrawing all those ceiling grids.** If you don't have a computer to draw your grid layers, draw sample ceiling grids once, have them printed as transparencies on "light-tac" stickyback (or use lightweight clear polyester prints for all-out paste-up work). Then never draw another grid. Lay your pre-printed grids over the floor plans, position as desired, and trim the edges. It's no problem to handle changes by picking up and resticking a grid, at least no problem when compared to re-drafting.

-- **Also, if not on CADD, the Ceiling Plan fixtures and symbols are especially suited to dry transfer** or pre-cut stickyback symbols.

-- **And do it all -- the preprinted grids and preprinted symbols -- on an overlay.** Between the appliques and overlay system, you'll do days of work in hours.

ROOF PLANS

-- **Avoid overly specific and detailed roofing notes.** Drafters tend to write out specification information regarding roofing and flashing. **Don't do it!** Unless it's a simple small building or an owner-builder job, keep the notes plain and generic. Leave the elaboration to the specs. This will save lawsuits as well as drafting time. Roof plans are also good candidates for keynotes, general notes, and/or standard strip notes.

-- **Roof plans sometimes look "empty" to the drafters** so they lay on textures and other extraneous data that just add time without adding clarity. Keep textures minimal, as noted before.

-- **Don't include roof elevation heights**

that either duplicate or contradict the structural drawings. For example, it's common to see parapets or different parts of the slopes on a roof plan noted as something like 56' 10-1/2" above the ground. These heights are properly shown in exterior elevations and/or structural cross sections.

-- **Have the roof framing plan and the upper floor plan printed as a composite screened** image in combination with the roof plan. That will verify the proper positioning and coordination of roof structural elements, roof drains, and roof openings.

-- **If roofing work consists of repair or remodeling, use photo drafting.** Roofing-related construction and renovation are very hard to draw and specify. Photos can explain all sorts of complex work in a flash.

-- **Notes and symbols again** ... those "slope" and "vent" notes and other repeats can be handled by keynotes. If not on CADD, skylights, vents, and detail keys can be handled with simple preprinted stickybacks or rub-ons. These are not only fast to apply, but easy to change during the inevitable rush revisions towards the end of the job.

EXTERIOR ELEVATIONS

-- **Keep door and window symbols off the elevations.** Such symbols are often included on elevations and it's pointless. Doors and windows are identified on the **plans.** Every duplication leads to contradictions. There will be a conflict between plan and elevation and someone will have to clear it up with a change order. "It doesn't hurt," argue some drafters. It doesn't hurt any more than having horizontal building dimensions on the elevations. It's just unnecessary and potentially hazardous in terms of creating errors and confusion on the jobsite.

-- **The same rule applies to door and window detail keys on elevations.** They'll either duplicate or contradict other drawings. The plan shows door and window symbols which guide the contractor to the door and window schedule which includes the frame schedule, and THAT is where the detail reference keys belong. (Yes there may be exceptions for one-of-a-kind conditions on small jobs, but the rule is a good one. It shouldn't be violated without good reason.)

-- **Avoid double-line/triple-line/ quadruple-line frames around doors and windows.** This pastime is popular for making drawings "look better." Meanwhile more arduous and important work is put aside in the process.

-- **Use pasteups or partial drawings and "typical" notes for repetitive exterior features such as masses of windows.** Working drawings for the World Trade Center show every window on every identical elevation of both buildings. People say: "That's really dumb," not realizing that the same amount of pointless drafting is done in their offices, too. It's just spread out over a lot of smaller projects.

WALL SECTIONS / CROSS SECTIONS

-- **The worst timewaster in cross sections: drawing the "back wall."** Drafters draw the back wall, then they add doors and windows, then door and window symbols . . . . The original and rather simple purpose of the cross section drawing disappears in the clutter.

-- **The worst timewaster in wall sections: duplication of data that's already shown in the overall cross sections and/or in the wall section details.** Very often, the end walls of the overall cross section are ample as detail reference drawings and no intermediate size wall sections are necessary. Where they are necessary, draw only those portions that are unique and use copies of the repeat portions of the walls. (Some sets of drawings are loaded with wall section after wall section, all drawn from foundation to parapet, and all are 80 to 90% identical. They're identical except for a spandrel variation, a balcony, a special window head, etc.)

To avoid weeks of wasted work on larger projects: Show the repeated portions of wall data as screened background and show only the variable information in solid line.
For ideal construction drawing coordination by means of paste-up or CADD, follow these steps:

1) **Decide on building systems and construction and choose your construction details.**

2) **Combine reduced-sized versions of the details into wall sections.** (Details may be 3" scale, for example; reduce them on computer or reprographically to 3/4" or 1/2" scale.)

3) **Combine reduced-sized versions of wall sections into the end walls of building cross sections.**

4) **Use the cross sections as base sheets for structural drawings and as screened background images for exterior elevations.**

WORKING DRAWING EFFICIENCY REFORMS IN A LARGE OFFICE

Large firms don't usually take well to office-wide reforms.

There are too many people with their own traditional ways of doing things. And, typically, there's insufficient top management interest or support to implement and enforce long-term changes.

So an office like SMP in San Francisco was a rare exception. SMP is a long established firm that does institutional, commercial, and transportation facilities. They're well known for their medical buildings, hospitals, and research facilities.

Morry Wexler, Technical Director at SMP, got the support he needed and proceded to do an amazing job.

A large part of his success was due to thorough education of staff members so they all shared a common understanding of the principles and techniques of systematic production methods. (Morry's teaching methods are described later in this book in the chapter on SUPERVISING AND EDUCATING THE TROOPS.)

Among the notable achievements of this firm:

--**Greater quality control** which has been seen in whole projects completed without a single extra or change order.

--**Superior documents** which have result-ed in tighter, lower bids, as well as the to-be-expected reductions in production time and cost.

Here's a menu of efficient practices well established at SMP.

FOUR MAIN VARIATIONS OF EACH FLOOR PLAN

Floor plans are created using Overlay Drafting so that four separate prints of each plan can be provided for bidding and construction.

1) Plans of walls and partitions with architectural data such as room names, door types, partition types, finishes, detail and keynote references, etc.

2) Floor plans with dimensions only.

3) Plans of equipment with walls and partitions shown as background.

4) Reflected ceiling plans. By using Overlay Drafting, relevant plan data is printed as screened background image.

Plans are 1/8" scale. This avoids the need to use potentially confusing "match-lines." Portions of plans that require more detailed attention, such as those showing installation of equipment, are reprographically enlarged.

Each type of plan serves a single specific function. The focused content of each plan reduces clutter and clarifies the construction process instead of confusing it. With such reduced content, the drawings are much cleaner and easier to read.

INTERIOR ELEVATIONS ARE KEPT TO A MINIMUM

Interior elevations aren't allowed unless there's something special to see on those walls. Anything else is just a duplication (or contradiction) of information that's on the plans. Fixtures are shown on separate fixture mounting height schedules.

FREEHAND DETAILS

Drafters find it fast and enjoyable. Morry tells of a project manager who asked for a drafter to do the finish drafting of a large bunch of details he had sketched out. *"Phil,"* Morry said, *"You don't need a drafter. They're already finished."* The sketched copies were cleaned up, mounted on a carrying sheet and copied onto reproducible polyester with a vacuum frame. (They could also have been copied onto drafting media with a large format engineering copier.) Grid tracing paper and long scales are used in the best Freehand Drafting to speed up and sharpen the details.

KEYNOTES

Drafters and checkers rough out their notes on the side as they go along, then notes are typed as keynote lists in large, clear uppercase letter style. There are only about 20 to 30 notes per floor plan even for complex buildings, so keynote lists are not unwieldy.

DESIGN DRAWINGS
BECOME WORKING DRAWINGS

The "Law of the Land" and an absolute requirement in the SMP office: From the first design concept, project managers have to think about the requirements of the construction documents. The plot plan starts right off with the precision of a construction document. Plans and elevations are done on base sheets in the right scale and positioning for use later in working drawings. Notes or rendering elements are done on overlays so plans and elevations are clean and ready for easy reusability.

FULL-SIZE MOCKUPS ARE THE CURE FOR A MAJOR PRODUCTION HEADACHE

"We completed working drawings for an $11,000,000 university building in two weeks." That was Arizona architect Michael Goodwin's summary of the results of a remarkably efficient, common-sensical, yet radical way of managing working drawing production.

Goodwin and other A/E's around the country have created a general solution to a near universal problem. In the process, they've created what may become a new national standard in working drawing production.

The problem in three parts:

1) Building designs are completed without adequate knowledge of building code and construction limitations. Then the working drawing process becomes a war zone between the competing needs of production staff and the designer. Drawings are changed so many times in some projects that manually drafted tracings become worn through and have to be redrawn from scratch.

2) Design changes are carried into working drawings, through addendum in bidding, and through change orders throughout construction. The building is essentially never "designed" until the tenants move in. **Guaranteed: Everybody loses their shirts in the process.**

3) Detailing is so excessive that bids go out of sight and out of budget. While the design and production team are just trying to clarify complex construction, the detailing makes the building look more complex and the architects harder to please, so the contractors crank up the prices.

Do these sound familiar? They should. They're happening to some degree at this minute in **thousand**s of offices across the nation.

As stated above, a few offices have discovered a method of production that solves these and many related problems, including problems of staffing, overall coordination with consulting engineers, and working drawing quality control.

We'll call the method "Full-Scale Working Drawing Mockup."

We all know about doing mini-mockups or "cartoons" of working drawing sets as aids to planning, costing, and delegating project work. This is full-size mockup. Most sheets in the working drawing set are designed and drawn completely in freehand at full scale until the design decisions have been settled, the details designed, and all the consultant coordination is completed.

Then, miracle of miracles, it takes only two weeks to redo the mockups as finished working drawings for $11,000,000 (and larger) projects. And the total amount of time spent on design AND production comes out weeks or months less than it would have normally.

A small-office architect in Berkeley introduced "Full-Scale Working Drawing Mockups" to us years ago as a lifesaver when he got an unexpectedly large hotel project. He prefers a small office, likes doing most of the work himself, but it looked like he'd have to hire six to eight drafters to get the job done on time. Instead he went to full-size sketch and paste-up sheets and assembled the entire project in accurate but roughly drawn form before going to finish drafting. Changes were easy to make during the freehand mockup process--far easier than they would be after hard-line drafting. After the coordinations were complete and the client signed off on the design, he "traced" the mockups and produced the complete set of drawings in tool-lined drafting.

That final step took only five days.

Here three similar examples:

--A Dallas architect/developer gets "half price" services from his engineers. They do sketches and mockups of their work, he does the final CADD inputting and/or manual hardlining. Typically, they charge half the usual fees in return for having their drafting done for them.

--The Design Collaborative in Tucson gets similar double-duty by doing freehand design drawings which become mockups and are often reusable for paste-up in the final working drawings.

--A structural engineer showed us the same timesaving technique in his one-person Orlando, Florida office. As principal designer for many industrial buildings and multi-family housing, he always sketches full-size working drawing sheets on large "no-print" blue grid tracing paper. He tapes yellow flimsy paper detail sketches to the sheets plus preprinted keynotes, and anything else that does the job. His special wrinkle is that he often doesn't have to redraw the mockups--he just has them photocopied onto tracing paper with a Xerox 1020 or other large-format copier at one of the local blueprint shops. Then he touches up the tracing paper reproducibles with pencil and has diazo bid sets printed in the usual manner.

Most A/E's voice the same objections when they first hear about the Full-Scale Mockup system:

"Clients don't want to see empty sheets or mockups, they expect to see real progress prints. If we don't have solid lines on paper the clients won't believe we're doing their work and we won't get paid."

Goodwin in Arizona handled client concerns nicely by bringing clients into the project planning process. He showed clients the differences between the traditional and new methods of design development-production. He showed them the steps and advantages of the new process and showed why they can't rely on normal "progress prints" as an accurate measure of job progress.

Then he conducted intensive full-day or multi-day coordination meetings with clients and consultants. Here are details of the main steps involved as the architect mocks up the job:

--Consulting engineers firm up their early schematic and design development work in one-to-one problem solving with one another, the architect, and the client.

--They make and record their decisions on full-size diagrammatic sheets.

--Some meetings that go to more than a day are separated by two-day breaks so consultants can proceed on their own to solve problems that need specialized attention.

--Exception to the freehand work: Floor plans are hardlined rather than sketched, to provide accurate base sheets for everyone's overlay work.

--The design partner proceeds to other finished work: fills in finish schedules and door/window schedules, and selects and organizes details from the standard detail library.

That's all Phase One--Essentially it's the wrap up of Design Development. Then there's Phase Two: **Graphic Documentation** in hardline of all the decisions and sketch work.

The freehand mockups go extremely fast because while accurate and to scale, they are essentially single line diagrams. With the decisions being made face-to-face by the key players, problems get solved in minutes instead of through weeks of memos, checkprints, and phone calls. Once everything falls into place, all that remains is plain non-stop drafting by hand, paste-up, or **CADD.**

The final documents are banged out by in-house and contract drafters. When staffing is small, final documents are done by moonlighters and small architectural firm contractors with the aid of a reprographics shop. All the outside drafters are given comprehensive cartoon sketches of details and **additional** special components that need to be drawn.

13

In the Goodwin example, his staff did the final assembly from pieces provided by the hired hands. Details came back on small sheets for paste up as did lists of keynotes and other parts.

The quality control finale: Final documents done with this process are so complete and well coordinated that the clients can expect about $500 worth of change orders due to A/E error for each $1,000,000 in construction.

The Full-Scale Mockup System makes working drawing design a part of the design development process. Traditionally, it's been the other way around: to make design development part of working drawings -- by default, whether anyone wanted it to be that way or not. And that's been a substantial part of the problem.

Much of design and working drawing data has to be changed during a project as previous decisions are reconsidered in the light of new documentation. Things that look right early in the job obviously need change when seen during later phases of work. The problem is that those early items have been drawn hardline as finish work when they were really just guides for later decision making.

The mockup system puts everything in proper perspective -- design decision making is done with sketches, the final recording and presentation of those decisions are done as a swift separate process unto itself.

CHAPTER TWO

PRODUCTION MANAGEMENT THAT REALLY PRODUCES

THE LOST SCIENCE OF EFFICIENCY

Two generations ago "efficiency" was a subject that was taught as a university course of study, like math and English.

There were campus Departments of Efficiency such as that at the University of Denver. There were textbooks such as "Fundamental Sources of Efficiency," published in 1914 by Dr. Fletcher Durell. There was an Efficiency Society complete with journal and conventions, and professional Efficiency Experts using techniques, such as the Taylor System, for analyzing and organizing the workplace.

One product of all this was employee-managed efficiency control. The Deming System is one of these efficiency methods. It was adopted by Japanese industry in the late 1940's, and is still one of the great "secrets" of Japanese management.

Unfortunately, the subject was much misunderstood and maligned. While efficiency experts invented ways to make workers more productive, labor unions and the popular press saw them as trying to exploit employees by making them work harder and faster. The irony was that "harder and faster" was the old low-production system the efficiency experts were trying to get rid of. They knew that what made a worker more productive was intelligence applied to the workplace, not sweat.

Pioneering efficiency experts discovered two processes that were fundamental to human effectiveness: Transformation and reuse. Then they went about finding faster ways to do both.

Efficiency can be defined as maximum output with minimum input. True productive efficiency is where the value of the output exceeds the value of the input. So a design office might put $10,000 worth of labor into a project and earn $11,000 for their trouble.

How's that? How can you get more out of something than you put into it? It's not that there's more taken out than put in, it's just that the raw materials are transformed to human use. The raw material of a computer chip is worth a thousandth of a fraction of a cent, for example. Transformed, it enhances human work and life and its value is multiplied enormously.

Transformation of materials or information is the primary source of wealth, but the primary tool to lower cost and save time is REUSE. Reuse in itself doesn't do the job; it has to be in combination with transformation.

For example, the only standard construction detail systems that work are those that are constantly improved. Transformation has to be applied in terms of job-site feedback, simplifications, elaborations, etc.

The primary job of the designers and drafters is, as always, to create new and better details. The difference is that they have an office database to start with and add to, so that everyone's hard work doesn't just disappear in the job files.

This is where most offices fall down: They can't improve their design process or technical resources on an office-wide basis because they have no way of storing such data in the first place. If it isn't stored, it can't be accessed. If not accessible, it can't be improved.

They don't have storage, access, and improvement for reuse, because they have nobody in charge of such things. There's no such job in most A/E firms as "Design and Technical Database Manager." And the person who used to do that job, the "Chief Draftsman," is no longer with us except sometimes under the title of "Production Manager."

One reason there aren't many positions like "Production Manager" is that it doesn't count well in the commonly used financial management statistics. The managers in most firms want to see a maximum of hours be directly billable, and a minimum attached to overhead. Otherwise their "Chargeable Ratio" or "Utilization Rate" doesn't look so hot.

It's an unfortunate misuse of financial management data. A statistic like Utilization Rate is intended to help spot inefficiencies, not create new ones. And where there is no one in charge of production, you can be sure there are loads of inefficiencies.

For example:

--No consistency in drafting, production and CADD systems. Every designer, drafter, and project team does things in essentially their own ways, most of which are grossly inefficient and error prone.

--No central source of quality control checking of all drawings. Some drawings get checked properly while many are not. Most documents are checked more for completeness than for Value Analysis or failures prevention. And, since most staff members are untrained in the latest and most effective checking systems, the checking is as inefficient as the production.

--No central source of education for the staff. If nobody is in charge, nobody is going to spend time distributing articles, arranging construction site visits, leading lunch-time seminars and group critiques of drawings.

But what if there is a Production Manager and he or she isn't well-informed?

That happens all too often.

Sometimes one of the older office principals is in charge of technical aspects of the office and insists on doing things in tried and true (read old and costly) ways. Combine that with an abrasive personality (the two often seem to go together) and it's a lost cause.

What are some of the qualifications of a Production Manager? We've known dozens who have done fabulous jobs for their offices and they've tended to have these qualities:

1)   Supportive personality. There are those who come on tough, pompous, and looking for trouble so they can insult employees rather than help them, and punish rather than teach. They've given quality control and effi

ciency systems a nasty reputation. The supportive managers, on the other hand, convey by word and deed that they care about the education and competence of the employees and that their primary mission is to help everyone do their best work.

2)   Open to all systems and able to apply each special technique to highest and best use. Someone only expert in CADD or only Overlay Drafting or only paste-up systems will lose money as he or she forces every job into the same mold. The worst offenders right now are CADD managers who don't understand production and reprographics. For example, they'll let people measure and digitize exterior and interior elevations of an existing building from scratch, with no understanding or use of photodrafting and/or video input to CADD.

3)   Full understanding and wide application of the database concept such as Standard details; standard notation; standard and simplified finish (and other) schedules; operational checklists; etc. In other words, the kind of person who reads systems-oriented material regularly.

4)   Enthusiasm for finding newer and better systems. They truly enjoy finding and inventing new and better ways to do things. It's an unusual talent and sometimes irritating to others who don't share such enthusiasm.

In summary:

Efficiency translates into using every means possible to expedite transformation of design data and finding more ways to continuously upgrade and reuse that data. Your best in-house efficiency expert is that one person in a hundred who actually cares about such things.

WHY PROJECTS
DON'T GET DONE

Everybody knows that projects usually take twice as long as their initial time estimates.

But not many project managers actually schedule their work that way. They're under pressure to meet a client deadline or an office budget. If their initial time and cost estimates were realistic for most projects, the offices wouldn't start them to begin with.

Fees are too low for too many projects. We all know that. Or if they're not too low, budgets get chewed up in the design process. Then bosses want to make up the difference in production.

At the outset, a project cannot be clearly seen in all its final complexity. It can be estimated as so many sheets and so many hours on average, per sheet. What isn't estimated at this early stage are the inevitable drafter errors, client information delays, regulatory problems, and designer changes.

A few other factors conspire to encourage inaccuracy in scheduling:

1) There's pressure to show a project as being complete to a certain percentage in order to mark a milestone on the timeline and/or to justify a phase payment.

2) The rules of thumb commonly used in estimating drafting or CADD schedules are too low -- at least half the time according to our working drawing time and cost surveys.

3) Drawings often LOOK half done when they're really only a third done.

As work progresses, the number of components on each drawing increases. So every addition and change effects more and more elements.

The early stage of a completed floor plan that shows walls, doors, and windows, can be very misleading in appearance. There's a lot of linework, the basic building is there, and the drawing may look 20% completed. But it isn't. The later layers of information -- dimensions, door and window symbols, notation -- will be increasingly detailed and increasingly time-consuming.

Our observations of the working drawing process show that what looks like 20% completion of a complex drawing is often closer to 5% of the final drawing and research time. That's one reason why the "last 5% or 10% of the job" so often seems to drag on inexplicably and take two to three times longer than anyone expects.

As the number of components on each drawing increases, so inevitably do the number of items that aren't "quite right."

Checking, corrections and rechecking go on and on, well after most of the work appears to be finished to most everyone. The fussier the participants, the slower the process grinds.

Here's a summary of some of the best methods around for making more accurate job schedules at the outset and for meeting deadlines, no matter what.

--A project schedule increases in accuracy, in direct proportion to the degree of detailed planning. If a project manager writes a list of drawings that includes "Window Details and Schedule," he or she can estimate "one detail sheet, 60 hours" without arousing too much doubt or guilt. *"A good detailer should be able to do a window sheet in about eight days,"* is the rationale.

Suppose the planning process goes further and includes a list of the items to go on the window sheet: How many window details? How many from the catalog? How many custom? How many variations in wall construction? Suppose the count is 25 semi-custom details and the guess is *three hours per detail -- design, coordinating with the manufacturer, drafting, and checking."* Then the time estimate goes to 75 hours and that's not including the Window Schedule. The Window Schedule will be at least a day and a half, plus checking and corrections, so the estimate is now 90+ hours, rising, and clearly more on target.

--Accuracy improves if lists of drawing components are included with sketches on mini-mockup sheets of the working drawing set. It puts the whole project into perspective, leads to production timesavers, and raises questions and cautionary notes such as: *"What are the chances that this exterior wall system might be changed?"* If the chances prove to be fair that the walls will change, that could affect the floor plans, exterior elevations, wall sections and details, exterior door and window details. In which case, none of that work should be hardlined without a definitive decision on the wall system.

--Pert and CPM diagrams help accuracy because the charts require a listing of finite details of the operation. The more items and steps listed in the working drawing plan and schedule, the more time that will have to be allocated to do the work.

Perts and CPMs also help because they identify the coordination and checking nodes that will be downtime for the production crew. If the downtime isn't anticipated and other work prepared, then it's time and money spent on the project even though no real work is getting done.

--Add 20% contingency in every phase for client, designer, or regulatory agency changes. That's the minimum time lag that can be expected, and if it doesn't happen, you've bought a little insurance for other unanticipated problems.

Here are two of the best front-end and back-end schedule makers we know:

1) Do the whole project as a mockup. Do the job as sketches, rough pasteups, freehand and semi-freehand throughout. When the job is finally checked, clear, and no major changes likely, bring in a full crew to "trace in hardline" all the work.

2) Use the best senior people you have to finish off projects in a hurry. For years, one well-known San Francisco firm pulled projects off the boards at about 90% completion and turned them over to one powerhouse who did nothing but such finish-up work. That cut through most of the nit picky delays and corrections. It eliminated the slowdowns due to boredom and exhaustion that add to the delays in the final weeks of most projects.

In summary:

The more job items you list, sketch, and mock up, the more accurate your time schedule estimates will be. And if a realistic schedule isn't fast enough, push for the timesavers that are the subject matter of this book.

THE OFFICE AS A "TIME MACHINE" AND THE REAL STRATA OF JOB ROLES

The monetary value of someone's job is dependent on the time frame he or she works in, according to Elliot Jaques, a British sociologist and management expert.

For example:

--**The lowest paid A/E office workers are those responsible for immediate results** such as doing job prints, filing, and maintenance.

--**Those responsible for longer range tasks with objectives set for days, weeks, or months away, will earn more.**

--**Those whose project goals extend beyond a three-month time span will usually earn managers' salaries.**

--**And whoever's projects and planning extend the most years into the future will be the highest paid.**

Jaques says:

"Within this complex and delicately balanced time machine, everybody carries out tasks or assignments linked to one another in time. One important aspect of the tasks of managers, therefore, is to set target completion times for getting jobs done, and in such a way that everything meshes together. It is what they get paid for."

Time-based measurements can be used to compare any two jobs, regardless of extreme differences in job types or company types.

Job size or degree of responsibility is measurable by its maximum time span. Mr.

Jaques says: *"A person may still have many short daily tasks but the longest tasks take longer and longer as he works his way to the top."*

In evaluating a job, the longest length tasks are the critical ones to measure. Jaques observes that workers' perceptions of the fairness of their pay coincides closely with the relationship of pay to their job's maximum project time spans.

A related observation: People will not willingly accept instructions or supervision from anyone whose job project time span isn't distinctly longer than their own.

Jaques' research shows there are fairly distinct divisions in maximum task time spans. He finds job strata appear at intervals of three months, a year, two years, five years, etc. From that he has evolved a set of **Job Strata** which mark these job role time spans as follows:

--**Stratum One,** "Shop and Office Floor," starts at the task time-span of up to three months.

--**Stratum Two** covers the span of three months to a year -- the work of first line managers.

--**Stratum Three,** the second line managers work in a time-span bracket of one to two years.

--**Stratum Four,** the General Manager's project time span, is two to five years.

--**Stratum Five,** a subsidiary CEO's time-span bracket is five to ten years.

Higher strata follow according to a top span of twenty years and beyond.

The Stratum identifications provide a means of categorizing job roles in terms of comparative compensation and of structuring work flow. Here are two helpful observations that follow from this type of analysis:

--**In any office line of command there should only be one person in a job slot at each stratum.** Otherwise there's trouble. The chain of command becomes too long -- bureaucratic.

--Those sharing the same stratum are hamstrung by a lack of clearly exclusive territory of responsibility. This will eat up time, lead inevitably to conflicts and problems of delegation among lower stratum staff, and if it's not resolved it will eventually lead to one party or the other leaving the office.

He says:

"One of the things that all jobs have in common is time. Any experienced manager, whatever his or her job, takes the great importance of time for granted. Things have to be done on time, planning is done about time, and organizing is done to achieve things in time. Sales efforts have to mesh with production capacity and programs; invoices and accounts have to be sent out and payments received on schedule . . . and so on for every function in the organization."

THE HIGH COST OF LOW-FEE WORKING DRAWINGS

Four to six errors per working drawing sheet. That's the average error rate we've described elsewhere in this book.

But what are the error rates for the lowest cost working drawings? What's the real cost of cheap services?

Last year we looked at a representative sampling of sixty sets of working drawings. They were a mix of building types, equally divided according to the project fees received: Lowest fees (2 to 4% of construction cost), Median fees (6-9% of construction cost), and Highest (11-16%).

The twenty lowest fee working drawing sets had an average of 17 "significant" errors per sheet. "Significant" is defined as an error that misrepresents some aspect of design, or is a contradiction with another part of the drawings; mainly anything that might cause confusion or errors in bidding or during construction. Misspellings, trivial omissions, and other minor items were not counted.

Some large complex drawings had up to 35 significant errors.

The mid-level fee projects had three to six significant errors per sheet -- mainly cross-coordination errors or lapses among architectural and consultants' drawings.

The highest fee projects had an average of zero to three errors per working drawing sheet. The highest fee project drawings were the cleanest, easiest to read, and while quite comprehensive and thoughtfully laid out, were rarely overdrafted.

Many mid-fee drawings were overdrafted with data overlaps and redundancies, such as door and window key symbols shown on exterior and interior elevations as well as plans.

The lowest fee, hand drafted drawings showed many traces of smudging, indicating much erasing along the way, and mixed drafting styles on the same drawings, usually with sloppy notation and faint or fuzzy line work.

Twenty sets of drawings were partly or wholly done on CADD -- mainly mid- and upper fee jobs. As you might expect, they were sharper and easier to read than the hand drafting jobs. But the CADD use had no noticeable bearing, one way or another, in error rates. Plotted output tended to be somewhat over-drawn, probably due to the common A/E belief that extra repeat data done on computer doesn't matter because repeats are automated.

We couldn't get statistics on the ultimate cost of the errors in low-fee jobs -- bidder problems, construction extras, failures, claims, etc.

There were ample horror stories, however.

--An architect described how a low-fee, non-paying client cut off a project prematurely, avoided roofing details or specs, and refused to pay for a day-long observation of the roofing application. The roof failed in the first season and caused $78,000 damage. The savings in detailing and site observation fees was about $740, **a hundred-to-one loss.**

--Another architect said his low-fee client had brought in design development drawings from another firm that needed to be "finished up." Then the client cut him off half-way through production and took the drawings to a contractor's drafting service for detailing. When roofing and other building components failed, he was included in all the claims.

--A developer client said he accepted a low-ball estimate from a design firm and found the final drawings totally out of whack. The site plan misstated the survey of property boundary lines. Grading had to be redone after construction because a number of drainage slopes on the site plan were opposite from what was required. Structural construction details had no relationship to the building and apparently were scissored whole from another job.

--Officials of a community center who had accepted the lowest design fee commented that their original architect defaulted and closed his office. They hired another to re-do most of the working drawings.

In no instance did a recipient of below-average fee service report satisfactory results. Nor were any of those who performed low-level work, and were willing to discuss it, happy with the outcome.

FINAL NOTE: None of these jobs represented tightly managed CADD, Systems Drafting, or any efficiency systems that might have allowed time for competent work at a low fee.

USE THE PROJECT MANAGER'S EXPERTISE WHEN ESTIMATING FEES

Insufficient fees are sometimes negotiated by a front office crew that brings in the work without regard for costs.

They toss the package into the project manager's lap, saying in effect:

"We got the job; doing it is your problem."

The front office may think it has to settle for a low or "break-even" fee, but the project manager or production manager can provide ammunition for higher fees.

Experienced production managers can explain to a client the potential costs of an inadequate set of documents, particularly in terms of change orders, extras, and claims.

Also, the more detailed the breakdown of a plan for working drawings and specs, the more realistic the time estimates. The production or project managers are best qualified to make such estimates.

Designer estimates of time and cost are often inadequate and to add insult to injury, when design time goes over budget, the production crew is expected to make up the difference.

Above all, use the fee negotiator's best friend -- the Scope of Services list. The production phase of a Scope of Work list will usually be the largest and most detailed of any phase, and after all, it's services ultimately that have to be negotiated in "fee" negotiations, not fees.

HOW DO YOU MAKE PRODUCTION IMPROVEMENTS BILLABLE?

The needs of financial management may conflict with the long-term production needs of your office.

For example, good financial management wants to minimize overhead and maximize billable work. **A high overhead rate spells inefficiency and trouble,** and clients certainly don't want to pay for it.

But the need for low overhead works against investments in non-billable new procedures and equipment. New procedures and equipment may save the office time and money, and improve quality control, but those savings don't translate into simple bookkeeping language. If the investment adds to overhead, it can be read as a negative in the financial reports.

As a result, most A/E offices can't justify the crucial job role of Production Manager or Technical Director to watch over the overall office production system.

If someone wants to improve the office's production systems, they often have to do it on their own time. Time is meted out during lulls between projects to keep the technical library and samples room up to date, to clip out articles on how to avoid roofing failures, to give lunchtime technical training to staff, etc.

That's why most office standard detail systems deteriorate from misuse and disuse. CADD systems meet the same fate unless they're managed by dedicated senior personnel and project managers. Keynoting, simplified standards, streamlined drafting, and all the other timesavers that fill this book, come and go in an office depending on who is hired, or who has left. There's little office-wide procedural consistency in any firm.

Such consistency has been found mainly where a partner or senior associate acted as the production manager. Top-level personnel aren't expected to have more than 30% to 40% of their time billable to projects so they can spend time watching out for the long-term needs of the office.

There's a related barrier to production that has penalized more than one forward-looking office: Namely, if you're more efficient, and if you charge by the hour, then you don't get paid as much.

Never mind that profit margins may be way up; if someone can point to a job being produced 120 hours sooner than it would have been otherwise, they can say: "That's 120 x our billing rate of $50 per hour. We just lost $6,000!"

It's a false view of things, but so's the other one about overhead rates, and they're both hard to argue against.

Here's a financial management-oriented solution, in two parts:

1) Restructure the fee system. Nearly 40% of projects these days are charged by a negotiated fixed fee, about 20% are charged as a percentage of construction cost, 20% hourly, and nearly 15% as a multiple of personnel expense. It's the hourly and multiplier fees that discourage efficiency. Since the most unpredictable aspects of A/E service in terms of design are design and construction administration, many offices are now charging hourly for the front-end and back-end services. And they're charging fixed fees on the largest and most predictable part of service, working drawing production. Most clients prefer a fixed fee and at the same time they recognize the uncertainties of design development, particularly where decisions have to be made by a client group or where there are regulatory design review impediments.

Advanced efficiency procedures are least used in design and construction administration, so there's little loss if those phases are charged hourly. If working drawings are charged to a client as a fixed fee at the going rates, or even slightly under, the high-speed production office can translate 30% to 40% of their revenue for production phases into pure profit.

2) **The second approach is to charge production services to project managers, just like reprographic services are charged.** CADD services are still often charged this way in sums ranging from $20 to $75 per hour; $40 per hour is about average. Some clients resist, but most go along with the charges because they gain added benefit from the creation of a CADD database and they recognize the very visible extra investment in hardware, software, supplies, and training time that come with CADD.

For over 15 years, Systems Drafting offices have charged time and cost of camera and vacuum-frame work for pasteup and Overlay Drafting as reimbursable services, just like CADD and for the same reasons. These days the charges go to defray investment and operating costs of large format engineering copiers, enlargement-reduction copiers, cameras, etc.

Another part of the bookkeeping solution is to let the production manager sell his or her services to the project managers:

--**The training of new employees by the Technical Director is billed to the project managers as technical consulting.**

--**Standard details chosen from the master Standard Detail File are "sold" to project managers at somewhat less than the cost of equivalent drafting.** (Morry Wexler would sell "mix and match" details at $40 or so per module on a debit-credit account, or "ready to wear" full 30" x 42" sheets of standard details for $1,600.)

--**Reprographic services for photo surveys of existing conditions, paste-up and overlay work are billed as from any outside reprographic service, as a reimbursable.**

The billing and overhead problem only's pertains to part of the Technical Director or Production Manager's time. He or she will also do a substantial amount of project-specific work, such as quality control checking of working drawings. So it doesn't take that much ingenuity to get his or her billable time up to a level that would be fully acceptable to any financial manager.

Other approaches to the problem:

--**Some offices have established separate business entities to handle all technical and production-related work.** Massive amounts of drafting at Edward Stone's office years ago was contracted through a drafting service owned by Stone's youngest son. Such design offices contract out their production services in the same way they hire consulting engineers, and no longer worry about details of production. (Design offices can reduce their liability this way too, since the drafting or production service can contract to bear most of the liability risk and, since efficiency systems go hand in hand with quality control, they're far less prone to claims.)

--**In some cities, production or drafting services are offered by well-qualified experts.** There, the design firms can turn their work over to the service bureau as they would to any outside service and bill the clients at "pass through" charges plus administrative markup. (Service bureaus are often ill-managed and short-lived; the main criteria in choosing one should be the architectural and construction document expertise and resources of the operator(s) of the service, not the equipment.)

--**Even some very small firms create production and reprographic subsidiaries.** The design office in one room and the reprographic service in the next room are two separate entities. Some small firms provide printing and related services to other offices in the same building or near-by, and thus can afford investments in large engineering copiers, diazo and laser printers, and full time repro personnel who would otherwise be out of reach.

--**When you get in-house Production Management services under control, offer what you've got to other design firms.** It can help them avoid overtime when they're rushed with spill-over work, and smooth out some of the peaks and valleys in your own practice. If you're good at drawing checking, that's a related service that's increasingly needed by firms wanting someone who can look at their documents with a fresh eye.

WHEN OFFICES ARE PAID BY THE HOUR, IT PAYS TO GO SLOW

"How can we be more efficient without being penalized for it?"

A lack of good answers to this question has cost A/E's dearly every time they've used a significant production timesaver.

--Whenever a job was negotiated on an hourly basis, more efficient production would cost the office money.

--Whenever an employee tried to set up a standard detail system or other long-term improvement, the time wasn't "billable" and it looked bad on the financial reports.

--When creating more useful and efficient graphics, the drawings were often cleaner and smaller and so didn't "look" like as much work had been done as did traditional crowded, oversized sheets.

CADD hasn't suffered similar problems for several reasons, mainly:

--For most users, it hasn't cut much time off the work. In many cases it's added to billable project time.

--CADD time can be charged separately from operator time and still is, in typical CADD offices -- generally at anywhere from $25 to $50 per hour, $35 per hour being average for small to mid-size firms.

Many offices have changed their fee schedule to charge for less predictable work on an hourly basis -- namely design, permits and approvals, and construction administration. They charge for the more predictable production phase by a fixed fee. With a fixed fee and a budget to meet, efficiency and faster production is rewarded with higher profits.

Since many timesavers can be applied to design drawings too, the question is: **Does an hourly design fee hurt the firm that does faster design documentation?**

Most such offices say NO. The E & B Partnership in Michigan is typical, and they say:

"We use whatever cushion the faster graphics give us to do more design options and more finished design work before moving into production. The hours get spent on solving more problems and catching loose ends in coordination rather than pencil pushing."

These more systematic design firms refuse to penalize themselves. **They set a budget based on realistic expectations of how long design should take and use every bit of it. That helps assure that they can do the best possible design and planning without hurting themselves financially.** (By contrast, about 14% of the design time for projects in the average architectural office is never billed -- it's just given away to the clients.)

How about the overhead problem? It takes time to manage systems, train employees, maintain up-to-date detail files, create CADD management systems, run reprographic equipment, etc., and if it can't be charged to project work, it's seemingly not paying its way.

One solution that's been a big success for top Systems Drafting and CADD firms: **Create a production consulting and drafting company as a separate company.** See the next section for more on this idea.

About that small and "empty" drawings problem . . . Big sheets and lots of drawings with more lines give the impression of more service, so the incentive to do inefficient, oversized, overdrawn graphics remains a persistent problem. Here are several ways to add to the bulk of drawings without losing time AND improve their content in the process:

1) Make the specs part of the drawings rather than printed as separate books. This works on small and medium sized jobs. It adds some weight to the set of drawings, and it's certainly more convenient for reference. Typically spec printout is laser printed in columns at 2/3

normal old typewritten size and pasted up in
strips on the working drawing sheets for reproduction.

2) Add extra specialized floor plans.
Some offices find building floor plan information is much more clearly conveyed in multiple
sheets instead of one:

--Floor plans of walls and partitions with
architectural data such as room names, door
types, partitions, finishes, detail and keynote
references, etc.

--Floor plans with framing, wall, and partition dimensions only.

--Plans of equipment and built-ins (as for
hospitals, labs, etc.)

**3) Provide several versions of the site
plan with varied trades in solid line on each
of their plans, indicating other trades to be
coordinated in a screened background image.**
This takes no more drafting, just judicious use
of overlays to beef up the set, clarify the work,
and help prevent confusion, conflicts, and duplication among the sitework contractors.

4) Provide added visuals, such as photocopies of specified fixtures and equipment from
catalogs, site photos, building code excerpts --
information that can be recorded by camera or
copier, doesn't require much time-consuming
hand work, and truly augments the usefulness
of the documents.

A THREE-TIER SCOPE OF WORK FOR WORKING DRAWING MANAGEMENT

We all know about overdrawing; about half of all sets of working drawings are either overdrawn or incomplete relative to the project.

Not so well known or appreciated is the time and cost of overly comprehensive services. Overdrawing, in fact, often results from a lack of knowledge of how much work is enough at all levels -- starting with design and including the whole project management plan.

Overly comprehensive services means simply: Time and attention are given to details of production service all out of proportion to the project budget.

Excessive services relative to project budget occurs in about 30% of projects we've surveyed. This wouldn't necessarily be a bad thing in itself and some degree of extra service is expected in any professional work. But when it's extreme, it means money is lost on one project which has to be made up on another. Then more important projects that deserve maximum time and attention get short changed.

The excessive work often starts with misunderstandings between design firm and client. If the Scope of Work of the project doesn't specify what work is to be done for what fees, there will inevitably be clients who honestly expect more work than they properly should get for the fee they pay. Then the design firm will overdo things to some degree and lose money, and the client may still feel cheated.

We've studied hundreds of working drawings for content relative to fees and find that **project time and cost budgets can be estimated reasonably well within three tiers:**

1) Comprehensive Services. These are for sophisticated buildings requiring considerable custom detailing and which are designed at top fees.

2) Median Services. These are your typical good quality, mid-range buildings which will be built at average construction costs and designed at design fees that are average for the building type.

3) Baseline Services. These are low-fee, minimal construction cost projects; usually of utilitarian construction such as low-budget, non-bid or owner-built developer buildings.

To avoid misunderstandings with clients and to minimize excessive work relative to fees received, use the **WORKING DRAWING PRODUCTION SERVICES SCOPE OF WORK** checklist in the APPENDIX of this book as a guide in negotiating fees and in working drawing planning.

WHAT MAKES A MANAGER GOOD, BAD OR AVERAGE?

How would you go about identifying the management personality traits that get the best results? The scientific approach would be to study the performance records of a large group of managers. Sort them out as low, average and high-level performers. Give them personality tests to identify dominant traits. Then see what, if any, sets of traits go with certain performance levels.

That's what social-industrial psychologist Jay Hall did. Here's what he found out:

First, take a look at low-level performance. Low-level performers have a conflicting desire for safety and ego boosting. They want to feel and appear important without taking risks. They tend to be a little pompous, defensive and always ready to pass the buck or block new ideas. They direct their time use toward negative rather than positive goals. They stop action rather than initiate it and criticize ideas rather than create them.

Average performance managers are less concerned with safety, but have strong ego-status needs. They tend to be more concerned with appearances. They'll often spend time trying to look--rather than be--effective.

High-production managers are primarily motivated by a desire for self-actualization. They're out for the experience of achievement. Their attitude profoundly affects their relationship with others and with their work.

They enjoy work. Such managers are not only the most productive kind; their fellow workers rank them as the best to work with. They don't look down on others or try to keep them down, in fact they help promote their peers and subordinates to higher positions.

Hall comments: *"The good manager, it seems, needs to find meaning in his work and strives to give the same meaning to others. What he does flows from his view that work is both a challenge and an opportunity for self-expansion. He looks upon innovation as an opportunity rather than a threat and is therefore willing to take risks. He believes that to be successful he must work with people under him and create opportunities for them to succeed. He is, in a phrase, an apostle of enlightened self-interest."*

Hall offers this provocative message about success: *"As a behaviorist, I believe strongly that we are what we do. This is fortunate for most of us because it means that the success of an executive doesn't depend on personal traits or extraordinary skills limited to a few outstanding individuals. It depends rather, on how we behave in our work, on the values we hold about personal and interpersonal potentials. All of these can be learned. The key to becoming a good manager--a success, if you will--is to learn to behave like one."*

29

THE PRODUCTION MANAGER'S JOB DESCRIPTION

"You keep harping on how important the Production Manager's job is, but you haven't published any real job description for the position," complains Connecticut architect H. Kaufmann.

Consider the oversight rectified.

PRODUCTION MANAGER, a special role comparable to what used to be called "Chief Draftsman" -- **an overall technical supervisor and coordinator of the design studio and drafting room.**

Capabilities and duties include those common to PROJECT MANAGERS plus:

__ Quality control checking of all documents for permits or bidding

__ Management and maintenance of the standard detail system

__ Management and maintenance of the office manuals and/or checklist system

__ Enforcement of office graphic standards

__ Management of the office master specifications (unless handled by a professional specifications writer)

__ Management of CADD standards, layering systems, production files, etc.

__ Overseeing or conducting cross-checking and coordination of specifications and working drawings

__ Production personnel allocation and employee reviews

__ Technical training and continuing education of design and drafting staff members

__ Continuing technical education of job captains and project managers

__ Enforcement of production standards among job captains and project managers

__ Maintenance of technical reference library

__ Maintenance of building code library and distribution and enforcement of the most important code rules

__ Management and coordination of in-office reprographic systems -- production computers, printers, copiers, etc.

__ Coordination of outside reprographic and CADD services

CHAPTER THREE

SUPERVISING AND EDUCATING THE TROOPS

THE BEST DRAFTING STUDIO MANAGEMENT

MOST DESIGNER / DRAFTER LABOR IS
MAKING MOCK-UPS TO THROW AWAY
. . . BUT WHO TELLS THEM THAT?

VOICES FROM THE "BACK ROOM" JUST BLOWING
IN THE WIND

YOUR BEST RESOURCE FOR EFFICIENCY SUGGESTIONS:
YOUR EMPLOYEES

HOW TO MAKE YOUR EMPLOYEES INTO
WHIZ BANG EXPERTS

A HIGH-PROFIT MANAGEMENT AND
TRAINING SYSTEM FOR DRAFTING STAFF

REINFORCING THE TRAINING WITH JOBSITE PHOTOS

IN-HOUSE SEMINARS GET EVERYONE ON THE SAME
WAVELENGTH

MBWA -- "MANAGEMENT BY WALKING AROUND"

A SIMPLE FACT OF LIFE THAT DEVASTATES
OFFICE PRODUCTIVITY EVERYWHERE

THE VALUE OF THE "GRAPHIC WORK ASSISTANT"

THE WRONG MESSAGE HAS BEEN DELIVERED TO
THE DRAFTING STAFF

YOUR OFFICE GUIDE TO FASTER, BETTER
WORKING DRAWINGS

THE BEST DRAFTING STUDIO MANAGEMENT

Two of the best drafting room manage-ment systems we've seen may seem contra-dictory. One is the most close-up, hands-on su-pervision you could ask for, and the other leaves some of the major supervisory lessons up to the employees. They both work and some firms have adopted them as a complementary set.

HANDS-ON TO THE MAX

An outstanding architectural manager, David Leslie, finished his projects con-sistently on time and on budget, while main-taining high personnel moral. Here's a brief de-scription of his system and its benefits:

First he met privately with every new drafter who was assigned to his projects. He would say: "I know you want to do your best work and you want to work efficiently. The worst problem you might have is getting the management support you need to do your best work. My job is to give you that support. I'll check with you about twice a day to see what you need, what problems there might be, and see if you have any questions."

Then he followed up by making the rounds about mid-morning, stopping to check each drafting station to see if everyone had everything they needed to do their work. After hours he stopped again at each station to pick up question notes or requests and do some quick checking of the work. The late day visits wer-en't snooping; he just wanted to respond to any questions or problems that might slow anyone down when they came back to work.

The result:

David always knew what everyone on the team was doing, how well they were doing, and exactly whether the project was on time from day to day. That's an extremely high level of awareness for any project manager.

David's drafters commented that open communication and information flow was the key to success. They said that information was always scarce when they worked on other jobs. They couldn't find references or get answers to technical questions or get a clue as to what de-sign changes were being made.

When information shortages become chronic, the work slows to a snail's pace. People have to slow down because they have no certainty that the information they're drafting is correct or up to date.

DO-IT-YOURSELF SUPERVISION --
THE FRIDAY AFTERNOON CLUB

A seemingly contrary management ap-proach lets drafters compare their work with one another, with little supervisory interven-tion. The method is the Employee Working Drawing Coordination Review. We've inves-tigated several variations on the system and can report in much more depth on how it works and its extended potential benefits.

It goes like this:

It's show-and-tell every Friday afternoon (when work usually slows down anyway). The project teams grab some conference room space, or a corner of the drafting room, and everyone on each team pins up the work they did during the week for review.

Each drafter takes five or ten minutes to show and describe each drawing: what the as-signments were, what they did, changes that were made, problems or questions that came up, etc.

Several positive things happen as a re-sult:

1)   Everyone sees how his or her work relates to the others' work. That's an item of curiosity in every drafting room and a cause of a lot of board hopping and conversations. By see-

ing how their work relates to one another, employees can catch contradictory information, or excessive duplication of information, and iron out the problems among themselves.

2) If somebody is doing unnecessary work, someone else will usually ask about it. So if interior elevations are being drawn that could be handled by a simple Fixture Heights Schedule, or the same floor plan data is redrawn completely at several different scales, and so on, it will be talked about. After awhile, if people want to do things in less efficient ways they learn they have to be prepared to explain why.

3) When somebody isn't carrying their load, it becomes obvious for all to see. Nonproductive drafters try to hide their lapses by shuffling drawings around and working on more than one project at a time. When they have to pin their work up for all to see for a few weeks in a row, there's no more hiding out. The motivation to be productive increases accordingly.

4) Supervisory problems show up too. Drafters often don't get the data they need quickly enough to keep making progress. And often there are too many uncoordinated changes in mid-job. When each week's review shows mainly a bunch of changes made from the previous week's work and no real progress, there can be no denying the situation. For this reason it pays for an office principal to sit in at times as an observer of these sessions. (Normally the boss is the last to know about such problems.)

5) Drafters come to see the total project as a whole and take a supervisory attitude towards the job. Drafters who get stuck on restroom plans and detail sheets for months at a time lose all sense of perspective -- and time. The weekly review brings the big picture back into focus.

6) Younger drafters learn from the more experienced ones. There are countless tips and techniques that most drafters "pick up" in random conversations and encounters. This weekly coordination review accelerates and enhances the learning process.

One essential rule to make this work, and say it out loud: NO NEGATIVE CRITICISM.

The objective is to trade data, compare notes, and improve coordination, not tear one another down. If critical personalities intrude themselves, the meetings will turn into nonproductive, defensive, adversary sessions.

And a supervisor must NEVER criticize someone's work in front of others. No matter how deserved it might be; it just plain doesn't work. There is no surer way of destroying morale, undercutting the motivation to work, encouraging sabotage of a project, and creating life-long enemies among staff. If a drafter needs some criticism, don't forget about it; just do it privately.

We have observed many variations of the "Hands-On to the Max" and the "Semi-Self-Supervision" Friday meetings for years. They've become especially useful in recent years, with the advent of CADD and the difficulty of monitoring an individual CADD drafter's work.

Both methods have been widely tested across the U.S. and Canada. Both will work for you, if you follow the rules, and remember that the whole point is to help employees do what they truly want to do: **mainly, their best.**

Has anyone applied both methods at once? Not that we know of, but it should make a great combination. Give either or both a try and watch morale and productivity move up simultaneously.

MOST DESIGNER & DRAFTER WORK IS MAKING MOCK-UPS TO THROW AWAY ...

BUT WHO TELLS THEM THAT?

What senior personnel usually want from junior personnel are sets of options:

--Rough sketches of alternative design features.

--Rough study models to work out spatial or massing problems.

--Rough detail drawings from which to choose and refine final details.

What junior personnel often deliver are:

--Finish design presentation drawings (or as finished as they can get in the time limits).

--Precision cut pasteboard models.

--Finish hardline drafting of data that will probably never be used as is.

The juniors complain that much of what they do has to be erased and done over. The seniors complain that they can't get fast and simple options. "I want some 'talking sketches' and they give me renderings!"

What compounds the confusion is that while complaining about time and cost, the senior personnel often LIKE the overly finished work. It LOOKS better, so they don't get down to basics with staff and make it clear exactly what they truly need and expect in terms of design or production studies.

The youngest personnel have only the examples of school to go by, where their work was created to be judged by design juries. They treat their assignments the same way.

Let them know that the objective in these assignments is to SEE, roughly and in general terms, how something looks. The point of seeing how something looks is so that it can then be changed to something else. Designers, clients, senior drafters, etc., want to see if something is too large, too small, in the wrong place, etc. So they don't want finished work, they want something quick and expendable -- like the rough draft of a piece of writing.

Teach by illustration and example: When giving assignments, show block models instead of finish presentation models, freehand sketches instead of hardline drafting, paste-up mockups from old jobs or other ready made sources --NOT new work that just duplicates what already exists.

VOICES FROM THE "BACK ROOM" JUST BLOWING IN THE WIND

What makes Japanese managers most different from US managers is that they LISTEN to employee suggestions.

The employees are closest to the front line problems, most acutely aware of shortcomings in service and quality control, and most painfully aware of management problems. Japanese management has learned to appreciate and use employees as a gold mine of ideas and information.

To help pass the word, here are remarks I have heard from drafters across the nation. In response to my question:

"What have been the worst barriers to production you've experienced in your drafting rooms?" I got quite an earfull.

1) Communication And Instruction

"There is never enough clarity or instruction. It's like they want us to draw up something and then they see if it's right or not. It could be done a lot faster with sketches or mockups instead of doing finish drafting and then deciding whether it's right or not."

"There aren't enough technical references and of those we have, we don't know which are supposed to be right."

"There is no instruction in how to detail or what the point of details really are. We just get old sets of prints to copy from. Then they call us 'dangerous' because we don't know enough about construction."

"No systematic instruction . . . no checklists. We have to guess what to do."

2) Sticking To Design Decisions

"They continue design into working drawings and every time they notice something they think can be improved, they make us change it. They don't have any idea how much time it takes to change something and then change all the other things that the first changes effect."

3) Lack of clarity as to who is in charge

"The most frustrating thing I get is contradictions in who's the boss on a job. The designer says one thing, then the boss contradicts her, then she contradicts the boss. Finally everybody slows down to wait for them to sort out their problems."

4) Thoughtless Work Environment

"Drafts, bad light, noise, lack of the right equipment."

"Second hand drafting furniture. Flat boards that break your back. Nothing is in the right place."

"If they don't care about us getting what we need to work efficiently, why should we care?"

5) Lack Of Simple Feedback Systems

"Some way of telling the boss about problems, or getting anything fixed? NO WAY! All the other problems we have in getting anything done seem to start from that one."

YOUR BEST RESOURCE FOR EFFICIENCY SUGGESTIONS: YOUR EMPLOYEES

"We've always gotten our best production ideas from our firm's employees,'" says Delaware architect J. Evans.

Their firm has an ongoing Suggestion System, but they get most of their input with annual surveys of staff and management. The survey is anonymous. Everyone is required to fill out the survey form, but it's collected and compiled by a local secretarial service so that nobody need fear recrimination for their remarks.

Over the years they've learned about dozens of problems such as poorly coordinated work assignments, drafting room distractions, and money losses in consultant services.

"People suggest good ideas with or without a survey." says Evans. *"But I think many others would not, or ideas would come up when we're worrying about deadlines and don't have time to think about improvements."*

Here is suggested text for an employee survey; use or adapt it in any way you see fit:

Our employees are our greatest asset and we want to find out more about what we can do to help you do your job.

Please fill out this questionnaire as completely as possible. Your answers will be strictly confidential.

The results will be analyzed by an outside company and a summary provided to office principals. No one in this office will see any of your original questionnaires so you can be completely candid.

We can't guarantee that we'll act on all recommendations that might be provided, but all will be given very serious attention.

Questions:

What office change or changes do you think might best improve services to the clients? (Use back of page whenever necessary.)

How do you rate these specific aspects of working with this office on a scale of one to ten: (Please add any comments you think will clarify your remarks.)

--Wages and Benefits ___

--Management support/responsiveness of supervisors ___

--Communications within the office ___

--Resources and tools to assist you in your work ___

--Co-workers ___

--Support staff ___

--Scheduling and disbursement of work ___

--Overall, how would you rate our office and services on a scale of one to ten? ___

Answer these questions as completely and candidly as possible.

--What change would be most helpful in terms of helping you do your best work?

--How do you think we could best improve our services to our clients?

--What do you consider to be the best aspect of working with our firm?

--What do you believe is most in need of improvement?

HOW TO MAKE YOUR EMPLOYEES INTO WHIZ BANG EXPERTS

After suffering a costly arbitration settlement last year, the technical partner of a well-known New York City office stood in the center of the drafting room and shouted:

From now on you are all going to become experts on joints and sealants. Read this article carefully and read everything else that comes through the office about joints, sealants, leaks, and waterproofing."

He quickly dispensed an arm-load of copies of an article on the subject.

Just as quickly, the articles disappeared into drawers, into binders, and onto shelves. It was back to business as usual. The only one who read the article with care was the spec writer and he already knew the material.

It was a flash in the pan, doomed from the start.

Two thousand miles away in a Dallas office of similar size, the production manager recently got a bonus and a raise for setting a new office record for quality control. He, too, started out with the intent of turning his technical staff members into experts and he succeeded.

Why the difference?

"I set individual responsibility," he told us in a phone interview. *"I don't ask everyone to become an expert on everything, but I did ask the staff for volunteers. Who would volunteer to become the office expert on wood failures, on sealants, roofing, fire codes, etc.?*

"Fifteen people volunteered to learn more about these subjects. I gave ten of the most ex-perienced people copies of data from my files and told them where else they could get additional information. Then I gave each one a deadline to learn the subject well enough to give a lunchtime seminar on causes and prevention of failures."

THAT was the difference -- individuals held to account for learning the material. First came an expression of interest by the staff members and then a direct and specific assignment to learn the data well enough to do a mini-presentation on it. Henceforth, each individual involved knew he or she had to come up with something useful and convey it intelligently to the group.

"We do a new six-week series twice a year. We have one lunch hour a week reserved for presentations. We make video tapes for those who miss it for any reason. We've updated the office manual to say that employees are expected to participate in continuing education programs in and out of the office. We have a large office but I would do the same thing if I had only three people. These days you can no longer afford to have people on the boards who don't understand construction."

A HIGH-PROFIT MANAGEMENT AND TRAINING SYSTEM FOR DRAFTING STAFF

Most drafting rooms are not managed.
There is no overall plan, no overall goal setting, no monitoring of results, no systematic training of staff, and hence, virtually no real progress or improvement in production in most A/E offices. **Most offices are as inefficient now as they were ten or twenty years ago.**

By contrast, projects within the drafting studio are most certainly managed; there's heck to pay when they're not. And there's almost always a full-time or part-time front office manager. While there's management of the overall operation and management of individual projects, there's a serious gap in between.

That's why the job role of Production Manager needs so much attention. Without such a person, there is no one to enforce significant, sustained improvement and efficiency in office production. Thousands of dollars will go down the drain year after year as designers, drafters, and project managers repeat the same obsolete working drawing practices and make the same mistakes as their predecessors.

CADD makes matters worse. Many CADD offices have a CADD manager, but that person is usually a computer system manager not a production manager. So many of the most wasteful practices in working drawing production are made part of the CADD process . . . and the best practices are not.

It's not just efficiency that's at stake.
A production manager is the one who educates employees in quality control in design, drafting, and construction. The traditional practices that slow down production are the same ones that lead to conflicts, bad morale, buck-passing, rushed schedules, and oversights and errors throughout the documents.

Meanwhile new drafting staff members enter the offices with no thought of production as an important part of their careers. The subject isn't taken very seriously despite the fact that drafting and production issues are likely to dominate most of their work days for years to come.

Technical Director Morry Wexler wrestled with these problems for many years at the Stone Marraccini & Patterson office (SMP) in San Francisco. He thought of many excellent solutions, tested them in practice, and made them better and better. Since retiring from SMP, Morry has been sharing his expertise in articles, lectures, and consultation.

Morry Wexler says:

"How do we instill the appreciation that drafting is not merely a ladder-rung on the way to design, but is an essential part of design -- that each ladder rung is a fusion of design and detail?

"Then, how do we train our young architects to be not only good designers, but to be good technical detailers? . . . With this in mind, several years ago we at Stone, Marraccini and Patterson opted for a three-phase program:

1) Explore and utilize available technological advances in drafting.

2) Research and create our own drafting and administrative procedures and techniques.

3) Develop in-house, on-going training seminars to teach our staff the new procedures and techniques.

"Hence, we developed our own version of systems drafting. . . . Our system, as taught in our seminars, touches on nearly all aspects of architecture, starting with the planning of a project, the budget, and ending with the notice of completion and post-occupancy evaluation.

"We believe our employees should have knowledge beyond the drafting board and into the board room. But the main and most important emphasis of these seminars is drafting -- error-free, lawsuit-resistant, on-budget, on-time drafting."

That's getting to the heart of things: *"error-free, lawsuit resistant, on-budget, on-time drafting."*

Regarding the role of Production Manager or Technical Director:

"Seek out the one person who is receptive to new ideas, has the imagination and skill to try them, and who likes to supervise, teach, persuade, ... That person's task should be, in part, to maintain and improve drafting standards, techniques, and practices; promote advanced procedures to effect quality, accuracy, speed, and efficiency; coordinate and integrate manual drafting with computer aided drafting; integrate consultants into the procedures, and control the quality of in-house reprographics work. Simple."

REINFORCING THE TRAINING WITH JOBSITE PHOTOS

Every day in every office there are lessons that can enlighten staff members. For example, there'll be errors, questions, challenges, and change orders from the job sites. The waste is compounded when these examples are allowed to slip by without being saved and reused . . . as teaching lessons, as examples of what not to do.

That's the value of job-site feedback photos. Photos are taken of problems during and after construction -- everything from a sagging counter top to a column in front of a door. Photos are taped onto 8-1/2 x 11 sheets to be photocopied, with explanatory (non-blaming) text. They are then routed among the staff and eventually end up in the Main Technical Library for future reference.

IN-HOUSE SEMINARS GET EVERYONE ON THE SAME WAVELENGTH

The operational center of Morry Wexler's "Systems Method Procedures" is a training seminar program.

"It's surprising how easy and how little time it takes to put together a seminar. Basically the Guru walks around the office and looks to see what's on the boards. He or she sees something that illustrates some principle of drafting, it is momentarily pulled off the boards, and a copy is made and put aside until enough material is gathered to cover a particular subject.

"When a project is finished, some original overlay drawings, paste-ups, schedules, etc., are similarly set aide. The material is then organized, assembled, mounted as flip-charts, and voile, you have a seminar."

The SMP series consists of eleven flip charts, mainly showing enlarged copies of drafting samples right off the boards. Each flip chart series covers an aspect of drafting in great detail. The entire series only takes about six to seven hours to present. **The training series modules include:**

A -- INTRODUCTION TO SYSTEMS METHOD PROCEDURES

B -- FLOOR PLANS

C -- PLANNING, SCHEDULING AND BUDGETING

D -- DETAILING

E -- RESOURCES

F -- INTEGRATING THE CONSULTANTS

G -- ADDENDA AND REVISIONS

H -- COLOR DESIGN (layers)

I -- ADVANCED PASTEUPS

J -- TRICKS AND SHORTCUTS

K -- ANATOMY OF A PROJECT

About the seminars:

Series A -- INTRODUCTION TO SYSTEMS METHOD PROCEDURES

"This is our introduction to what many think is a strange new world. . . .We talk about making a profit. We talk about errors, and conflicts, and omissions. That leads us to talk about lawsuits. We show a whole bunch of newspaper articles reporting how architects got sued for one thing or another.

"We talk about how Systems Method Procedures will help us make good drawings, and how it will alleviate the perceived drudgery of drafting, and even help make drafting interesting and fun."

Series B -- FLOOR PLANS

"By real example, taken right off the boards, we describe how we make drawings in three dimensions by means of layers, which information goes on which layer, and how the layers are registered and composited into a single drawing. . . . We describe how some schematic drawings are easily converted to design development drawings and how nearly all design development drawings are converted to working drawings."

Series C -- PLANNING, SCHEDULING AND BUDGETING

"We show a drawing count with the associated hours per drawing. ('You mean this one drawing costs $2,000? Wow!') . . . We show how mini-drawings of each anticipated drawing helps us visualize and plan final drawings . . . and budget the hours."

40

Series D -- DETAILING

"We describe the module, paste-up procedures, our detail and drawing numbering system, drawing blackouts, phased details, free-hand details, . . . composition and layout, schedules, how to integrate word processing with detailing and the role of CADD in detailing."

"We show good practices, sloppy details, subtle errors, redundant details, useless information, and over drawing and rendered details (Mona Lisas). We give a short course on how to coordinate the drawings with the specifications and vice versa."

Series E -- RESOURCES

In this session, Morry explains the Main Library, where things are, and how to use them.

Series F -- INTEGRATING THE CONSULTANTS

This explains the roles and relationships of consultants. Morry has a special separate two-hour seminar for consultants to help them mesh with the prime design firm.

Series G -- ADDENDA AND REVISIONS

"This series . . . is given on a project-by-project basis, just prior to a project being let for bid. It describes how to revise the drawings and specifications for addenda during the bidding period, and how to revise them for bulletins during the construction period. Specially designed revision methods result in post-construction record drawings which are as clean as the original drawings, accomplished with minimal effort and expense."

More next month on this and the remaining aspects of Morry's training program. We'll also include an interview with Morry on his general philosophy of Technical Management.

Series H -- COLOR DESIGN

"This series describes how to create color design drawings by using the original base architectural floor plans with overlays for the color work.."

Series I -- ADVANCED PASTEUPS

"Here we go into the fancier paste-up techniques . . . symmetrical designs, common reusable elements, key plans, etc."

Series J -- TRICKS OF THE TRADE

"We demonstrate easy, fun and effective ways of converting working drawings to presentation drawings by applying overlayed entourage -- people, trees, cars, clouds, etc. -- to certain layers of previously prepared drawings. We also demonstrate phased drawings, montages, and other techniques."

Series K -- ANATOMY OF A PROJECT

About the final seminar of the series:

". . . We go through a typical project step by step and summarize the seminar on Systems Method Procedures. If we're short on time, we do a question-answer period which is often more effective anyhow.

"Now comes the most important part. It's about four o'clock and a lot of information has been pushed into everybody's head. What should have taken a year or so in school has taken only a few hours of one day. In a few days what will remain in everybody's head will be vague impressions of what transpired at the seminar. But the ideas are there. The most important part now is the follow-up.

"The principle duty of the Guru, the Chief Draftsperson, the Manager of Production, the Technical Director, or whatever, is to walk around the office every day, visit everybody working on the boards, and teach, explain, demonstrate, instruct, give direction, and heap praise. One more thing: Answer questions. No matter what you are doing, stop in your tracks . . . and answer the question."

MBWA --
"MANAGEMENT BY WALKING AROUND"

"MBWA" means "Management by Walking Around." It's an old idea that's been repopularized in recent years by some of the better management books. The phrase refers to managers and supervisors who truly know what's going on in the office; **managers who pitch in and help people do productive work as opposed to mangers who keep their noses stuck in paperwork.**

The most productive design and drafting studio manager we know has brought project after project in on time and on budget with MBWA. His method is so simple and powerful it seems crazy that everybody doesn't use it. Here's how it works, step-by-step:

1) He takes every new employee to lunch and lets him or her know that the supervisor's job is to act as SUPPORT to the employees. He says he knows the employees want to do their best work and they mainly need to have management remove barriers that might be in the way of working efficiently. That means:

--Quick management response to
    employee questions.

--Ready employee access to references,
    drawings, technical references and fob
    records.

--Plentiful and convenient tools,
    materials, and office equipment as
    needed to do the job.

--Continuous communication of what's to
    be done--long-term as well as
    short-term.

--Feedback on how employees are doing
    in work quality and quantity.

--Information on how one's work fits into
    the overall larger projects.

2) The manager has thus made a support contract with the employees. He follows up with twice-daily contact at each workstation.

--He visits the boards and workstations after work hours--a 30-minute round to glance over the day's production, write instructions, and check questions and request notes that are left for him.

--He does follow-up visits to his production team members at mid-morning to review work in more detail and respond to verbal questions and requests.

His supervisory contact is for superior to the norm while the net amount of time he spends on it is way less than normal. The reason is that **he's preventing problems instead of correcting them**. He initiates contact when it's most convenient for him instead of when cornered by staff.

3) He conducts team meetings on late Friday afternoons ("when not much gets done anyway") for employees to display their week's work. This lets employees see how their work relates to others, and to allow them to compare their work output. (This has negative impact only on those few who perform well below the norm and refuse to improve themselves.)

Proof of the system: Besides making deadlines, his teams' work rarely require much in the way of addenda during biding or change orders during construction. **Employee morale is high when he's in charge, and clients who know him always ask that he be in charge of their future work.**

A SIMPLE FACT OF LIFE THAT DEVASTATES OFFICE PRODUCTIVITY EVERYWHERE

A Fact of Life is one of those things everyone knows about, that causes all kinds of problems, and that nobody talks about. The culprit in this case costs millions of office workers an average of one hour a day in slow productivity. It's especially evident in drafting rooms.

For the evidence, visit any A/E drafting room before the mid-afternoon break and observe the work flow. It's a time when some employees are doodling and almost snoozing at their workstations, shuffling drawing sheets, slipping out on errands, shooting the breeze on the phone -- whatever they're doing, it's slow motion.

The main culprit, say nutritionists who do corporate consulting, is lunch. Virtually anything other than a simple salad or plain meat dish will send the blood sugar crashing in the early afternoon and as far as productivity and mental agility goes, it's the Night of the Living Dead.

Even more glaring, the post-dinner slump for those working overtime. Overtime work has a limited cost effectiveness anyway, and those who are on charette who do a "regular" dinner may as well not even go back to the office. Architect David Deppin, at a large San Francisco office, observed that *"When half the team just had a simple meat dish like salmon, and the other half had the usual pastas or mixed carbohydrate meals, the others could hardly work.. The contrast was startling."*

It used to be worse in the days when people ate jelly doughnuts and did heavy re-charges of morning coffee. Many of those addicts lost all their afternoons, not just an hour or two, and did so day after day for years. For more, just check any nutrition book on the impact of diet on blood sugar and daily energy cycles.

THE VALUE OF THE "GRAPHIC WORK ASSISTANT"

The primary role of the office secretary is to expedite communication. He or she takes messages, transcribes notes and gets correspondence out, routes mail, places calls . . . it's all communication and everyone knows how important the job is.

There's another job role that's equally important to the A/E office and it's virtually unknown; that's the role of the Graphic Work Assistant. This is the "Assistant To" applied to the drafting and graphics realm.

There are dozens of jobs that a GWA can handle expeditiously that otherwise will be done by higher priced drafters, designers, managers and office principals.

The Graphic Work Assistant watches over the production infrastructure: Equipment; the repairs and warranties; drafting, copier, FAX, and printer supplies.

Paste-up work is a natural for the GWA. The more you use the large format engineering copiers, the more paste-up work everyone will do, and much of that is repetitive and best delegated to the Graphic Work Assistant.

Offices that use CADD have even more day-to-day GWA needs. Pen-and-ink CADD plotters need almost constant maintenance, as do pens in general in offices that use ink drafting extensively. **Other chores in the GWA job description:** There's field photography to be done, rough building measurements to be made, direct courier service required. Much of this job is in the vein of what used to be called an "Office Boy" or "Print Boy."

Some offices use part-timers, especially students, in this role. It's also been a re-entry job position for mothers and housewives who are getting back into the work force. Several offices have had great success hiring retired people as part-timers. One firm got a lady who retired from publishing and now does proofreading and office library maintenance along with other basic GWA chores.

Small offices have enlarged the secretarial job to include GWA; it helps keep the secretary on hand full time where a low work load wouldn't otherwise justify it. Warning: don't try to foist extensive GWA duties on a secretary who is already working full time . . . it won't usually go over even if you think the secretary is presently underutilized.

ANOTHER ROLE FOR ASSISTANTS -- FOUR-HANDED DRAFTING

Extensive paste-up drafting is best done in teams according to California drafting room supervisor, George Abronski.

"You get a lot more done with four hands on the same table especially if you're doing a lot of repetitive work, like multi-unit housing or hospital plan paste-ups," says Abronski.

We saw the same technique used in plain drafting many years ago in New York where an architect told us he was inspired by the "Four-Hand Sit Down" method introduced in dentistry. Dentists said that having easier patient access in a recliner and an assistant working at all times was a great expediter for most dental chores.

Four hands were better in drafting, said the New York architect who subdivided drawing sheets into units, usually multiples of 8" x 10", and assigned different assistants to work with him on different parts of the drawing. They all worked around one large work table. He was able to give quick response to questions and had others to do side chores for him which sped up his work. He said that working at a group drafting station was much faster and more interesting than working alone at a traditional single-person workstation.

THE WRONG MESSAGE HAS BEEN DELIVERED TO THE DRAFTING STAFF

Most new A/E staff members don't think of drafting as an essential and important part of their careers.

Why should they? Most design schools don't provide working drawing courses. Such courses are labeled as "training" instead of "education" and are left to vocational or technical schools. The implications are unmistakable and the results have been disastrous for the profession.

Morry Wexler, who dealt with this problem for many years at the Stone Marraccini & Patterson office (SMP) in San Francisco, described the elements of a top-notch staff orientation and training program designed to undo the negative aspects of design education. Here are notes from a Question-Answer interview with Morry:

What's the current status of "Systems" in design offices?

"Lots of firms know about Overlay Drafting and you see pin bars on the drafting boards. But after you walk around the office for awhile you realize nobody is using the pin bars.

"Some offices have all the equipment they need, but nobody knows how to use it or the staff is not encouraged to use it. So they're stuck with stickybacks.

"Many offices have made attempts at creating a library of Standard Details and Reference Details, but the details are often not organized or too difficult to use, and therefore ineffective.

"Not long ago I was doing some consulting work on a hospital addition and I asked for the office's standard casework details. There were no casework details! Drawings that should have taken an hour to assemble from standards took two days to draft from scratch."

What's the main problem in keeping a design office on a Systems track?

"There's no Technical Director (an overall manager of documentation and production). There's no one person assigned to establish and manage their drafting system, to visit the boards, answer questions, and help employees find information. That person has to be dedicated to this role and be faithful about visiting the drafting boards twice daily.

"It doesn't matter what equipment or systems you have; if the employees aren't trained and comfortable about using the office resources, they won't use them."

The problem extends to CADD. Morry observes that many offices have computers and CADD these days but the equipment is used in haphazard ways without a systematic overall approach to design and working drawings.

And the problem boils down, once more, to having someone in charge -- a manager of the systems -- **and support from the top** to enforce the best procedures.

YOUR OFFICE GUIDE TO FASTER, BETTER WORKING DRAWINGS

One of the best production timesavers we've seen in years is the "prototype" or generic instructional set of working drawings.

A "prototype" or "instruction" set of drawings shows the drafters all the best practices. It shows where typical data goes and how it should be sized and structured. Instead of **telling** drafters what's right and wrong, prototype sample sheets **show** them the most efficient and clearest formats.

Here's the simplest way to make a prototype set, in four steps:

1)   **Find prints of past working drawings of building types common to the office.**

2)   **Find the best examples of particular drawing types** -- A good General Information Sheet, the best Site Plans, Floor Plans, Elevations, Finish Schedules, etc. Add handwritten notes to explain why elements in the drawings are especially good or not so good.

3)   **Cut and paste, where appropriate, to combine the best examples.** It doesn't matter if good sample parts of drawings are from different projects.

4)   **Make large-format photocopies of the marked and pasted up sample sheets.** These will now be the office generic guides for each kind of drawing for your common building types.

Henceforth, when a job is under way, the drafters will have the best possible examples to refer to. They'll see the data that should be on a particular drawing type, the preferred scales for various drawings, the office's best ex-

amples of streamlined and functional drafting. The sets can also include clearly marked examples of what not to do, such as undersized lettering, overdrawing, over-elaborate schedules, etc.

Some production managers spend a little more time on their prototype working drawings and do the following:

1)   **Select, review and mark up old prints, as above.**

2)   **Choose an all-purpose working drawing module if the office doesn't already have one.** (We recommend any module close to 6" wide x 5-3/4" high as a unit that is large enough to encompass virtually any detail, and one that can fit evenly within all standard A/E drawing sizes. The half-module of 3" width is suitable for title blocks and keynote strips.)

3)   **Have a drafter trace the outline around key elements on each sheet.** Then the finished prototype sheet just shows outlines of drawing components such as site plans, window details, etc., all positioned within the modular drawing system. The idea is to show outlined blocks of **types** of common working drawing data without confusing the issue by showing the particular job data itself. These drawings show the location and format of window details, for example, but not the details themselves.

4)   **Add a generic keynote list suited to each type of drawing.** This is a list of working drawing contents which will do double duty as a prototype notation checklist. In other words, it's a reminder to drafters of items to include in the drawings, and it's a guide to the final drawing notation, keynotes, and specifications and detail coordination. (The spec and detail coordination only works if key-notes are clustered by CSI divisions and include their CSI coordination numbers.)

The best prototype working drawing sheets have indexes of recommended content and are enhanced with timesaving do's and don'ts.

GENERAL INFORMATION SHEET

__ **Drawing Index**
    __ Use large, legible letter style as per sample.
    __ Use reduced sized photocopies or laser-printed output as mini-indexes on remainder of drawing sheets, as per sample.

__ **Vicinity Map**
    __ Photocopy or scan from road map.

__ **Lexicon & Abbreviations**
    __ Copy from Office Standards Files.

__ **Symbols & Materials Indications**
    __ Copy from Office Standards Files.

__ **Names/addresses/phones of consultants & building officials**
    __ Copy from project manager's project manual.

SITE PLAN

__ **Grades & Contours**
    Have original survey done in architectural scale.
    __ Scan or reproduce survey as screened shadow print background image.
    __ Check the site to confirm the accuracy of critical survey dimensions, elevations, and site features.

__ **Aerial Photo**
    __ Combine background image with survey to check survey accuracy.

__ **Datum and dimensioning start points**
    __ Establish construction dimensioning start point, preferably at surveyors benchmark.
    __ Establish construction grade datum for first level finish floor as 0'-0".

__ **North and Reference North arrow**
    __ Copy from office standard.

__ **Construction setback lines**
    __ Confirm compliance with the latest governing ordinances and property easements.

__ **Work to keep and to remove**
    __ Use solid lines for site elements to remain, dash lines for those to be removed, and solid lines with notation or keynotes as "new" for new work.

The same kinds of guides are provided for Floor Plans, Exterior Elevations, Sections, Roof Plans, etc. Similar guidance is provided for the consulting engineering drawings in the most comprehensive prototype working drawing sets.

Prototype working drawing sample sets also include guidelines for drafting construction details. Many younger drafters have extreme difficulty with detailing and need all the guidance they can get. Guidance checklists with prototype drawing samples can be fairly extensive, as per the following list of recommended practices:

CONSTRUCTION DETAIL LAYOUT

__ Draw in sequence from the most general to the most particular.

__ Draw in "layered" phases so that a detail is substantially visually complete at every stage.

__ Keep the exterior face of construction facing to the left and the interior face to the right. (This is a general rule that works well for consistency and readability most of the time. Abandon the rule whenever it fails to add clarity.)

__ Keep most notation as a list down the right-hand side of a detail. Place other notes as appropriate for maximum clarity.

__ Add thicker "profile lines" around the most important portion(s) of construction being shown.

__ When showing a manufactured product such as extruded aluminum windows or storefronts, show the outline or profile of the object without drawing unnecessary details of the interior extrusions.

DETAIL NOTATION

__ Most notes should be simple names of materials or parts.

__ If more information is required, add the data as an assembly or reference note after the material or part name.

__ Notes should provide information in a consistent sequence.

__ Size of the material or part, where size is relevant and not duplicated by a dimension.

__ The name of the material or part--generic, not specific names are usually preferred. (No product brand names unless your drawing and specifications are one and the same.)

__ Position or spacing of parts unless they are dimensioned.

__ Arrow leader lines from notes follow a consistent office standard. Recommended: a short straight line starting horizontally from the note and breaking at an angle to the designated material or part.

SCALES FOR DETAIL DRAWINGS

__ Use 3/4" = 1'-0" scale only for the most simple light framing, landscaping, and cabinet details.

__ 1-1/2" = 1'-0" scale is widely used for simpler construction components in roofs, walls, floors, ceilings, etc., but 3" is often better for clarity.

__ 3" = 1'-0" scale is mainly used for doors and windows, wall fixtures and connectors, *and more elaborate* components of the walls, floors, ceilings, etc. When in doubt, the larger scale is preferable.

__ One-half size and full-size details are only for shop drawings or to show the smallest of construction components. Extra-large details take up space and time all out of proportion to their usefulness, and are rarely justified.

DIMENSIONING

__ Avoid fractions in dimensions as much as possible. The smallest practical fraction in most dimensioning is 1/4".

__ Dimension lines should connect only to lead lines extended from the faces of materials, and should not connect directly to material profile lines themselves.

__ Don't duplicate the same dimension on two sides of a detail.

__ Use consistent and simple dimension connection symbols -- arrows, slashes, or dots. (Ignore this rule if using "Modular Drafting," where each dimension symbol has a special meaning.)

DRAFTING STANDARDS

__ Small lettering is a problem on any job. All notation lettering and dimension numbers should be at least 1/8" high, preferably computer-printed or typed.

__ Light linework tends to fade away, so lines should be consistently black, more differentiated by line width than "darkness" or "lightness."

__ Small symbols tend to clog up in reproduction so symbols should be large and open. That includes arrowheads, circles, triangles, etc.

__ Crosshatch patterns tend to run together when reproduced, so line patterns should be spaced at least 1/16" apart.

__ Poche made by using grey tone drawing or graphite dust does not reproduce well, so use dot or Zipatone-type patterns to achieve "grey" background effects.

__ Numbers and lettering should not touch linework or they will tend to flow together in reproduction and lose clarity.

STANDARD DETAIL SHEET FORMAT

Since details have to be drawn anyway, they may as well be drawn for potential reuse in the office standard detail library. To expedite standardization, instructions like these may be added to the office's detail sheet samples and/or checklists:

__ Follow a consistent size for the detail "cutout window."

__ 6" wide x 5-3/4" high as a size and shape that accommodates the largest number of details of different types, scales, and sizes, and still fits evenly within most working drawing sheet sizes.

__ Establish standard sizes and positions for:

__ Notation strings or keynote code strings
    __ Dimensions
    __ Detail Title and Key
    __ Scale
    __ Detail File Number

__ Create a standard detail format sheet that incorporates the "cutout window" with markers for standard positioning of break lines, notes, and dimension lines.

SEE THE CHAPTER ON **FAST-TRACK CONSTRUCTION DETAILS** FOR MORE INFORMATION ON RECOMMENDED DETAIL FORMAT AND CONTENT.

CHAPTER FOUR

PERSONNEL MANAGEMENT

JOB DESCRIPTIONS IN THE DRAFTING ROOM

Job descriptions used by design firms vary considerably. A "Junior Drafter" in one firm is a non-university graduate with less than six months' work experience. In another office a Junior is defined as "A graduate of an accredited architectural school who has from zero to three years' working experience."

Similar wide differences are found in various firms for all other job categories. To help bring order to the situation, we recommend you follow the standards used in this survey. They represent the averages of the firms in our national fees and production surveys.

Intern drafter or Designer Interns are most often described as employees who are still in school and working part-time, such as in a work-study program, or gaining work experience through summer employment. Their pay is typically at the low end of the scale for Entry Level employees.

Sometimes there is no pay for those who are extremely inexperienced or who are receiving a special educational experience at the office. This is the equivalent of the entry-level apprentice position of a couple of generations back. In rare cases where an office or studio offers a dedicated educational program, the apprentice may pay the employer.

ENTRY-LEVEL DRAFTER / DESIGNER

Entry-level means fresh out of high school, technical school, community college, or university. University graduates are often less employable than those with two-year college training, but they are expected to learn fast and become productive within three to six months of hiring. Their nominal responsibilities are essentially the same as those listed below for Junior Level Designers or Drafters.

JUNIOR-LEVEL DRAFTER / DESIGNER

Some firms use the designation "Architect I" or "Drafter I" for this job slot. It typically refers to someone who has up to two years' working experience. A Junior with school experience would be expected to move to Intermediate or Architect or Engineer II status within a year. Those with little or no schooling, or who are taking night courses, may remain in this category for a couple of years.

Nominal responsibilities of Junior or Intern status in most offices may include:

__ Design and planning of simple structures

__ Translation of engineers' site surveys into architectural plot plans

__ Transposition of sketched details, sections, and notes to working-drawing sheets

__ Layout of floor plan drawings and elevations from preliminaries

__ Hatching, texturing, and finish-up entourage of presentation drawings

__ Model building

__ Measurement (usually with a more senior person) of existing conditions

__ Directed CADD inputting as a CADD operator

__ Site and building photography

__ Clerical tasks such as maintaining technical and job files

__ Diazo printing and photocopy

__ Paste-up drafting

__ Word processing assistance

__ Office operations, maintenance

__ Errands

INTERMEDIATE DRAFTER / DESIGNER

Unlicensed, with two to four years of cumulative working experience. An employee who had a substantial amount of technical training in school and part-time work experience during school may have Intermediate status within a few months of employment. A graduate of a five- or six-year program might qualify as an Intermediate upon graduation depending on work experience accumulated during schooling. A graduate of a four or five-year program with little experience might require two to four years office work to qualify.

Nominal responsibilities of Intermediates in most offices include:

__ All phases of design and planning

__ Completion of site, foundation and floor plans, elevations, sections and schedules for smaller buildings

__ Transposition of sketched details, sections, and notes onto working drawing sheets

__ Near completion of site, floor plans, etc. for larger buildings

__ Detail design for simpler, smaller building construction situations

__ Some site construction observation and construction administration

__ Assistance at presentation meetings and some degree of direct client relationship

__ Coordination of architectural and engineering drawings for smaller buildings

__ Self-directed CADD input and checking

__ Word processing

__ Outline specification writing

__ Shop drawing review with close supervision

All work of the Intermediate is subject to thorough higher level review and checking.

SENIOR DRAFTER OR DESIGNER

May be licensed but usually not until the latter time in this job category. After licensing, the job role is often renamed as "Staff Architect," "Project Architect," etc.

Five to eight years of cumulative education and working experience. A graduate of a six-year program with little office experience would probably require two years of comprehensive office experience to reach senior status. A fifth year graduate might need two to three years' experience to qualify.

An employee with a substantial technical training and part-time work experience during school may have Senior status within a year or less of full-time employment.

Nominal responsibilities in most offices include:

__ Response to Request for Proposal for smaller and medium-size projects

__ Predesign project planning and programming

__ Schematic design for projects of most any size and complexity

__ Complete original building design and client presentation

__ Financing documents for small to medium-size projects

__ Coordination of permits and approvals process for small and medium-size projects

__ Complete self-directed original construction detail design and drafting

__ Drafting and supervision of working drawings for projects of virtually any size

__ Check-print phase checking for progress, errors, and omissions

__ Comprehensive engineering drawing coordination and checking

__ Comprehensive shop drawing review and
coordination

__ Specification writing for non-specialized
building projects

__ Bid administration for small and medium-
size projects

__ All phases of construction site observation
and administration

__ Client contact and liaison

__ Post-construction reviews / post-
occupancy surveys

__ Production and design office
administration for smaller offices

PROJECT MANAGER, JOB CAPTAIN, PROJECT ARCHITECT OR PROJECT ENGINEER

**Typically seven to ten years combined
school and work experience.** An energetic employee who worked summers and part-time
throughout school and who aggressively pur-
sued work and continuing education after graduation from a four or five year program might
qualify to be a project manager within a year or
two of graduation.

The responsibilities are much as follows.
These responsibilities typically require eight
years combined work experience and education
and completion of all or most of the licensing
exams:

__ All phases of marketing, presentation, and
client relationships

__ All phases of permits and approvals
acquisition

__ Site planning and some degree of skill in
landscape architecture and interior
furnishings selection

__ All phases of design for buildings of
virtually any size or complexity

__ Supervision of one or more design and/or
production teams

__ Complete design and production of
original construction details

__ Drafting and supervision of working
drawings for projects of virtually any
complexity

__ Check print phase checking for progress,
errors, and omission

__ Comprehensive engineering coordination
and drawing checking

__ Comprehensive shop drawing review and
coordination

__ Complete specification writing

__ Bid administration

__ Supervision of CADD teams or
departments

__ All phases of construction administration

__ Post construction reviews and
post-occupancy surveys

__ Production and design office
administration

53

BRINGING JUSTICE TO THE EMPLOYEE EVALUATION SYSTEM

Employees are too often the victims of snap judgments.

--They do something especially well that happens to be seen by the boss and bang, instant acclaim.

--They do something wrong at the wrong time and place, and without warning they're canned.

All in complete disregard of a total employment record -- it's not fair.

Also not fair: Some marginal employees keep output and quality to a minimum, but never obviously minimal enough that it seems reasonable to fire them. And when they are fired, the office may have problems justifying it if they fight back.

Evaluation charts help a lot but they're surprisingly rare. These are like charts often used to compare job applicants. Desired qualities are listed and the evaluator writes scores from 1 to 10.

During a new employee's probation period, evaluations should be done monthly by at least two observers. In a larger office the observers would be the immediate supervisor, a production manager or office principal. After probation, the evaluations should be repeated quarterly or semi-annually. The results of the evaluations can be presented if the employee has to be fired, or later, as the basis for employee performance reviews and raises.

Here are the primary points covered by several firms that use numerical employee evaluation charts:

__ Reliability: meets deadlines with completed work

__ Quality of drawing

__ Technical knowledge and skill

__ Initiative

__ Cooperation with supervisor

__ Creativity

__ Cooperation with other employees

__ Leadership potential

__ Speed

__ Accuracy, low error rate

__ Other . . .

Some offices include items such as marketing skills, efforts at continuing education, community service, etc.

Since some factors will be considered more important than others, they should receive numerical "weights." "Weight" -- or importance of a factor being rated -- may be 3 for "Very Important," 2 for "Important," and 1 for "Desirable." Then the rating numbers will be multiplied by their weights and the whole list tallied for a final score.

An employee's score only has meaning relative to other employees. Unless the score is extreme one way or the other, it shouldn't be a basis for action until all employees are rated.

FLEXTIME -- THE LOW-COST (OR NO COST) PRODUCTION BOOSTER

Flextime definitely improves productivity, say about half the employees, supervisors, and bosses involved. Twelve studies cited in HARVARD BUSINESS REVIEW articles over the years suggest a 12% median improvement in worker output as a result of Flextime.

It's a major low-cost employee benefit. Sometimes there's no discernible cost at all--a lot of return with no investment.

Like it sounds, Flextime means the office has some flexibility built into its daily work hours. It meets the needs of staff members who inevitably have important family or personal needs that conflict with regular office schedules. Workers need time to give spouses rides to work; to deliver and pick up children from day care and school; to see doctors, care for family members, etc.

A general midday "core time" is established when everyone is supposed to be on the job. And everyone is expected to put in the normal total work hours each week. The only difference is that employees choose whether to come early and leave early or come late and leave late. A typical Flextime office opens at 7 am and closes at 7 pm, but it isn't necessarily open to calls or visitors except during the normal "9 to 5" standard office hours.

Nobody expected the production gains. Flextime was originally introduced to improve employee benefits and attract top quality help during periods when help is hard to get. The notable improvement in work flow came as a bonus.

Why the productivity improvement? Management consultants say Flextime creates a situation that allows staff members to use many of the best time management techniques--techniques that are usually not open to employees because of typical rigid work schedules. For example:

--By choosing non-rush hour arrival and departure times, suburban employees cut back drastically on commute time--sometimes saving at total of one or two hours a day. That means less stress, fewer problems at home, and more energy for work.

--Staff members set work hours to match their personal productivity cycles. Many people are tops in the morning, the earlier the better for them. Others are worthless in the first hours of the normal workday but perform quite well if they have the option to come late and stay late.

--People create automatic "Quiet Hours" at the start and end of each day. Quite hours, widely used by design firms across the country, are highly productive periods free of distractions and interruptions from others. Interruptions, sidetracks, and the more trivial tasks peak during the midday's core hours. Early and later hours are comparatively quiet and allow people to do their most concerted work.

--Employees almost universally want to keep the Flextime benefit and tend to work a little harder to help make sure it works for the office. They normally take extra care to coordinate their assignment deadlines, meetings, absences, etc. so that the Flextime system doesn't cause project management problems.

Some Flextime offices are extraordinarily flexible. A large consulting engineering firm in Atlanta asks only that employees put in a total of 80 hours within each two-week work period. When they work is a matter of individual choice within the constraints of team coordination and project deadlines. Some people come in as early as 5 am; many arrive late and work evenings every day. Everyone has to take special care that their output is coordinated with what the project managers and other team members need. Office morale is unusually high. The chief engineers say they haven't had any particular scheduling or supervision problems even during rush jobs or during especially complex large projects.

THE FOUR DAY WEEK -- HOW PRACTICAL FOR DRAFTING DEPARTMENTS?

Like Flextime, it works . . . when people work at it. The basic idea is a simple trade-off: People want three-day weekends, so they do 40-hour weeks in four days instead of five.

Problems: Getting 40 hours work out of four days isn't all that easy. Ten-hour days are tiring. Employees in big cities find they can't make commute connections late in the day. Most clients and contractors don't have four-day workweeks, and they have to be accommodated.

Many offices have found ways around the problems after a little give and take between staff and management. A lot depends on motivation. If key staff members AND the bosses really want three-day weekends, they'll figure out the details, solve the problems, and come up with additional advantages that make the idea more attractive to everyone.

Here's how one A/E firm successfully implemented the changeover, step by step:

1)   The start point was a basic 40-hour work week. Work days were 8 a.m. to 5 p.m. with a one-hour lunch and two 15-minute coffee breaks. The first trade off was the coffee breaks, which were eliminated in favor of coffee any time at the board (already standard in many offices). This provided an extra half hour per day. (Most "15-minute" breaks actually run longer anyway.)

2)   The lunch hour was reduced to 30 minutes. This was resisted by some staff, and welcomed by others, but it's not that unusual. An extra 30 minutes was added on pay-days to allow for group lunches and errands.

3)   The 8 to 5 schedule was modified and changed to run from 7:45 a.m. to 5:15 p.m. This adds another 30 minutes to each day without getting into extremes of early or late hours.

4)   At this point one and a half hours have been added to the work day. Four days of this means an extra six hours is added, almost enough to allow the fifth day off each week.

5)   Eight paid holidays were traded by the employees in return for the routine of three-day weekends. If a holiday falls on Monday, the work week goes from Tuesday to Friday. Eight paid holidays total 64 paid hours per employee per year, or an average of sightly over one hour weekly in a 50-week year.

6)   The vacation week accumulated time was refigured on the basis of calendar weeks rather than five-day weeks. This freed another one to two days added work time per tenured employee. Finally, their previous allowance of ten days' paid sick time per year was reduced to six days. The five to six days added another 40 hours to the work year and completed the adjustment needed to get forty hours of work into four workdays with minimum disruptions.

An added selling point to employees was that travel time to work and back, averaging one and one-half hours per employee daily, was eliminated for the extra day off. That's a free-time gain for employees of an average of 75 hours per year.

Finally a rotation system was worked out so that someone would be on hand on Fridays to take messages and deal with copable problems for those who were out of the office.

As you see, the four day week doesn't require vastly longer work days and it can take any number of forms. Play around with the trade-offs and study the possible advantages/disadvantages. Many employees are extremely happy with the idea, **overhead is reduced** because paid holidays and some sick days are eliminated and, usually, **absenteeism and tardiness decline dramatically.**

FROM EMPLOYEES TO INDEPENDENT CONTRACTORS

A/E's reduce personnel management problems by farming out work to independent production consultants and service bureaus. Dropping the traditional employer/employee relationship benefits everyone. Independent contractor designer/drafters gain work hour flexibility and tax advantages. Offices contracting with them slash high in-house personnel overhead costs. Here's the picture:

Employees are more costly than ever -- but it's not due to straight wages. Employee-related taxes and benefits equal over 30% of base wages in most cases. Government paperwork and regulatory hassles are another added cost. And there's plain overhead; support services for employees have doubled in cost in the past ten years.

There's another high-cost factor that most people don't like to talk about. Architectural employees tell us they actually produce work only 5 to 7 hours out of an 8-hour day-- usually because of management problems. Whatever the causes, that's a 12.5 to 37% cost factor on top of straight wages. When all costs are tallied, farm-out contracted drafting becomes very attractive.

With contract arrangements, offices pay only for accepted work -- not just for hours spent in a drafting room. Fees may be by bid but are often negotiated.

When consultant drafters push a job through, they often earn a substantial profit. Some portion of payment may be held back to assure there aren't any major glitches in the drawings, and payments are normally contingent on client approvals and payments.

Jobs have to be planned with extra care before handing them to outside consultants. That's a plus for the office. The extra time spent planning the work reduces errors, redundancies and contradictions in the final drawings. Job supervision time and costs are drastically reduced. Delivery time is usually cut, partly because of the extra time given to job planning.

There are tax benefits for former wage earners who are now independent contractors. Many expenses that are not tax deductible for employees are proper business deductions for independent consultants. Some further advantages, tax and otherwise:

--The consultant must have permanent office space at home. Part of the mortgage or rent is then deductible, as are portions of the utility, insurance, phone, and other home-related expenses. (A consultant's home office should be the principle place of business, or used to meet clients, or be a separate structure. "Part-time" office space doesn't cut it legally.)

-- Costs of business travel and entertainment, tools, materials, office furnishings, subscriptions, continuing education, etc., become similarly deductible.

-- Consultants receive full payments--no income tax withholding. That means they don't lose hundreds of dollars of interest on tax money withheld, as salaried employees do. The independent consultant can establish tax saving investment pension plans that are not available to salaried employees.

-- Consulting is ideal for people who have special needs or problems with their working hours. Some skilled students do it, as do young mothers who are stuck at home for a while. Couples or teams sometimes share part-time consulting so they can pursue personal business that might otherwise conflict with an office routine.

Tax advantages are so attractive that the IRS sets strict rules against sham arrangements. For example, you can't fire someone and keep him or her at the board as a "consultant" -- even with mutual consent. The tests of a valid consulting arrangement must be met. These test are fairly simple and easy to meet by anyone entering into a genuine independent consulting practice.

57

The primary legal test is whether the consultant is truly running an independent business. The IRS wants to know:

--Does the consultant have all the trappings of an independent office such as business name, business license, business bank account, business telephone, stationery, advertising, charge accounts with suppliers, a billing and accounting system, contracts with contracting clients, files Schedule C forms with the IRS, etc.?

--Is the consultant or contractor responsible for the means of implementing the work? You specify final results; the consultant determines method.

--Does the consultant or contractor work exclusively for one firm? This isn't mandatory -- some consultants will contract with a single firm for six months or a year at a time -- but it's considered part of the picture.

--Is work actively sought from other firms?

--Does the consultant work in his or her own office, clearly separately maintained from the contracting office?

There is a long list of other tests used by the IRS and the courts in borderline cases, but these listed are the most important. For details from the IRS, ask for Circular C, "Employers Tax Guide," from your nearest IRS office.

WHO WORKS THE HARDEST?

The Institute for Social Research at the University of Michigan has spent years pursuing the question of who puts out the most effort on the job. There may be some surprises for you in the results below:

--The average female employee puts in more hours and more effort on the job despite lower pay, than male counterparts.

--The average male spends 52 minutes each day on breaks, trips to the rest room, etc.

--The average female employee spends 35 minutes in rest breaks.

--Researchers measure the effort people give to work by an elaborate "work effort scale." It shows women giving 112% the effort of that given by men.

--Craftspeople, people with high monthly incomes, and young males spend the most time in breaks, scheduled and unscheduled.

--Among women, the unmarried ones take the least break time and score highest in the work effort scale. Aside from unmarried women, part-time workers, professionals, and union members put out the most effort per hour at their jobs.

CHAPTER FIVE

THE IDEAL DRAFTING STUDIO

HOW TO CREATE A GREAT DRAFTING ENVIRONMENT

Question: What makes a great drafting environment?

Answer: A learning environment.

A few top production rooms are specifically designed and managed as learning laboratories. The employees treat working there as a privilege and behave accordingly, with unbeatable, high-energy productivity.

It takes a trivial dollar investment to turn a design and drafting studio into a learning lab; just a little extra attention by those in charge.

The ironic thing is that most design offices are schools anyway. They're just not run that way. In most offices, the advantages of the total learning opportunity that are already in place for staff and management are completely overlooked.

Look at all the parallels between an office and a school setup: There are supervisors (instructors), a library, assignments which require research and learning by staff members, A/V equipment and seminar (conference) rooms and, usually, a large percentage of young people in the process of learning their trade.

Education in the office is usually considered a headache for the employers. "Why do WE have to be responsible for teaching these kids?" complains the boss. But it's a headache only if it isn't seen in a more positive light, as a distinct opportunity to have much higher productivity and quality control than other offices.

Here are the most visible features of the best design and production studios in the country, in ten lessons:

1) The office physical environment is used as an instruction tool.

--Feet and inch dimensions of the office, including doorways and windows, are painted as large-size measurements on the wall so drafters learn to eyeball real dimension sizes more accurately.

--Office working drawings are exhibited on the walls in the same way renderings and design drawings are usually shown. (It's almost shocking to see an office that's so conscientious about their working drawings that they put them on display.)

--Sometimes the actual framing behind the walls and ceilings is drawn or painted on those surfaces for decorative as well as instructive purposes. A few design firms and schools have deliberately left parts of rough construction exposed, for visual impact and to show how the place was built.

--Several firms use changeable wall panels and/or posters to illustrate the best practices of construction and detailing. They make giant blowups of pages of books and manuals at photo poster stores to use as displays. The exhibits are changed monthly and follow themes such as Concrete, Masonry, Wood, etc. The blowups show basic components, sizes, detail practices, and top examples of office work. Enlarged illustrations are sometimes taken from office drafting standards and drafting books such as **Wakita & Linde, Ramsey/Sleeper, Systems Drafting, Systems Graphics,** and the **Northern California AIA Production Office Procedures Manual.**

2) They give a helping hand to employee professional development.

--It's common for A/E firms to encourage employees to take the licensing exams by paying exam fees. Many go further and pay for in-house seminars or paid courses with exam preparation tutors.

--Some offices provide a complete exam study library: Handbooks, practice exams,

NCARB books, and all the basic references, such as the Steel-Wood-Concrete-Masonry handbooks and building codes. And all are tabbed as the most relevant sections for easy reference

--A few firms let employees schedule evening study hours in their libraries. Knowing that many people study best with others, they encourage employees to create after-hours study groups. Some foot the bill for take-out lunches or dinners when employees do in-house seminars, study sessions, or lectures.

--Many firms buy, rent or borrow continuing education course materials from their professional societies. The CSI, AIA, and engineer societies all offer valuable courses and/or course materials such as manuals, audio tapes, and video tapes.

--They send employees to professional society study courses that come to town. The CSI spec writing certification course is a favorite, for example.

--They send employees out of town to university continuing education programs, such as the extremely successful short-course programs on specs, construction failures, project management, and production techniques offered by the University of Wisconsin Extension in Madison. Most smaller firms can't afford to send employees to such sessions, but some provide paid time off, share the cost in some equitable way, or loan the money at no interest.

3) They run their library as a support service for employees.

--A few offices encourage employees to contribute books to their libraries and they allow check-out for home and lunch-time reading.

--They make photocopies of the best current journal articles and hand them out as giveaways.

--They circulate professional newsletters and do so fast enough that people can read them while they're still current. (An alternative is to post newsletters and have them checked out for quick reading, or just photocopy them for in-office circulation.)

4) They hold in-house workshops with star speakers.

--Virtually all the best-known A/E writers and speakers have been hired to do in-house half-day, full-day, and even multi-day workshops or trainings. I've done it for more firms than I can count, as has Stu Rose on his highly personalized and effective marketing methods, Ernest Burden on design presentation, Weld Coxe on management, Gerre Jones on cold-call marketing, Chief Boyd on microcomputer CADD management, and so on. Offices may pay $1,000 to $2,000 in fees for a day's program but can split the cost with other firms or even charge others. (Several small firms who "couldn't afford" a full presentation have paid speaker's fees plus expenses and ended up making money by charging others to attend.)

5) They give unusually attentive management support. That, too, promotes learning.

--The "Buddy System" prevents a lot of problems common to newcomers in larger firms. In essence, a senior member of the firm is assigned to be mentor for any newcomer to the office: Introduce him or her around the office, have a lunch to explain office policies and expectations, and then just be available to answer questions and provide counseling from time to time. This level of support often makes or breaks the office experience for younger staff members. It helps break down the all-too-common "Us vs. Them" barriers between lower echelon staff and management, and does incalculable good for overall morale and employee support of the office.

--There's a confidential ombudsman system in a few larger offices. The bosses may say: "My door is always open," but most employees will never cross the threshold. The problems are often too sticky to tell anyone but the spouse and outside friends: Racial or sexual harassment; a project architect with a drinking prob-

61

lem who is running a project into the ground; favoritism; . . . the kinds of things that employees think the boss really doesn't want to hear about.

6)   Some use an Answer Box, a sort of in-office "Information Please." "How wide should the toilet plumbing chase be for a double-loaded chase?" "Is thicker sheet-metal flashing better or not?" "How much shim space should we allow for wood frame window rough openings?" These are the kinds of questions that get bandied about every drafting room, but too often there is no ready answer. So drafters guess, or spend time looking in the wrong handbooks, or ask the wrong person and learn the wrong answers. Solutions:

--Some supervisors ask that all such questions be posted on yellow Post-It notes on the edges of drafting boards or CADD terminals. They pick them up and get the answers back ASAP.

--One office kept an "Answer Box" for awhile for employees to drop off the most common technical questions that came up. The production manager compiled the questions and answers in an office version of Graphic Standards.

--Some firms have always had their own versions of Graphic Standards. They collect the technical reference data that particularly deals with the projects, building systems, and products they use most often (such as planning standards for schools and hospitals, Sweet's product data), photocopy the information, and distribute copies for employees to keep in three-ring binders. It cuts way back on employee search and research time.

7)   Many now use an Office Data Base and Jobsite Feedback System. All offices have some sort of Corporate or Collective Memory, it's just usually kept inside people's brains. When those people aren't around, or when they leave the office permanently, that part of the office memory goes with them. So some firms attempt to salvage the best of office wisdom with devices such as the Design Database and Management by Checklist systems described in this book. They also all happen to be excellent re-

sources for staff education. **For example:**

--Standard Details are the best instruction drafters can have in the do's and don'ts of detailing. (Of course some offices teach the don'ts unintentionally with bad details. But with a detail file, at least people can learn from the mistakes. Without a detail file that's open to the whole office, bad details just remain buried in old working drawings, like time bombs that will inevitably resurface to do more damage.)

--Standard detail systems come alive when an office insists that all visitors to jobsites take pictures of any trouble spots they see and/or provide some notes. Then whoever is in charge of the system has to take those pictures and notes and make sure details are upgraded to match.

--Construction and postconstruction trouble spots are a further resource when they're photocopied and made available in the office manual: "The headroom at the landing of this stair is so low that it'll give free lobotomies to users for years to come." "These cracks in the pavement are trying to help us make up for the movement joints we didn't design." Remarks like that lighten the subject and make the lessons more vivid and easier to remember.

--Standard notation teaches in the same way as standard details do. And like the details, sometimes by bad examples as well as good.

--Office design, production, and construction administration checklists are part of the act, too. We've said plenty about those as supervisory and management tools. What's not said is that they are teaching tools. They teach the management process to subordinates by breaking all such processes down into discrete, easily understood chunks, and listing them in the most desirable sequence of action.

8)   Computers and microcomputer CADD are self-taught in quite a few offices.

--When offices cannot afford extensive CADD or computer trainings, they sometimes offer small computers (Macs or Compacs) for employees to checkout for weekend practice.

--All manner of instructional videos, audio guides, and self-instructing software are available so that most of what newcomers need to know about computer use can be self-taught. A few offices stock these in their in-office libraries for employees to check out.

--Computer lecture tapes from professional societies and conferences, such as A/E/C SYSTEMS, are much used by management for learning while driving. A few pass on their "used" tapes to the office library so the staff can gain the same benefits, as noted in the next item.

9) Audio and video tapes for management, spec writing, construction, project management. The best office libraries now include instructional tapes of all sorts--from handyman instructions for home remodeling (which are **great** for showing inexperienced staff the rudiments of construction practice) to time management. Staff and management are encouraged in some firms to donate their instructional tapes to the pool since, once used, they tend to gather dust.

10) Objectification of the promotion process promotes learning, too.

--Many firms now have specific job titles, openly publicize their pay ranges for those titles, and publish the experience and responsibilities that go with the titles. No one pretends it's easy to manage this kind of openness but it helps employees to no end and is very encouraging to self development and continuing education. (A Dallas A/E firm has published a list of jobs and a comprehensive list of what they expect employees to know to qualify for the jobs. Once an employee demonstrates skill and knowledge adequate for a formal proposal, they only have to wait until the job slot opens.)

--In-office promotions are infinitely preferable to frequent hiring of new personnel to fill higher job slots. The latter practice is typical of high-turnover, high-overhead, and low-morale firms. When an office keeps personnel in the dark as to what's necessary to move up, and then as soon as a new project opens up, starts to hire new people in better positions . . . it poisons the entire office atmosphere. Published job slots and in-office promotion fit hand-in-hand with in-office education systems and commonly go together in the most desirable offices.

REPLANNING YOUR DRAFTING STUDIO

"Work is mainly moving things from one place to another."

"Efficiency is moving the right things to the right places faster."

Those descriptions point directly to what we need to do to simplify and speed up our work efforts. Consider **Webster's New World Dictionary** which shows that the word "work" comes from various ancient European terms ("werk," "weorc," etc.) which refer to "doing" and "action." "Doing and "acting" refer to moving, putting, setting, or placing.

"Moving things around" isn't just manual labor; it's also abstract, high-level activity. "Things" can be words, ideas, attributes. Teachers move data from their source to the minds of the students. Writers compile ideas, compare them with other ideas and move them around until they click together. Designers and drafters are in the idea and information business and they're valued not only for how good their output is, but how fast.

If work and efficient work involve moving things from place to place, then efficiency problems are mainly placement problems -- problems of things being in the right place at the right time.

With this in mind, consider what most design studios and drafting rooms are like:

--**Workstations, drafting boards, and desks are a barrier to work, not a help.** There are no table-top holders for commonly used supplies and tools -- they're piled in drawers and subject to constant rummaging and upheaval.

--**Reference data is out of reach.** When shared with many staff, it's out of place or just "out." There aren't enough individually accessible reference sources, or they're not the best of their kind, or staff members don't know about them.

--**Drafting room floor plans feature flat files at one end of the office and print machines at another.** There's constant foot traffic back and forth just to get reference prints.

--**The job files and other storage areas don't group similar items together and none of it is close to the people who need it most.** Most drafting rooms are planned at one time or another but their current arrangement is usually a random accumulation of numerous unrelated changes over time.

--**Phones are not at the drafting or CADD workstations.**

--**Key people who share project work are separated from one another.**

--**Lighting is at the ceiling --** maximum distance from the work surfaces it illuminates.

Conversely, some unwanted facilities and items are in close proximity to drafters. The most common complaints about drafting station locations are that they're too close to air outlets or intakes; or too close to doors, restrooms, halls, print machines, transformer closets, common phones, and other noisy or much used distractions.

Recommended: Make a spatial interaction diagram of the key elements in your design and production studio. List all the major components and plan those that have the most interactions to be closest together. List those which have negative relationships so that they can be separated.

The best general rule: Items used the most by the most people should either be at every workstation or be most centralized in the workspace. Thus flat files and quick print facilities should usually be at the center of a drafting room, not at one far end or the other.

THE MOST COMMON WORK PLACE PROBLEMS ARE EASY TO FIX

A Lou Harris work place survey has found that 74% of office employees are hamstrung by easily correctable working conditions.

The most common and troublesome barriers to getting work done according to employees surveyed:

___ Inadequate instructions for work assignments.

___ Poor lighting.

___ Inadequate conversational privacy.

___ Inadequate work surface space and storage.

___ Poor access to other areas and departments.

___ Lack of convenient access to tools, equipment and materials.

These simple items block workers at every turn. They start tasks with tight deadlines and then discover management is playing hide-and-seek with vital information and tools. They make errors and redo work because of poor communication, inadequate lighting, or improper work tools. The pace of work is stop-and-go, with little chance to get up momentum because of cramped work spaces and too many distractions and interruptions from others.

If typical work places are that bad, why doesn't management change them? Mainly because of the dominant problem: communication. In the Harris survey, 63% of the executives thought employee satisfaction had improved over the previous five years. Seventy percent of the employees said otherwise.

Even bosses who seek to change the work place get on the wrong track. They consult everyone but the workers and decide that remodeling and refurbishing will enhance working conditions. The employees disagree once again. **They say they don't mind amenities, but what they need most are the items listed in the survey:** clear instructions, adequate work space, some privacy, and easy access to the people, the tools, and the materials necessary to do their jobs.

OPEN DRAFTING STUDIO OR SEMI-PRIVATE DRAFTING CUBICLES?

Researchers at the University of Tennessee did a clever piece on work stations. They first asked office personnel to rate their work spaces in terms of privacy and noise level.

The employees evaluated their satisfaction with their spaces, rating them as "good" or "not-so-good" places to work. Not surprisingly, semiprivate cubicle work stations got the highest rating.

Then supervisors were asked to rate the work performance of the various employees who had been surveyed. In general, the employees who had the semiprivate cubicles got the highest ratings in work performance.

Obvious conclusion: Good work stations that provide some privacy and distance from others improve worker productivity.

Supervisors say the same about drafting rooms: Cubicles work best. What's the best height? The consensus: 4-1/2' to 5' is the preferred height. That's high enough to provide privacy during work, and low enough to allow across-the-room visibility when standing up.

The best cubicle designs include controllable work surface task lighting, shelves, pin board, file drawer, electrical strip outlet, and telephone.

Nominal plan dimensions are: 3' wide and 5' to 5-1/2' long drafting space, with comparable size reference space beside, or behind, the drafting space.

The most noticeable difference between a cubicled drafting room and an open room is the comparative quietness. Employees do much less random socializing, and less board hopping. Those who want to get right to work can do so without having to stop and chat with co-workers. So the noise level is down, and the intensity of concentrated work is up.

Cubicle design can help with touchy personnel problems. People often get very upset when they have to be moved to a different work station. A few offices have unbelievable hassles and competition for favored spots. Some have dealt with the problem by designing completely identical and movable work cubicles. When personnel must be moved, they can bring their whole drafting station with them: the files, shelves, etc. they're used to. That doesn't eliminate all such conflicts, but it can drastically reduce them.

DRAFTING STUDIO CLUTTER AND "MAKING SPACE"

Drafting room clutter -- all office clutter -- has one primary cause: Horizontal work surfaces are used for storage. Horizontal storage is the most inefficient storage there is.

Here's what to do about it:

--Tag all horizontal surfaces as WORK surfaces. In the worst offices this includes the floor as well as tables and tops of bookshelves.

--Have someone tackle each pile on a work surface and categorize what's being stored flat -- letters and memos, magazines, drawings, photos, specs, materials samples.

--Install vertical file folder holders for desktop storage of all current basic office paperwork.

--Use the list of categories of small paper items -- letters, memos, etc. -- to make labels for vertical file folders. Use a color code for overall classifications of data -- current projects, personnel, financial, etc. -- so that when a folder is left lying about, it's quickly obvious where it should go.

--Obtain vertical file holders for drawings. Recommended: Plan-Hold spring friction files that keep drawings upright in very large folders. It takes far less space to store drawings this way than in flat files and the files are very easy to get to.

--Install shelves to vertically store all books and magazines that are currently stored horizontally. Books, manuals, specs, etc. stored horizontally on shelves can usually be better stored and accessed when reset vertically.

--Box all "semi-obsolete but not quite" items for "semi-long term storage" in a back storage space -- accessible but out of sight for now.

--Keep spare folders and tags right at hand so that it's as easy to start a new file for vertical storage as it would be to start piling up the material.

--If worried about losing track of boxed data in long-term storage, such as boxes loaded with project manuals, old books, etc., take Polaroid shots of the contents, and note the box number and location on the snapshot. Make a "storage" file folder to record where you've stashed such items.

A FIRST PRIORITY -- THE TECHNICAL LIBRARY

The first main task for any production manager is to create an office "Main Library" that goes beyond the usual a technical library. That's the advice of production consultant Morry Wexler.

This is in addition to the usual book shelf of manufacturers' catalogs and technical data. The new "MAIN LIBRARY" contains the Reference and Standard Detail Libraries, Operating Checklists, technical reference books, AIA Handbook, CSI specs, etc. Furthermore, according to Morry:

"The Main Library should also contain, preprinted on vellum, your office's Standard Symbols and Conventions; reference symbols, architectural material symbols . . . etc."

Interestingly, a lot of offices don't use a single set of symbols and conventions. The lapse leads to endless little inconsistencies, misunderstandings, and errors.

"The Main Library should contain, preprinted on vellum for hand drafting and on bond for typing, your office's standard modular paste-up SCHEDULES: 'Room Finish and Material Schedule,' 'Modular Casework Schedule,' 'Partition Schedule,' etc.

"The Main Library should also contain a collection of DRAWINGS OF PREVIOUS PROJECTS (schematic, design development, and working drawings) and SPECIFICATIONS for ready reference. 'What quarry did that marble we used on project 674 come from? Their rep was pretty good. Let's look it up in the specs.' It happens all the time.

"The Main Library should also contain SAMPLES -- not only carpets, paint chips, tiles, etc., but also the technical stuff that many of our young architects know very little about. I refer to nails, screws, bolts, anchors, door hardware,

metals of various types and finishes, expansion joints, control joints, seismic joints, roofing systems, insulation, etc. It's surprising what knowledge a sample library contains, and how useful it is in designing details."

Amen! We've visited numerous university architectural schools and only found one with a display of actual building components such as Morry describes. That was at the Polytechnical School in Barcelona, Spain. There must be more, but who knows where? Meanwhile, virtually none of the architectural offices I've ever visited have such a resource. It takes up space and that's a problem, but what a benefit!

Continuing:

"The material in these libraries, assembled and formalized into one central area, not only provides technical data to the staff, but makes the data easily and quickly available. If the data is not easily available, people may not take the trouble to look it up and may speculate instead.

"Result: error and/or omission. . . . All too often the same data are developed and drawn over and over again because similar data from previous projects was too hard to find, and once found, it was too hard to reproduce. Hence, ready to use, easily available, easily reproducible technical data are essential."

MANY OF THE BEST TECHNICAL REFERENCE SOURCES AREN'T WELL KNOWN

Most detail and drafting text books are either obsolete or stuffed with examples of the worst and least efficient drafting processes. And most of the large graphics and materials standards books are loaded with technical boilerplate that serves no purpose for the designer or drafter. A **few** books deal with specifics of drafting, design, and materials Here are some of the best of recent years. Inquire with your book dealer or contact the listed sources for latest editions and prices:

Construction Sealants and Adhesives, by Julian R. Panek and John P. Cook. John Wiley & Sons. ISBN 0-471-09360-2.

Construction Disasters--Design Failures, Causes & Prevention, by Steven S. Ross. McGraw-Hill Book Co. ISBN 0-07-053865-4.

The Professional Practice of Architectural Detailing, by Watika & Linde. John Wiley & Sons. ISBN 0-471-91715-X.

Roofing Concepts & Principles--A Practical Approach to Roofing. Paul Tente, Paul Tente Associates, Box 6819, Colorado Springs, CO 80934.

NRCA Roofing & Waterproofing Manual (Includes NRCA recommended details). National Roofing Contractors Association, 8600 Bryn Mawr Ave., Chicago, IL 60631.

Avoiding Liability in Architecture, Design & Construction, edited by Robert F. Cushman. John Wiley & Sons, Inc. Wiley-Interscience. ISBN 0-471-09579-6.

Structural & Foundation Failures, by Barry B. LePatner & Sidney M. Johnson. McGraw-Hill Book Co. ISBN 0-07-032584-7.

For all manner of construction design, drafting, and spec aids we highly recommend:

Directory of Publications of Loss Prevention Aids. (Free list of books, manuals, video and audio programs. These are helpful to all design professionals, not just the ASFE members.)

Association of Soil and Foundations Engineers:
811 Colesville Rd. Suite 225
Silver Spring, MD 20910

The CSI Catalog of Services and Publications (Free catalog of manuals, self-study aids, cassette tapes, and model specifications sections and technical aids. A priceless source of data.)

Construction Specifications Institute
601 Madison St.
Alexandria, VA 22314

BRP Publications "Reports for Sale"

National Academy Press
The Building Research Board
2101 Constitution Ave., N.W.
Washington, DC 20418

National Research Council of Canada "List of Publications" (Lots of extremely useful research data on construction of all kinds.)

Publications Section
Division of Building Research
National Research Council of Canada
Ottawa, Ontario, Canada, K1A 0R6

DRAFTING STUDIOS TURN ON THE LIGHTS

"Let's clean up the office. I want all those drafting lamps out of here." So said an architect to his technical staff after a new ultra-bright ceiling light system was installed.

After a couple of years, the desk lamps are back and the ceiling lights are out. The high-level fluorescent environment met accepted lighting standards, but it also created an all-pervasive brightness; a pasty, washed-out visual environment; veiling reflectance from computer screens, tracing paper, and drafting tools; and uncontrollable shadows on work surfaces. It drove the employees crazy.

Arguments for getting lights out of the ceiling are persuasive. Connecticut lighting consultant Sylvan Shemitz and the Interspace company have listed them:

1)   **A company's tax write-offs accelerate and investment credits may be permitted for non-integrated lighting fixtures.**

2)   **Total floor-to-floor space is reduced in new construction when the ceiling lighting plenum is eliminated.**

3)   **Electric lighting power consumption can be cut in half.**

4)   **Air-conditioning loads can be reduced by about 25%.**

The arguments aren't just speculation. Successful reforms were first made years ago in the widely publicized ARCO Philadelphia headquarters building. Now it's common for office lighting to be integrated with desks, work tables, filing cabinets and office partitions. Ceilings are left bare except for low-level background lighting. The result: Overall lighting is pleasantly subdued. Work surface lighting is concentrated and excellent. And despite the higher cost of incandescent lighting, there is a new energy cost savings because of overall reduced electrical and heat loads.

. . . AND TRASH THEIR FLUORESCENTS

Fluorescent lighting becomes a clearly noticeable nuisance as design offices install more computers. The fluorescent lights flicker, the computer screens flicker, and the combination is a major peripheral vision irritant. (Peripheral side vision is much more sensitive to fast movements such as flickering than head-on vision. The irritant is there but it's often not directly sensed, and that increases general eye and body stress.)

There are persistent reports of nervous reactions to prolonged exposure to fluorescent radiation. Filmed observations of school children show an extreme difference in hyperactivity among certain kids depending on room lighting.

If it's not practical to unhook the fluorescents entirely, at least they can be dimmed and combined with task lighting. This is common in design firms that rent space in buildings, especially in their CADD work areas and areas near perimeter windows.

If it's not possible to undo the fluorescent lamps, at least install "full-spectrum" tubes. They emit a range of color spectrum that approximates that of sunlight and reduces the washed-out quality that standard bulbs impose on environment.

One point that most top interior designers and environmental lighting consultants agree on these days: Give the individual as much personal control over his or her workstation lighting as possible. All people have unique visual sensitivity and they're the best judges of what they need in order to see and work effectively.

THE MYSTERY OF THE VERTICAL DRAFTING BOARDS

Many drafting rooms look like a scene from The Night of the Living Dead.

Bent and twisted bodies drag themselves across large dead-flat drafting boards. Some climb on stool rungs or crawl atop the stools themselves, stretching vainly to reach the furthermost portion of their drawings.

Those who have been drafting the longest show it the most -- rounded shoulders, hollow chests, hunched backs , . . . they look like coal miners who have spent their lives in under sized tunnels.

What's the justification for using flat built-in drafting tables? We asked several employers and managers in San Francisco offices.

"It's the look of the thing. I won't have the clutter of tables at all different heights and angles," said one architect. Another architect concurred: *"It looks nicer with everything uniform."* A structural engineer looked out over a sea of flat doors on saw horses and said: *"I don't know why. It's just always been this way."*

We don't have statistics on the health hazards of hunched-over flat board drafting. But drafters in non-adjustable board environments have plenty of anecdotal comment, mainly in the vein of: *"My back hurts . . . ." "My sides hurt . . . ." "Back here around my kidneys. . . ."*

Manufacturers of vertical drafting boards naturally have very positive statistics about happy drafters who can draft better but, being self-serving, those statistics are suspect. Suffice it to say that anyone who uses a flat board for very long starts to hurt, gets tired, and has to stretch, sit back, or walk around to get some relief.

A board doesn't have to be fully vertical, but some adjustability helps enormously. All users of angular or vertical boards we've talked to agree they're more comfortable and conducive to more efficient work. Some users didn't want to change but were compelled by doctor's orders.

And speaking of "all users" . . . virtually the entire world uses adjustable or vertical drafting boards -- Europe, Japan, Latin America . . . everywhere in Canada. The United States seems to be the only major non-third world nation that still uses mainly flat drafting tables.

The arguments against easier-to-use drafting boards (other than appearance)?

"The parallel bars slip." Answer: They don't slip unless you're using a high angle or vertical board. And a vertical board is even better for most drafting because you can use a bar drafting machine. *"Drafting machines aren't good for very long horizontal lines."* Answer to that one: A minute percentage of drafting involves long horizontal lines; most line work is short and is expedited by combining the drawing straightedge and scale into one single tool.

"They cost money." That's the capper for most A/E firms which traditionally have refused to look at the real cost of non-capital investment to improve productivity of their employees.

"I've seen those drafting tables in Canada and Europe, and they're overkill," says a New York City architect in reference to the powered pedestal drafting boards, the kind with foot pedals reminiscent of barber chairs that allow automatic adjustment of height and slope. He says: *"We made our own adjustable tables which allow plenty of slope by manual adjustment for 1/5th the cost of the electric boards."*

And about the drafting machines or vertical bar drafters . . .

Every fully professional drafter we've ever known has insisted on using them, sometimes in conjunction with an auxiliary drafting station with a parallel bar.

The magic of the vertical drafter or any drafting machine is in its integration of functions. Just combining the straightedge and scale saves several steps on every scaled line that's drawn. Instead of scaling to a start point, marking it, scaling to the end point, and then setting down the scale and shifting the parallel bar to draft it ... you skip most of those steps by just laying down the scaled straight edge at the start point and drawing the line. Users get added savings by not hunting and positioning triangles. It cuts out half or more of such moves. It even eliminates brushing of eraser crumbs. Small things, but they're small units that add up to much of the time-drag of all manual drafting.

We did our own statistical comparison of productivity of drafting machine or bar drafter vs. parallel bar and triangles years ago -- it's at least 8% faster for draftspeople on all routine drafting chores, more so for many individuals. Assuming wages plus perks for a full time draftsperson add up to $500/week, an 8% improvement saves $40 a week, or $2,000 during a 50-week work year.

Vertical drafters, drafting machines, and adjustable boards are usually highest in price at the local drafting supply stores. You can usually do better from mail order discount houses. Sometimes you'll find an off-brand model in art or drafting supply stores for a couple of hundred dollars. (For small manual drafting -- 11x17 sheets -- or small sheet work on the road or on jobsites, we recommend the **Otring** integrated portable drafting board from $40 to $60.)

Although U.S. firms have been the slowest to change, many are now at least thinking about it. CADD and non-CADD firms know that manual drafting will be with them for some years to come and while the flat boards may "look good" to some, draftspeople and designers who can sit up straight and walk, without looking like evolutionary throwbacks in a horror movie, look even better.

LIGHT TABLES THAT ARE INDISPENSABLE AND USELESS AT THE SAME TIME

An anonymous suggestion arrives in the mail:

"Light tables are great for cross checking engineer's drawings, paste-up, and CADD layer checking, but at our office they became useless because everyone piled old prints and Sweet's catalogs on them. It was easier to use a window than to unload all the trash off the light table."

"Since people were using the window anyway, we mounted the light table vertically too, on the wall. With clips and pin bar at the top, a ledge on the bottom, and a slight slope, it's better than before and now we have more file space."

Good move! In fact, that's how many graphic arts studios and print shops also place their light tables; and for the same reasons.

Two reminders:

1) For most efficient utilization, provide about one light table for every six design and drafting staff.

2) You don't have to shell out big bucks. A plywood box, cabinet fluorescent fixtures from the hardware store, and a glass top with spray or taped matte plastic under the glass to disperse light will do the trick for a couple of hundred dollars instead of the $1,000 plus that some suppliers charge.

NOT MUCH RESPECT FOR THE DRAFTING STUDIO

--If the drafting room is where life-and-death decisions are to be made, let it be treated as a place of importance, not the "back room."

--If the decisions and craft expected in working drawings are to be of the highest professional quality, let there be generous investment in the equipment and continuing education needed for people to do their best work.

What would you think of a profession that gave the **least** prestige and attention to the aspects of its work that required the most personnel, work space, time, money, and general office resources?

Or what of a profession that gave the **lowest** educational priority to aspects of its work that were most responsible for time and cost overruns, lawsuits, property damage, and even deaths?

Looney, eh?

But we're professionals whose leading architecture schools consider it a point of pride that they are NOT "trade schools." It's not just a matter of neglect or indifference; it's a matter of prestige for the top schools that they are NOT training students about how to prevent leaking roofs and how to keep their buildings standing upright.

If medical schools and law schools declined to teach the actual craft as well as the art of their professions it would be a national scandal. If the architectural profession continues its schizophrenia -- prestige and pride in appearances, and embarrassment over technical expertise -- it will become a public scandal for us. Our profession may pay a terrible price in terms of lost public trust.

Think it over:

--If drafters are to maintain enthusiasm for their jobs and their skills, let their work be granted full prestige and recognition from everyone in the office.

MORE ENHANCEMENTS FOR THE STUDIO

Here are some more steps offices have taken to upgrade technical staff recognition and working conditions. They all have high impact, way out of proportion to the minor costs involved.

--In-house seminars with production managers, consultants, spec writers, contractors, etc. to upgrade employee skills and knowledge.

--Wall displays of the office's best drafting, just as renderings are traditionally displayed.

--Acknowledgment of designers and drafters who complete a job with a roster of participants on the working drawing title sheet, and full names of drafters on the title blocks, not just their initials.

--Names of all tenured personnel on the office stationery. This is admittedly most practical for offices that can whip out new letterheads quickly via desktop publishing, or those with low turnover. Also increasingly common now: Business cards for all studio staff members.

--Public "thank yous" to project teams when they meet or beat deadlines and budgets. A few firms we know in New York City have always provided Friday lunches for teams to celebrate successful phase completions, and dinners to mark job conclusion.

--Periodic open-house receptions for employees' families and friends. It's a chance for spouses to meet boss and co-workers and see the work being done. It tells employees that they're appreciated.

--Small gifts, especially trophies for jobs well done, are great low-cost morale boosters.

--Personal home-delivered subscriptions to at least one technical-oriented journal.

--"Technical article" three-ring binders for all employees, followed with photocopies of the best technical articles that are published. And a customized office version of "Graphic Standards" as three-ring binders to employees, with inserts of handbook charts, code sections, product literature, etc. . . . whatever is most useful in day-to-day operations.

--Upgrades in the workstations, including simple touches such as a name plaque at each station.

--Field trips to jobsites. There's always a big interest in how design features and details actually work out on the job. Field trips are one of the best forms of education possible and are another form of recognition.

CHAPTER SIX

USER-FRIENDLY WORKING DRAWINGS

CONVENIENCES AND AMENITIES TO ENHANCE YOUR WORKING DRAWINGS

As sure as night follows day:

First, bidders discover they have to turn a set of drawings inside out to find information.

Second, they give up after awhile and start writing a list of potential extras.

It can be simple things they can't find, like which way is North? What's the project address?

Or bigger things, like how far is the building from the property line?

Here are some of the most common unanswered questions in working drawings:

--Are plan dimensions to framing or to finish surfaces?

--What does "North Elevation" on an exterior elevation drawing mean? Is the building facing North or is it the observer?

--When the drawing index says FLOOR PLAN, CROSS SECTIONS, etc., **what else is on those sheets?** What details or schedules are also included?

--When there's an isolated or unusual detail, where is the original reference to it in the broadscope drawings?

Here are uncommon examples of a few of the best convenience features we've seen added to drawings to help expedite plan checks, bidding, and construction:

--Single-purpose architectural floor plans such as plans with space ID and framing dimensions only. Another plan shows door and window symbols, finish keys, detail keys, construction notes, etc. In some cases, such as for medical buildings, another floor plan may be dedicated to identifying casework and special equipment. Each such plan is relatively uncluttered, more readable, can be of smaller scale, and follows the natural sequence of construction.

--Evaluation questionnaires and a return envelope with bid sets to get suggestions for improvements from bidders. The design firm gets free advice and the bidders are put on notice that the design firm actually cares about the clarity, organization, and usefulness of the documents.

More convenience features are illustrated on the pages that follow:

REFERENCE DATA TO ENHANCE WORKING DRAWING CLARITY

Below is an example of an architectural drawing index reduced and reproduced on every drawing. This cuts out the constant turning back and forth from one sheet of drawings to the index in order to find another sheet -- the user can just go direct.

DRAWING CROSS-REFERENCE INDEX

ARCHITECTURAL
A0.1	PERSPECTIVE
A0.2	GENERAL INFORMATION
A1.1	SITE PLAN
A1.2	IRRIGATION PLAN
A1.3	LANDSCAPING/PLANTING PLAN
A1.4	SITEWORK DETAILS
A2.1	FLOOR PLANS & FINISH SCHEDULE
A2.2	ROOF PLAN
A2.3	FLOOR PLANS, DOOR/WINDOWSCHEDULES
A2.4	DETAIL PLANS & INTERIOR ELEVATIONS
A3.1	EXTERIOR ELEVATIONS
A3.2	CROSS SECTIONS
A3.3	WALL SECTIONS
A4.1	EXTERIOR DETAILS
A4.2	EXTERIOR DETAILS
A5.1	REFLECTED CEILING PLANS & CLG. DETAILS
A5.2	STAIRS & MISC. DETAILS
A5.3	CABINETS & INTERIOR DETAILS

STRUCTURAL
S.1	FOUNDATION PLANS
S.2	FOOTING & SLAB DETAILS
S.3	FLOOR FRAMING PLAN & DETAILS
S.4	WALL FRAMING DETAILS
S.5	ROOF FRAMING PLAN & DETAILS
S.6	TRUSS ELEVATIONS
S.7	ROOF DETAILS

MECHANICAL
M1.1	SITE UTILITIES
M2.1	HVAC PLAN
M3.1	PLUMBING FLOOR PLAN
M3.2	PLUMBING ISOMETRICS

ELECTRICAL
E1.1	SITE & LANDSCAPING LIGHTING
E2.1	CEILING FIXTURE PLAN & DETAILS
E2.2	ELECTRICAL POWER PLAN

A key plan symbol printed on all relevant drawings to clearly show which face of the building is shown in exterior elevations, or what part of the floor plan is covered on a particular sheet.

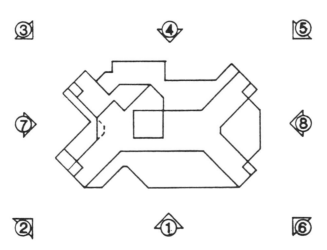

A clearly illustrated general note about floor plan dimensioning to framing.

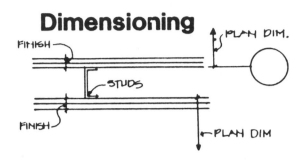

Dimensioning

ALL DIMENSIONS ARE FROM FACE OF STRUCTURE

Two examples of Ceiling Heights schedules. When these are included on reflected ceiling plans, the contractors don't have to search through cross sections to figure out the heights.

CEILING HEIGHTS

1	8'-0"
2	8'-6"
3	9"-0"
4	9'-4"
5	9'-4" W/8'-0" AT DOOR
6	9'-6"
7	10'-0"
8	BOTTOM OF JOIST
9	EXPOSED
10	MATCH EXISTING
11	7'-6"
12	VARIES
13	EXISTING TO REMAIN

CEILING HEIGHTS

A - EXPOSED

B - 7'-8"

C - 8'-0"

D - 8'-6"

E - 8'-9"

F - 9'-0"

G - 9'-6"

H - 10'-0"

ANOTHER LEVEL OF CONVENIENCE

SMALL-SCALE DRAWINGS THAT SHOW EXACTLY WHAT THE DETAIL DRAWINGS REFER TO

In this example, the wall section drawings are keyed to details that are on the same sheet. They unmistakably show the relationship of construction detail to the overall building

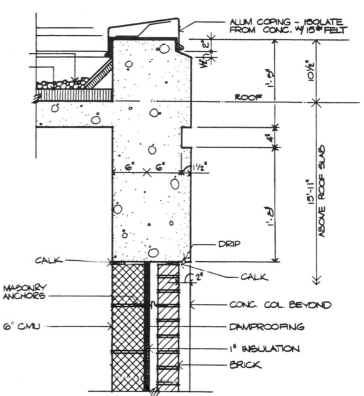

14 DETAIL PARAPET AT MECH ROOM

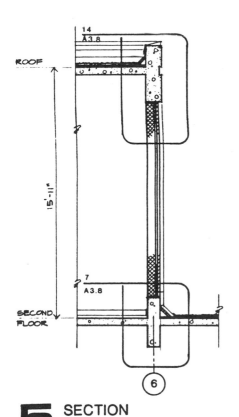

5 SECTION

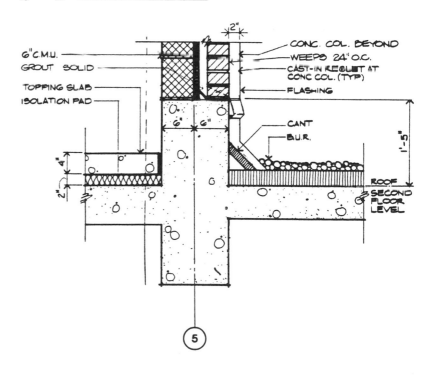

7 DETAIL

A small-scale stair section keyed to the larger detail. In this case it wouldn't have been necessary to show the reinforcing and concrete indications in the small-scale drawing since that just duplicates the larger scale drawings.

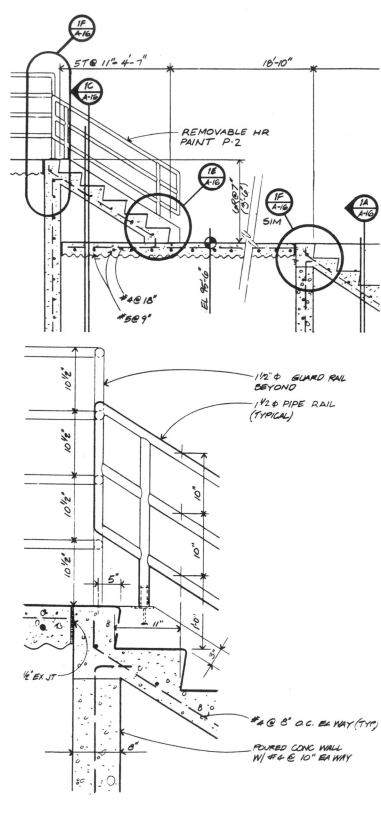

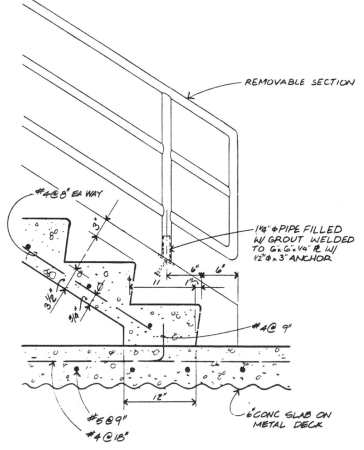

REMOVABLE SECTION

#4@8" EA WAY

1¼"⌀PIPE FILLED W/ GROUT WELDED TO 6"x6"x¼" PL W/ ½"⌀x3" ANCHOR

#4@9"

#5@9"

#4@18"

6"CONC SLAB ON METAL DECK

1½"⌀ GUARD RAIL BEYOND

1½⌀ PIPE RAIL (TYPICAL)

½"EX JT

#4 @ 8" O.C. EA WAY (TYP)

POURED CONC WALL W/ #4 @ 10" EA WAY

1E	**DETAIL**
	SCALE: ¾" = 1'-0"

1F	**DETAIL**
	SCALE: ¾" = 1'-0"

A plan section of complex wall construction is shown in the same manner as a vertical wall section and referenced directly on the same sheet to the larger scale detail.

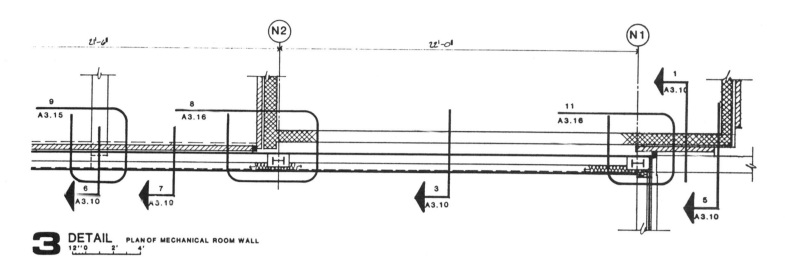

3 **DETAIL** PLAN OF MECHANICAL ROOM WALL
12''0   2'   4'

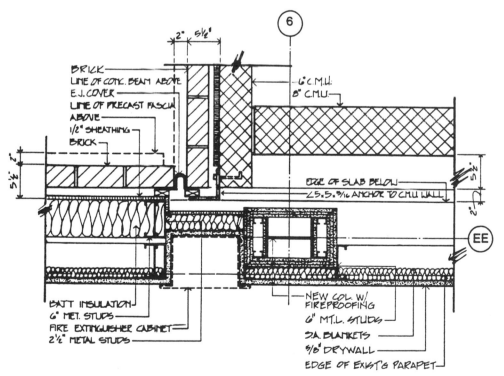

BRICK
LINE OF CONC. BEAM ABOVE
E.J. COVER
LINE OF PRECAST FASCIA ABOVE
1/2" SHEATHING
BRICK

6" C.M.U.
8" CMU.

EDGE OF SLAB BELOW
∠5.5.5/16 ANCHOR TO CMU WALL

EE

BATT INSULATION
6" MET. STUDS
FIRE EXTINGUISHER CABINET
2½" METAL STUDS

NEW COL. W/ FIREPROOFING
6" MTL. STUDS
2A BLANKETS
5/8" DRYWALL
EDGE OF EXIST'G PARAPET

SEE DETAIL 9 A3.16 FOR DETAIL 2 PRECAST FASCIA

8 **DETAIL** PLAN AT SECOND FLOOR
12''   6''   0   1'

81

ANOTHER CONVENIENCE: WALL TYPE SCHEDULES

Many firms use wall type keys and schedules to identify wall construction.

These are like detail keys and they can refer to details, but most often they are referenced to a wall legend like the one below.

Wall thickness dimensions are especially useful for checking plan dimensioning.

See the chapter on Standard Details for samples of standardized wall section details.

WALL TYPES

1. 4" BRICK, AIR SPACE, 2" RIGID INSULATION AND 12" CMU FOR AN ACTUAL THICKNESS OF 1'-6"
2. SAME AS 1 WITH VENEER PLASTER TO MATCH FACE OF ADJACENT WALL
3. 4" CMU, AIR SPACE, 2" RIGID INSULATION AND 12" CMU FOR AN ACTUAL THICKNESS OF 1'-6"
4. 4" CMU
4A. 4" LT WT CMU
5. 3 1/2" METAL STUDS @ 16" O.C. WITH 5/8 GPDW BOTH SIDES FOR A NOMINAL THICKNESS OF 5"
5A. SAME AS 5 WITH 3 1/2" SOUND BATTS
6. 6" CMU
6A. 6" LIGHT WEIGHT CMU
6B. 6" CMU WITH VENEER PLASTER
7. 4" COMMON BRICK OVER 4" CMU
8. 8" CMU
8A. 8" LIGHT WEIGHT CMU
8B. 8" CMU WITH PLASTER VENEER TO MATCH ADJACENT WALL
9. 6" METAL STUDS AT 16" O.C. WITH FIBERGLASS INSULATION R-19 AND (2) LAYERS 5/8" GPDW TYPE "X" BOTH SIDES
10. NOT USED
11. NOT USED
12. 12" CMU

WALL TYPES

SYMBOL	DESCRIPTION	ACTUAL THICKNESS
1	4" CMU	3-5/8"
2	6" CMU	5-5/8"
3	8" CMU	7-5/8"
4	SAME AS WALL TYPE 3, WITH SACMU AT THE HEIGHTS SHOWN IN SECTION 7/A8	
5	12" CMU	11-5/8"
6	3-1/2" METAL STUDS AT 16" O.C. WITH 5/8" GPDW BOTH SIDES	4-3/4"
7	SAME AS WALL TYPE 6, WITH SOUND ATTENUATION BLANKETS FOR ASSEMBLY STC RATING OF 47	
8	8" CMU WITH 3-1/2" MET. STUDS, BATT INSULATION (R=11), AND 1/2" GPDW ON FACE OF STUDS.	12"
9	8" CMU WITH 5/8" GPDW GLUED TO EAST SIDE	8-14"
10	6" CMU, 2" RIGID INSULATION (R=10.8), 4" FACE BRICK	12"
11	4" FACE BRICK, 4" CMU AND 4" FACE BRICK	11-5/8"
12	10" CMU	9-5/8"

NOTES FOR WALL TYPES

1. ALL INTERIOR WALLS ARE TYPE 2, UNLESS OTHERWISE SHOWN.

2. ALL EXTERIOR WALLS ARE TYPE 10, UNLESS OTHERWISE SHOWN.

3. WALL TYPES ARE TYPICALLY INDICATED ONLY FOR WALLS NOT OTHERWISE DETAILED.

4. ALL INTERIOR PARTITIONS SHALL EXTEND A MINIMUM OF 8" ABOVE CEILING, EXCEPT TO STRUCTURE WHERE REQUIRED FOR BEARING, STC RATING OR FIRE RATING.

5. PROVIDE TYPE-X GPDW AS REQUIRED FOR FIRE RATING.

ADD PICTURES OF ELECTRICAL, MECHANICAL, AND PLUMBING FIXTURES

Why not? They're in the manufacturers' catalogs anyway and easy to photocopy to stickybacks. It makes for a much more vivid, useful, and reliable set of Fixture Schedules. Put a key number beside each picture of a fixture and copy the key code numbers on the plans. You can include catalog descriptive text, too. Pictures too large in the catalogs? Use a reduction copier. Every downtown copier shop has a reduction copier, just in case your office doesn't.

And while you're at it, don't forget catalog and book pictures of hardware, furnishings, tile patterns, concrete textures, wood grain, cabinet work . . . wherever there are pictures in Sweet's, handbooks, and the magazines, there's easy-to-add clarity for your working drawings.

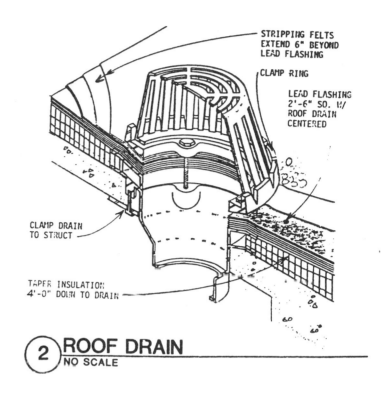

STRIPPING FELTS EXTEND 6" BEYOND LEAD FLASHING

CLAMP RING

LEAD FLASHING 2'-6" SQ. W/ ROOF DRAIN CENTERED

CLAMP DRAIN TO STRUCT

TAPER INSULATION 4'-0" DOWN TO DRAIN

② ROOF DRAIN NO SCALE

5

CLAMP FASTENER

light duty

High preloading — up to 90-lb. pull down pressure. Fastener actuated by finger-tip pressure on large wing-nut.

180° turn locks or unlocks.

Load-carrying capacity up to 300-lb. tension.

Wing-nut can be eliminated and a bolt or screwhead used for opening or closing.

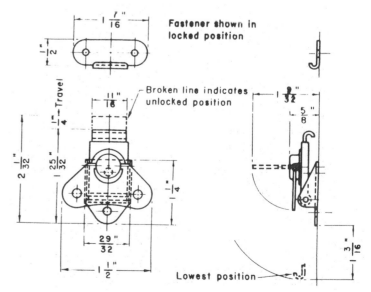

Fastener shown in locked position

Broken line indicates unlocked position

Lowest position

83

JUST THE RIGHT EMPHASIS ON DETAIL KEYS MAKES DRAWINGS MORE READABLE

In this sample, the office has drawn a very small scale roof plan, but it's still readable and especially so because of the bold outlines on the detail key symbols.

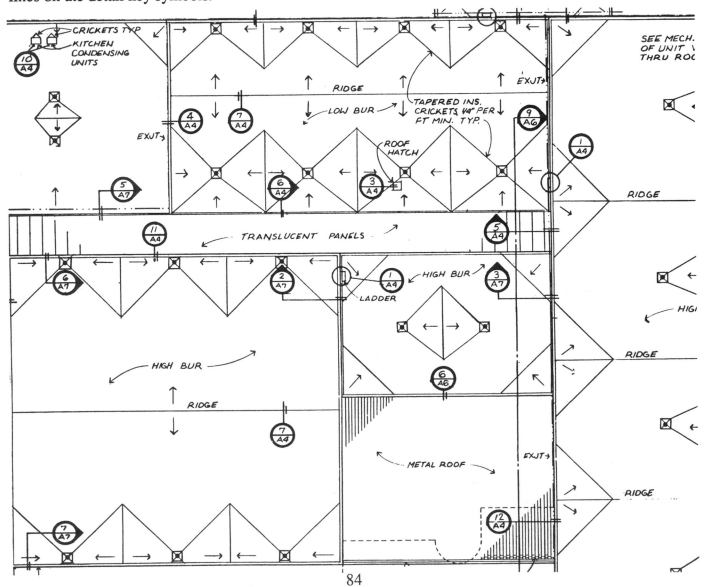

DIMENSIONING SIMPLIFIED WITH A MODULAR GRID

Modular grids are a design timesaver and also a big help in organizing plans and elevations in working drawings.

Frank Lloyd Wright's floor plans, for example usually had very few dimensions. Walls and partitions were referenced to an alphanumeric coordination grid similar to that shown below.

For more on the timesaving advantages of grids, see the section on A GREAT DESIGN AND DRAFTING TOOL . . . AND IT'S FREE FOR THE USING in the chapter on FAST-TRACK DESIGN & PRESENTATION.

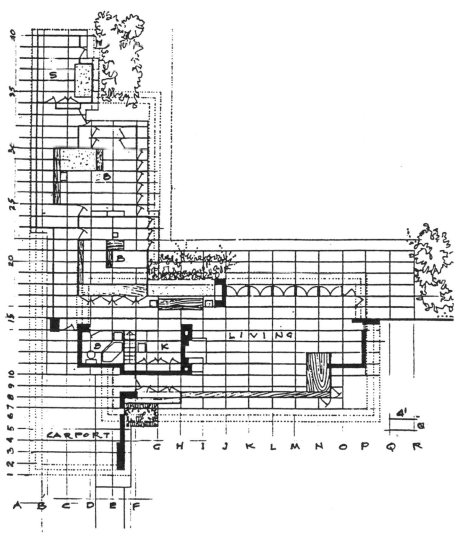

PLAN DIMENSION COORDINATION

Most dimensions can be replaced by a coordination grid and this general note:

THIS BUILDING IS LAID OUT ON A 4' x 4' GRID. UNLESS OTHERWISE NOTED OR DIMENSIONED, THE INSIDE FACE OF THE EXTERIOR WALL AND THE CENTER OF INTERIOR PARTITIONS AND MULLION ARE ON THE GRID LINES. ALL DIMENSIONS ARE GIVEN TO FACE OF STUDS OR CONCRETE AND TO THE CENTERS OF BEAMS AND COLUMNS.

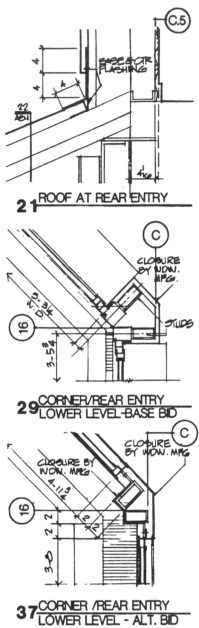

21 ROOF AT REAR ENTRY

29 CORNER/REAR ENTRY LOWER LEVEL-BASE BID

37 CORNER /REAR ENTRY LOWER LEVEL - ALT. BID

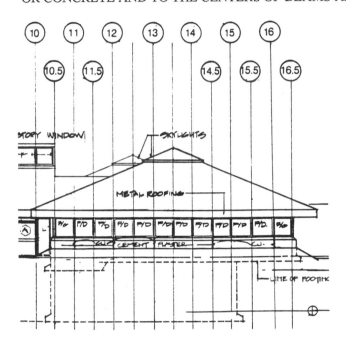

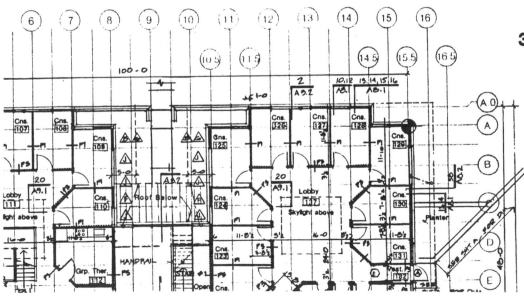

KEYNOTES AND KEYGUIDES

1) Photos of existing conditions are keynoted and directly linked to the plan of the building.

2) The keynotes are also keyed ("keyguides") to the screened perimeter floor plan of the building.

Here's an unusually well thought out drawing -- a combination of two clarifying techniques:

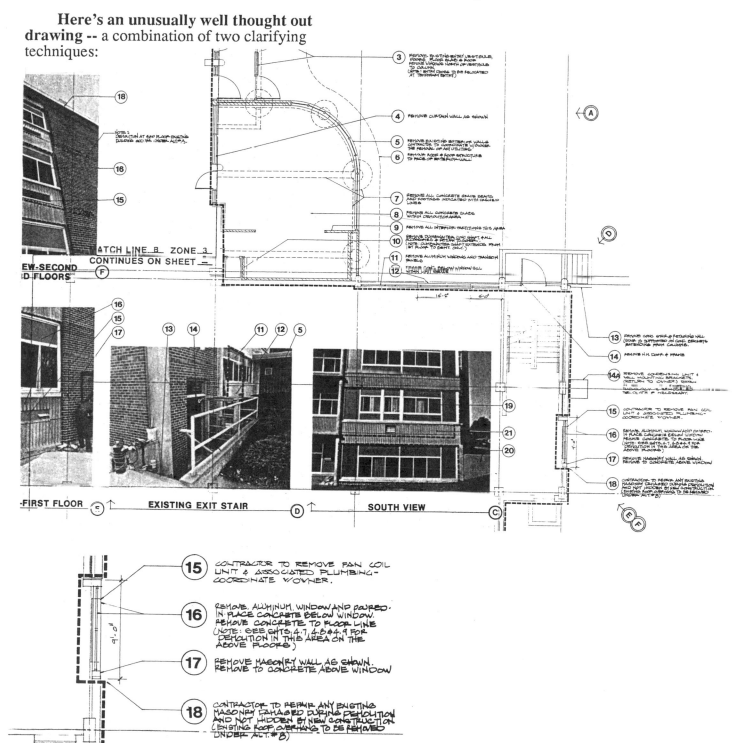

87

A FEW EXTRA LINES
GIVE 2-D
DETAILS 3-D
READABILITY

It's a common practice in manufacturer's catalogs -- a construction element is shown in traditional orthographic projection, but the face elevation is shown in a 30 degree extension to give a "3-D" look.

Here are some swimming pool curb details that illustrate the point -- how much clearer these are with "side views" added than they would be otherwise.

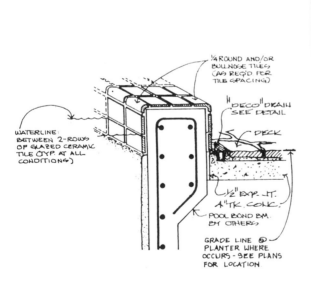

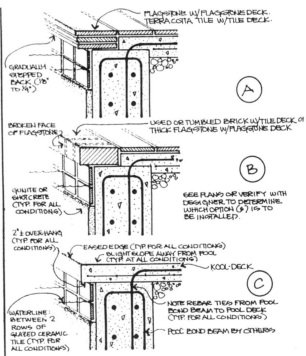

CHAPTER SEVEN

STREAMLINED DRAFTING TECHNIQUES

DRAFTING TIPS AND EXPERIMENTS FROM THE PROS

The Klipp Partnership in Denver has introduced many excellent improvements in manual, systems, and computer-aided drafting.

Larry Jenks, Vice President in the Partnership has been generous in sharing his expertise with the profession.

Jenks says:

"We have a very young office, and our people do not always have the drafting and lettering skills we would like to see. We took note of the fact that architects tend to be competitive, and that people's work improved when they were grouped together with other people who were similar in experience, but who had complementary skills.

"We took this idea a step further. We issued a direct challenge to everyone in the office to replicate an example of one of the partner's work. We offered Mont Blanc and Rotring pens as prizes, and we had three categories: conventional drafting, freehand drawing, and computer aided drawings. We had 3 entries in each category, and about 50% participation. Our people had fun competing against each other, and we were abel to get some fee attention to drafitng skills. . ."

They've also introduced a "Wall of Fame Frame" where work in categories other than design and presentation can be displayed. The display changes monthly and provides well-deserved public recognition for those who's best efforts are usually only seen in job prints and flat file drawers.

The partnership has made excellent use of Keynoting. Each drafter is given copies of a master list or "Master Menu" of drawing notation to guide them in making Keynotes.

A portion of the list looks like this:

SECTION 5.4

STRUCTURAL METAL FRAMING

5.4__ STEEL STUDS
    3-5/8" 18 ga. steel studs @ 24" o.c.,
    unless noted otherwise

5.4__ STEEL STUDS
    6" 18 ga. steel studs @ 24" o.c. unless
    noted otherwise

5.4__ STEEL "C" JOISTS
    See Structural

5.4__ . . . etc.

Each note selected for a job will have a letter added after it: "A," "B," etc.

The drafters yellow highlight notes that apply to their project and write in additional notes. Then the modified list will be revised on the word processor and a new reverse printout made on stickyback film which will be adhered on the back of each drawing sheet.

New notes added to the master list are reviewed periodically and those that are reusable are added to the office Master Menu of notation.

One client complained he couldn't remember the content of a note when looking between the keynote number on a drawing and the legend on the side of the sheet. So the Klipp Partnership has added a clarifier: the Title or Identification of each note--usually the name of a material--is printed in upper case letters on the keynote AND is added as part of the keynote number on the field of the drawing.

Adding a Title or Identification note with each keynote on a plan, elevation, or section is a good clarifier for any keynote system and a good double-check system to make sure the keynote numbers used are correct. **This level of keynoting provides multiple values to the office.**

For example:

-- **It's an excellent application of the concept of creating an Office Design Database.** The office first selected the best and most generally usable notation from past jobs to make up a Master Menu. Then the Menu is added to and improved upon by the experience gained in each new project.

--**It's a training tool for younger staff members** on what should and shouldn't be in drawing notes.

--**Keynotes eliminate clutter in drawings** which is an automatic clarifier. That in turn allows the use of smaller, easier to handle sets of drawings.

--**An office Master Menu ensures consistent office nomenclature** from drawing to drawing and job to job no matter who does the drafting.

--**The notes are automatically linked to specifications** and offer an excellent checklist of Specification content for the spec writer.

--**The use of standards** which are reused and re-examined from project to project **is a boost to quality control.**

--**And finally,** this intelligent use of keynoting is just a good, solid all-round timesaver.

Although this firm uses CADD as well as Systems Drafting techniques, much work, as ever, is still done by hand. So, rather than fight it, they do what they can to expedite it.

Three more simple, productive drafting tool improvements from the Klipp office:

--**Adjustable triangles with sun angles and roof pitches.** They applied matt, self-adhesive film to adjustable triangle to mark sun angles (for their 40degree N. latitude) and roof pitches. Although some commercial drafting tools come with similar guides, they're rarely useful because they don't fit the specific needs of an individual office. Making one's only such guides and templates is timesaver that pays off day after day after day.

--**They plotted lettering guidelines on the CADD system** and issued sheets to everyone to place under drawing sheets to reduce or eliminate the need to draw guidelines. They included some vertical guidelines too for lining up multiple lines of text.

--**They've etched lines in triangles as drafting guides.** According to Mr. Jenks the set one line in about 1/8" to draw guidelines for vertical lettering. A drafter draws the first guideline and then moves the etched line on top of it to draw the second line. It works for quickly drafting evenly spaced hatching and cross hatching too.

These kinds of ideas are thoughtful solutions to real day-to-day, sometimes minute-to-minute problems in the working world. They represent the kinds of often-ignored or unsung ideas easily save hundreds of work hours over a year's time in any medium size office. Even as we stand on the threshold of all-out CADD, these ideas still count and as long as they're used, they'll help make up for the still-common money-losing computer systems.

IT PAYS TO EXPERIMENT

Productive experiments in working drawings have a long history. For example:

Early in the century Architect Bernard Maybeck used a "magic lantern" (opaque projector) to project his sketched details and ornament to full size on a "drafting wall." The drawings could be refined and modified in real size on wall mounted sheets, then traced to make templates and molds to for the final construction.

Some engineers with the US Army Corp of Engineers used overlay drafting and reduced-size, full color offset printed working drawings in the early '50's.

In the 60's small office engineers and architects found a gold mine with their diazo print machines. They discovered they could make diazo paper prints of repetitive elements, such as repeat building rooms, fenestration, etc., Scotch tape them onto tracing paper, and make new sepia paper reproducible prints by slowly running the paste-ups through their diazo machines. This allowed owners of the smallest firms to quickly and economically do their own production work without hiring drafters.

Through the 50's and 60's various small-office architects experimented with such novelties as:

--**Working drawings in a book.** Book sized combinations of broad-scope working drawings, details, and specifications. Many offices around the country still produce substantial amounts of work at remarkably low cost by using variants of the book format, usually with 11 x 17 size sheets, photocopied on both sides, with plastic spiral bindings.

--**Isometric working drawings.** Every few years building designers rediscovered that you can show a one-story building plan, exterior elevations, sections, framing and interior eleva-tions, all in one single drawing by creating 30-60 degree isometric views. Two opposing views of the same building show just about everything there is to see and it's all in "solid modeling." They're much easier and faster to draw than you might think and they show an extremely clear image of the final building product. (With isometric options in the low-cost CADD programs, we expect to see renewed experiments along this line.)

--**Actual 3-D drawings, complete with glasses.** Graphic devices have been sold that allow drafters to use red and blue linework stereographic images and red and blue 3-D glasses to draw perspective or isometric architectural drawings and even structural engineering details. Although these haven't seemed likely to catch on, it is easier to make such drawings on some CADD systems than by hand, and they'll undoubtedly be done again.

Some timesaving experiments of those times which are still used and still worth considering by include:

--**Models as working drawings.** Scale models of building projects were photographed and reproduced on drafting media, then drafters added dimensions and notes. This was done for interiors as well as exteriors, particularly by a well-known interior design firm that did elaborate study models of their work.

--**Photo details** that combine photos of real construction with hand-drawn cut views of what's behind or underneath the finished construction work. A stair, for example, will have a cross section of the construction inset on the photo with notes and dimensions. A/E's first take photos of standard construction and have them reproduced in halftone as moist-erasable reproducibles on polyester (Mylar).

Most repro shops can handle this but **specify that you want the least expensive method.** Then they erase part of each photo and draw in what's holding that part of construction together. They make extremely clear standard construction details.

THE MOST TIME-CONSUMING PART OF DRAFTING

One of the most time-consuming parts of day-to-day drafting is plain old dimensioning.

Here are proven tips for improving and speeding the dimensioning process:

--**Use a drafting grid.** Probably the easiest to use and most beneficial timesaver in any manual drafting room is "no-print blue" gridded tracing paper. The average timesavings for using gridded paper on a typical drawing is about 3% to 4%. A speedy 40-hour per drawing job at straight drafting wage of $10/hour means $400 direct cost. A 4% savings equals $16. The gridded paper sheet cost for a 30" x 42" sheet is $1.50. (The actual employee cost and the real savings is usually considerably more, but this low estimated savings still illustrates the point.) A 1/8" grid is generally more useful than the commonly used 1/10" grid.

--**Use a building grid too.** When the building plan follows a consistent modular unit system, the drafters and construction workpeople can mainly count grids instead of getting into minute dimensions.

--**Keep dimensioning at 4' and 4" increments as much as possible.** Those modules are simplest to draft, easiest to read, and most compatible with building material sizes.

--**Use an extra long flat scale** to avoid end-over-end fussing with small scales for long dimensions. Using short scales on long dimensions is an invitation to error.

--**Mark your drafting tool scales to make them more readable.** Where a scale has dual dimensions, such as 1/8" from one end and 1/4" from another, rub a red pencil over one or another dimension sets to reduce the chances of misreading scale numbers.

--**Make underlay templates for the most commonly used dimensions of specialized drawings.** If you do a lot of housing, for example, mark the common building component heights for your section and elevation drawings -- 6'-8" door and window heights, 8' ceilings, and 3' high counters, for example. If doing brick and block masonry buildings, make a template of standard masonry courses for laying out wall sections that show sills, lintels, spandrels, etc.

--**Make or buy inch-to-feet conversion charts to keep at all drafting stations.** And buy the plan dimension and conversion calculators available at most print and drafting supply shops. They'll pay for themselves in a week or two of drafting.

FREEHAND DRAFTING, STILL THE FASTEST

A reader recently sent us some working drawing sheets and offered a dare: *"I dare you to find any technical difference between the freehand details and the ones drafted with tools,"* he wrote.

Not only could we find no difference in technical quality, we couldn't tell which details were freehand and which weren't.

The reason: He had reverted to the old system of drawing freehand at "double size" and then using a reduction copier to reduce to "half size." The photo-reduction smoothed out any wiggles in the freehand line work so that it took a very close look to see which details were which.

This harks back to the discovery made years ago during deadline emergencies. There wasn't time to take details that had been "sketch" drafted and have junior drafters redraw them in hardline. So the sketches were cleaned up a bit, pasted up, and photographically turned into working drawing sheets.

The funny thing was that most people who reused freehand work directly in working drawings only did it in emergencies. They thought it was wrong to draft in freehand unless there was no other alternative. Meanwhile, others realized they had hit on something important.

--An architect in San Diego lost his drafting staff to the flu for a week, and had to finish up a set of residential working drawings himself. As he poured over the previous freehand and "semi-freehand" plans, elevations, sections, and details, he realized that *"By God, they were perfectly good drawings."*

--A Dallas architect had been doing his structural engineering drafting for years -- the engineer submitted yellow tracing paper freehand versions which the architect redrew. The saving in engineering fees, he said, was about 50%. During a deadline rush, he realized the engineering drawings were actually quite precise -- part freehand and part tool drafting but perfectly adequate. So he started skipping the redrawing part.

--Our friend, Tucson architect Frank Mascia, who is as CADD oriented as anyone in the business, uses freehand too. Computer purists are offended when they see a sheet of drawings that combines CADD, regular tool drafting, and freehand . . . never mind that the hybrid approach gets the final product out twice as fast as the "pure" approach. But Frank and others keep saying: "Do what works." That means focus on the end result you want -- good, readable, accurate, fast documents -- and don't give priority attention to the means by which you get that result.

FREEHAND DRAFTING EXPERIMENTS PAY OFF YEAR AFTER YEAR

Freehand drafted working drawings have been discovered and rediscovered numerous times during high-pressure rush projects. The designers do properly scaled and well-drawn freehand or semi-freehand design sketches and someone realizes that they're good enough to use in the working drawings. It helps when such drawings are done with grid background guidelines and drawn at double size. When they're reproduced at half-size on a reduction copier they come out looking like straight-line tooled drafting.

Same-size freehand work can be perfectly acceptable too and has included designer elevations, cross sections, interior elevations, and even complete floor plans. (Same-size freehand drafting is usually done with the aid of grid sheets and long scales. The scales are used as straight edges for rapid drafting of longer lines.)

One of the most intriguing uses of freehand drafting: complete consultant participation in the planning and mockup of working drawings.

Here's how it works:

1)   After preliminary floor plans are essentially decided and tool drafted, or computer plotted, they're pinned on a conference room wall.

2)   Representative of the major consultants, especially structural and mechanical, are brought in to meet together and work directly on overlay sketch sheets. They and the architect have to work out all construction conflicts and interferences on the spot. This proves far faster than doing it long-distance, piecemeal, with back-and-forth checkprints during working drawing production.

3)   The architect and engineers then continue to semi-freehand draft their work throughout the project and meet to repeat the co-ordination process as the building is firmed up.

4)   When production is at 80-90% completion and all major decisions are finalized, the freehand drawings are assembled and all the consultants' plans and sections are finish drafted. The finish drafting is no more than copying and, since it's not held up by questions and decision making, it goes extremely fast. Smaller drawings, especially details, that have been freehand drawn are reused directly without redrafting when possible. If they're not good enough for reuse, they're copied by hand or on CADD, but such straightforward, uninterrupted copy-drafting goes at high speed.

There are two important secrets behind the success of this method:

1) MOST DRAFTING CONSISTS OF MAKING CHANGES.

2) DESIGN AND DECISION MAKING DRAWING IS TOTALLY DIFFERENT AS A PROCESS FROM DRAFTING. WHEN IT'S MIXED IN WITH DRAFTING, EVERYTHING GETS BOGGED DOWN.

When drafting or inputting graphic data is ONLY that, it can go like lightning. When design drawing is used for decision making only and not slowed by the rigors of drafting, it can go considerably faster than usual. So the object is to separate the two different processes and delay the finalization, the finish drafting, until the last phase of work.

We first learned of an office making full-size mockups of their working drawings in 1969. It was a small firm that had received a large hotel commission as a rush project. Three architectural and three engineering personnel completed the multi-million dollar project in six weeks. Three months would have been normal (with added personnel) using the draft-erase-draft and redraft system.

FREEHAND DRAFTING . . . GOOD FOR MORE THAN DETAILS

An example of a freehand cross section and exterior elevation used in working drawings.

The elevations and cross sections in this job were originally sketched on flimsy tracing paper and then cleaned up and reused directly in working drawings.

Freehand plans, site plans, door/window frame schedules and other freehand or sketch drawings have all been successfully incorporated in working drawings.

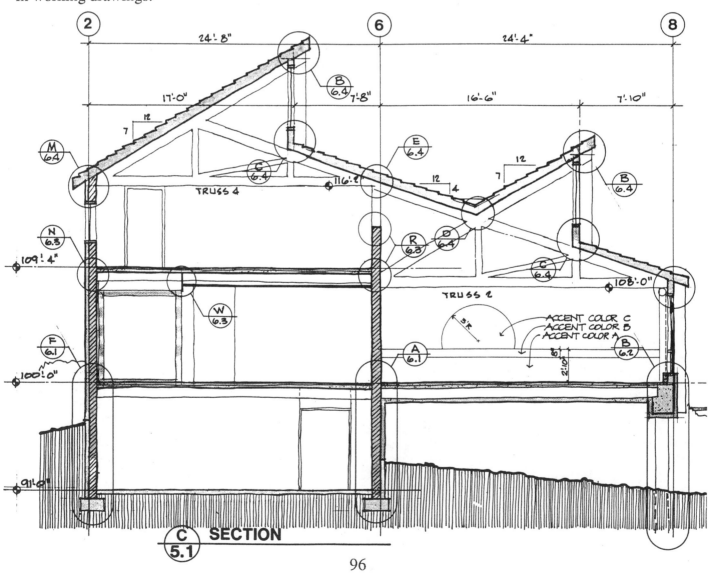

OLD-TIME TIMESAVERS ON THE OLD-TIME DRAFTING BOARD

You may have seen the reprinted edition of the 1890 Sears Roebuck catalog that was published a few years ago. It was amusing to see the buggy whips, butter churns, and bustles.

And it was a little shocking to see that virtually **the only items in that museum-piece catalog that are still in use today are our standard A/E drafting tools.**

CADD or no CADD, most drafters in most offices are still pushing pencils and using 19th century tools. And people still need every drafting timesaver they can lay their hands on.

Problem: It always takes time to place, align, and tape a drawing on the drafting board.

Solution: A strip of hard spring steel screwed into the upper left hand corner of a drafting board will let you secure a drawing in place and hold it down while you tape the other corners in place.

Problem: The drafting tape isn't always handy.

Solutions: Convert heavyweight tape dispensers sold for Scotch tape into drafting tape holders. . . . Or run strips of drafting tape on the parallel bar and cut them into the right size segments. . . . Or use the plastic or cardboard strip that drafters often place along the top of a drafting board to protect the top of the sheet from the parallel bar as an alignment and placement device. If the protective strip is placed in a straight line, it'll also act as a temporary sheet holder while you do the taping. . . . Use the overlay drafting five-hole punched sheets and pin bar even if not doing overlay drafting for automatic placement and alignment.

Problem: Drawings slip out of alignment when you're shifting and taping them down.

Solution: Use tic marks on the drafting board surface to match standard drawing sheet borders to show where you always want your drawing positioned. . . . A grid sheet underlay also solves that problem besides speeding up scaling and drafting in general.

Problem: Drafting tools get scattered across the board and out of reach.

Solutions: Cut slots for triangles and drill holes for lead holders in a wood or lucite block. Add any other grooves or special slots you want for erasers, scale, etc. . . . Add self adhesive hooks as sold in hardware stores to the drafting lamp to hang the most commonly used small triangles and erasing shield.

Problem: Some unusual small elements, such as odd-shaped planters or site furniture on plans have to be drawn again and again.

Solution: You can quickly make your own templates for repetitive shapes by cutting out a square of drafting polyester ("Mylar") and cut any special shape you want in it with an X-acto knife. It'll work as a remarkably efficient temporary symbol template.

Problem: Triangles and other flat tools smear graphite on the drawings and sometimes catch and tear the sheets.

Solution: Buy triangles with small nubs on them that keep them a fraction of an inch above the drawing surface, or as noted in the June, 1987 article on tool improvements, add your own plastic nubs as dots of Elmer's glue.

Problem: It's hard to grab triangular scales and hard to find the scale markings you need.

Solution: Add binder clip or tape as pick up handles, and code the scales on each side with color dots.

SYSTEMS FOR DRAWING EXISTING CONDITIONS

Time to measure an old building, so out comes the tape measure, flashlight, and clip board. Drive to the site and take measurements. Poke through the basement, climb through the attic and after a half day or day of this, return to the office to "draw up the building."

Oops, the walls don't seem to come together on the new drawing. Some dimensions are missing on the field sketches and notes aren't completely clear. Back out to the building . . . .

This ritual is repeated time and again on rehab and remodeling jobs by younger staff who don't know better.

What follows is a checklist for building inspection and measurement -- as you'll see, preparation is half the job.

Write a procedural CHECKLIST, the sequential steps for doing complete and efficient building surveys. Checklists are great reminders for the veterans and absolutely essential for beginners.

And make up THE SURVEY KIT. This is a kit of the tools of the trade -- ready to go whenever someone has to get out to the field. A checklist of recommended contents begins on the next page.

Here are recommended procedural steps of **Preparation, Surveying,** and **Redrawing.**

PREPARATION

__ **Obtain previous drawings from any possible source** -- current or previous owner, real estate broker, building department, loan appraiser, contractor, etc. Any drawing you can start with is better than nothing.

__ **If you find previous drawings,** make a cleaned-up photocopy of the plan and photo-enlargements of important rooms and elevations. OR scan the drawings into computer, convert to raster or object-oriented, and print out copies for use at the site.

__ **Make a quick operational checklist of tasks and standards unique to this project.** That's to prevent staff from getting overly-detailed in measurements if the job is just for minor alterations and space planning, or under-detailed if it's a historic rehab or preservation job.

__ **Double check that you have all the tools needed** for an efficient building survey and move on to the next phase.

THE SURVEY

__ **Two-person teams are more efficient than a survey party of one** and they provide better cross-checking. Three people, one to write dimensions while two others measure, is most efficient of all, especially for a large project. Measurement surveying is very tiring work and errors will increase in direct proportion to the hours spent. The last part of the work is always the most error prone and least checked.

__ **If there are no existing drawings,** do an overall on-site schematic floor plan of the building to be measured. Use grid paper under-layment (8 squares to an inch) to aid freehand drafting. If you need more accuracy, use a small portable drafting board with a mini-drafting machine on it, such as the "Rotring" sold at drafting and art supply stores.

__ **Draw in dimension lines and points to be dimensioned before doing the measurements;** it'll make much neater drawings.

__ **Don't bother showing feet ' and inch " marks.** A simple 9-3 says the same as 9'-3" and says it without the fly specks.

__ **Establish datum points** -- reliable start points for all subsequent plan and elevation measurements. Measure heights simply from 0'-0" not from surveyor heights above sea level.

___ **Walk through the rooms and keynote number each room on the sketch plan;** that is, write the room name, function, or number in a legend and use an I.D. code letter on the plan. Use keynote numbers for notes and comments to avoid cluttering up the plan drawings and obscuring dimensions. A tape recorder can be used for verbal notes to save time on the site.

___ **Do the overall cross dimensions of each room.** If the measurements need be accurate only to a couple of inches, use an electronic tape measure that talks or gives visual display of dimensions from wall to wall and floor to ceiling.

___ **Take long cumulative dimensions wherever possible;** they're likely to be more accurate than dimensions taken by moving the tape from point to point. Keep tapes on floors or flat against walls; sagging tapes or off-angle measurements are inaccurate.

___ **If detailed dimensioning is required,** proceed in a consistent direction starting to the right of the entry of each room. Proceed counterclockwise around the room.

___ **Use diagonal "triangulation" dimensions as well as regular point-to-point dimensions to check for squareness of rooms.** This is also a way to double check size and shape of odd-shaped spaces.

___ **Take every dimension twice as you go along,** whether using a tape measure or electronic range finder.

___ **Use levels and plumb lines** to make sure critical dimensions are taken.

THE KIT FOR FIELD SURVEYS

An architectural and interior design firm that hits measure jobs like a swat team keeps the following stock of survey tools:

___ Large clip boards or clip binder

___ Grid paper

___ Legal pad

___ Pencils, colored pencils, erasers, and sharpeners

___ Architectural scale (1' long)

___ Portable drafting board with mini-drafting machine or freehand drafting board

___ Battery powered reading lights to attach to clip boards when necessary

___ Chalk & felt tip marker pens for field marks

___ Steel tape measures, wide and rigid enough for vertical wall measurements (bring a spare)

___ Electronic tape for rough dimensions or preliminary dimensions

___ 50' & 100' tapes for long plan dimensions

___ Folding carpenter's rule with 1' marks highlighted to show scale in photos

___ Telescopic measuring rod for vertical measurements where there is no wall support

___ Levels:
    ___ Spirit level
    ___ Plumb bob
    ___ Flexible plastic tube water level

___ Hand-held tape recorder with extra tape cassettes

___ Camcorder

___ 35 mm camera (with wide angle and telescopic lenses where needed)

___ High speed film, 400 ASA minimum

___ Panorama box camera for exterior reference views

___ Panorama flash box camera

___ Hand-held flashlight

___ Lamp style flashlight

___ 100 watt trouble-light with extension cord

___ Electrical outlet tester

___ Pocket knife

___ Compass

___ Plumber's wrench

___ Pry bar

___ Nylon cord for setting levels and datum
line along walls or comparing heights of
sills, etc.

___ Duct tape for attaching notes, cord to walls

___ Surveyor's transit (if knowledgeable on
use) and heavy-duty supplies if needed,
such as ladders

___ Personal items such as:

   ___ Hard hats
   ___ Disposable coveralls
   ___ Hand cleaner and rags
   ___ Bottled water

LOW-TECH MATH, FAST AS A CALCULATOR

When a calculator isn't right at hand, here's an alternate way of adding large columns of figures by hand. Super fast and accurate, it's called "adding by tens." It works like this:

As you add down a column of numbers, make a mark at each point where there is a sub-sum of ten or more. Take each remainder over ten down and add it to the next number. Write the last remaining sub-total on the answer line and count the marks for the total number of "tens" and carry that to the next column.

Try this simple step-by-step example and you'll see immediately how it works: It's just as fast and foolproof for dealing with page-long columns of numbers.

7   7 plus 9 is 16;

9   make a mark under 9; carry over 6;

3   6 plus 3 is 9;

8   9 plus 8 is 17; make a mark and carry 7

4   7 plus 4 is 11; make a mark and carry
       over 1;

4   1 plus 4 is 5,

5   which is written in the answer space.

35   Count the marks; there are three and carry
       that number to the next column.

Try it once and any time you have to do manual addition you'll never do it the old way again.

INK DRAFTING
Ink drafting?
In this day and age?

Actually you see a lot of it, even in what are supposedly automated offices. Computer plotted output often has to be revised and it's often faster to do it by hand on the original plots than to replot whole new sheets.

For those who continue to do manual drafting, many have long known that ink drafting is FASTER than pencil drafting -- anywhere from 5% to 15% faster per drawing . . . depending on the drawing, the number of changes, the experience of the drafter, etc. But faster overall for sure.

Ink drafting is faster? That will be hard to swallow by those who never learned to use modern ink drafting tools and media or who remember using the old inking methods. Or you may know an old-time methodical inker who manages to take as long as management allows to get the linework done.

It's notably faster in the long run partly because erasing is so much cleaner and faster -- and corrections are a much larger part of drafting than most people appreciate.

The idea that making changes is faster with ink will also surprise some people, but it's true. You have to follow the rules, however, and you have to invest in the right materials.

Here's a refresher on ink drafting with some updates from our Systems Drafting and Systems Graphics books:

Ink drafting and polyester ("Mylar") drafting media go hand in hand. The plastic media isn't easy to draft on with graphite or plastic lead and ink doesn't go well with paper. But ink and plastic sheets are a great match.

To draft on Mylar, you must use jewel tip technical pens. Metal tips wear out in no time and no matter how much cheaper it seems to be to buy a non-jewel pen tip, you'll soon lose your pen tip, your cost savings, and the time you're supposed to be saving.

Invest in a small ultrasonic pen cleaner for quick inexpensive cleaning if and when pens do clog up. It'll pay for itself in no time.

Follow the pen manufacturers' instructions. Users typically ignore the instructions, miss out on savings, and create problems for themselves. For the novice, here's a run down on technique:

1) When you start, shake the pen slightly until you hear the little "click-click" that means the wire that helps keep the ink supply tube open is operating. If you don't hear the click, **never tap the tip against a surface to loosen it up;** just follow the manufacturer's instructions for checking the pen. Any difficulty in starting is usually a matter of inexperience.

2) Test a few lines on scrap paper to check the ink flow before you start.

3) Use underlayments or hard leads such as 7H for guidelines or construction lines for inking. If the guidelines are light you can ink right over them. Gradually, with practice, you'll need fewer and fewer preliminary guidelines.

4) Don't try to letter notes or dimensions with the technical pens. Use a narrow tip, water soluble ink, felt tip pen for hand lettering (better yet, of course, keynoting with computer printed standard notes).

5) When drawing, keep the pen moving slightly as you touch down on the drafting surface and continue the movement as you raise the pen. This helps avoid the slight enlargement or "bulbing" at the beginning and end points of lines.

6) Use a light touch. Don't press on the pen or indent the polyester. Don't use a hard grip. A major advantage of inking is that it's smooth and easy, less straining and fatiguing than pencil drafting. But some people apply pressure and grip the pen hard out of habit and stress and tire themselves unnecessarily.

7) Prepare the drafting surface by wiping it clean with a paper towel and rubbing alcohol. Dirt or light traces of oil or finger prints can affect the ink linework.

8) Don't use ponce or cleaning powder on the Mylar; it'll clog the pen.

Materials and supplies:

--For starting out, a basic set of the most often used tip sizes will serve fine: 00, 0, and 1. Add to that later if you find a need for more subtle variations in line widths.

--Drafting tools with nubs to keep them slightly above the drafting surface to avoid smearing fresh ink lines.

--Fast-drying ink; you'll be surprised at how little you have to raise the drafting tools off the board and how little time you have to wait to avoid smearing fresh lines.

--Plastic, erasing fluid imbibed erasers. Keep cotton swabs, tissue and alcohol around for special erasing or cleaning jobs. For plain erasing of small elements, use a plain plastic eraser; don't use an abrasive eraser or you'll grind away the matte surface.

--An ultrasonic pen cleaner, as already recommended.

--Electric erasers can only be used with extreme care. Even non-abrasive erasers in an electric eraser can burn off the drafting surface by heat friction.

The "Mylar" drafting media can be 3 to 4 mil thick. 3 mil is fine for 24" x 36" size sheets, 2 mil for smaller drawings, with drafting matte surface on one side only. There's no need for two-sided matte Mylar (unless you draw on the back, and that's an obsolete practice).

As for pen recommendations, the major brands of technical pens are fine but they're overpriced in most stores. Watch for sales or use a mail order discount supplier.

BIG TIME SAVINGS IN A SMALL PACKAGE: MINI-MOCKUP WORKING DRAWINGS

It's the simplest of tools that some of the oldest firms in the US and overseas have used for nearly 100 years. They wouldn't dream of doing otherwise. Same with some of the nation's most advanced CADD offices. Meanwhile, oddly enough, many others have never heard of it.

The tool is the old-time mini-mockup working drawing set and these days the old timer is more useful than ever.

The mockup set is usually drawn on 8-1/2 x 11 sheets -- one-quarter size rough sketches of site plan, floor plans, elevations, details, etc. -- every sheet that's anticipated in an upcoming working drawing set is included.

"Why bother?" says the "show-me" manager. "My people have done so many working drawings, they could do them blindfolded."

Blindfolded is the word. **Doing a set of drawings without a mockup guide is like driving in a strange city without a road map.** No matter how many times people have done working drawings, such drawings are still full of false starts, redundancies, errors, and blank spots that wouldn't be there if the job had been planned from the outset.

Like all good things, **the mockups have multiple values, any one of which would justify their use:**

--**They're invaluable for doing more precise project time and cost estimates.**

--**They help show the client the total Scope of Work,** and have been used successfully to raise fees during client negotiations.

--**They guide the drafters when the supervisors aren't around.** They keep drafters busy, help them draw the right work at the right scales in the right places, AND help them avoid duplicating or contradicting what others are doing.

--**They help guide the checkers during check print time.**

--**They keep CADD production under control** -- a situation where work is often forgotten or overdone because it's hard to see the whole job at once.

They have so many values that NONE of the top production offices we know ever start a set of drawings without a mini-mockup and a contents checklist to go with it.

Here's how the users make a good thing better:

--Some firms preprint their working drawing sheet formats long-way on oversized sheets, such as 11" x 17". They show the outline of the standard drawing size on the upper left corner, and then add a drawing content and to-do checklist on the right hand side.

--Some save time making the mockups by using reduction copiers to recycle reduced size prints of design development drawings as working drawing sheets.

--Some clarify their CADD layering or manual overlay systems by including sample mockups of base sheets and overlays to show how they actually relate.

--Some include a drafting grid background on mockup sheets to assist freehand sketching. Mockup sheets often also include the office standard drafting modules (such as the 5-3/4" x 6" detail module we recommend).

--The best organized firms use mockups to show the production enhancements that are supposed to be used on a job. If paste-up and photodrawing are to be used, for example, that's what's shown in the mockup. If a fixtures height schedule is used instead of endless interior elevations, that's what's specified. The mockup can be a sort of specification of graphic standards as well as a "working drawing" for the working drawings.

--Some supervisors use a copy of the mockup to keep track of how many hours are spent on each working drawing sheet. That helps refine the office time and cost database so future jobs can be estimated more accurately.

Mockups are especially fast and easy to make if you take advantage of the office enlargement-reduction copier. Since some elements will repeat themselves on the mockup, just as they will in the drawings, there's no need to resketch them over and over. Just use photocopier pasteups.

Don't forget to distribute copies of the mockups as a guide to all drafting staff on a job. And send out new copies when there are major changes in the composition of the drawings. Some mockups are created for planning and then disappear into a file. They lose half their value by lack of use throughout the project.

Some cons with the pros. All tools get abused, and the best ones get beaten to death. Example: when a drafter is assigned to do a mockup and does it with all the excessive overdrawing typical in architectural drafting -- a supervision problem -- not a failure of the system.

See the pages that follow for illustrations of the appropriate level of completion of sketched mini-mockup working drawings.

MINI-MOCKUP EXAMPLES

These drawings show the limited amount of detail required for mini-mockups. Enough to show the intent without getting too fussy.

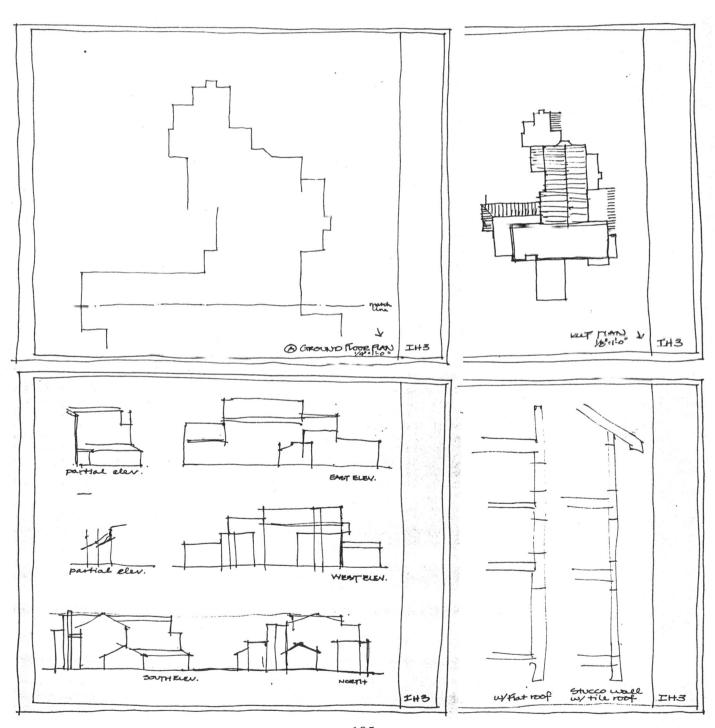

**Another partial example of a simple,
efficient, mini-mockup drawing.**

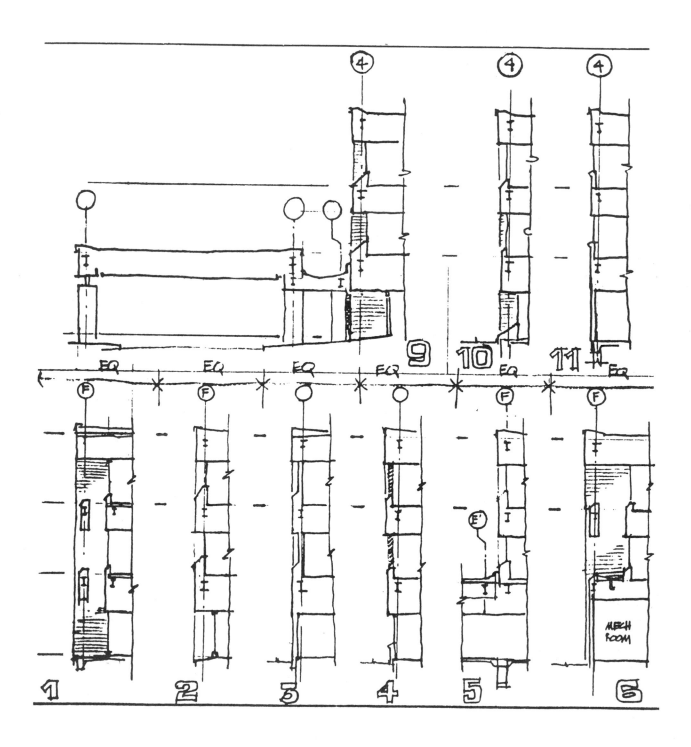

CHAPTER EIGHT

FAST-TRACK DESIGN & PRESENTATION

WHAT HAPPENS WHEN DESIGN IS CHANGED DURING
WORKING DRAWINGS

A GREAT DESIGN AND DRAFTING TOOL . . .
AND IT'S FREE FOR THE USING

DESIGN TIPS THAT HELP SPEED UP WORKING
DRAWING PRODUCTION

HOW TO GAIN A 20% HEAD START ON EVERY SET
OF WORKING DRAWINGS

"DON'T TALK TO US ABOUT 'SYSTEMS,'
WE DON'T DO REPETITIVE WORK"

WHY WORKING DRAWING PROBLEMS OFTEN START
IN THE DESIGN STUDIO

WHAT DO THE DESIGNERS NEED FROM THE CLIENTS?
 WHAT DO THE DRAFTERS NEED FROM THE DESIGNERS?
 WHAT DO THE ENGINEERS NEED FROM THE ARCHITECTS?

HOW CAN YOU PREVENT EXCESSIVE LAST-MINUTE
DESIGN CHANGES?

A SPECIAL HELPER: THE ROOM-BY-ROOM
DATA-CAPTURE FORM

WHAT HAPPENS WHEN DESIGN IS CHANGED DURING WORKING DRAWINGS

It's insidious how a few seemingly minor design changes during working drawings can destroy a project's budget and time schedule.

In some cases, after changes start to multiply, it would be best to scrap the working drawings and start over. People only realize that after the job is over, it's never obvious at the time. So drawings are erased and scrubbed out part by part, again and again. Eventually large areas of some sheets are redrawn twice, three times, even four times over.

What misleads people the most is that simple changes made during advanced design development or working drawings are, in themselves, no problem. Changing a room or moving a door is nothing. It's the unexpected side effects that are the killers. Change begets changes. And those beget other changes, none of which are obvious right off the bat.

Here's one instance, a classic case study: It started when the principal in charge of design returned after an illness. He didn't know the reasons for design decisions made during his absence and ordered minor changes in the plan of a corporate training center building. "Move that door so it lines up with the one across the way," he said, pointing at a meeting room. Moving the door would require moving a built-in corner slide projector cabinet. But that seemed easy to fix, so no one objected to the change.

Then the domino effect took over. The projector cabinet was moved from the corner of the room and had to be redesigned and redetailed. The room had to be enlarged slightly to maintain its required occupant load and projection screen visibility.

One change forced other changes; those changes caused a few more. For example, the change in the length of the conference room required that an adjacent room, which could not be reduced in size, be shifted to match the wall shift in the first room. That adjacent room was bounded on the opposite side by a restroom. When it moved, the restroom plumbing wall had to be moved, which meant redesigning and redrawing the restroom and its plumbing fixtures in 1/8" and 1/4" scales.

That work was minor compared to the reverberation of changes in reflected ceiling, structural, electrical, plumbing, HVAC, and sprinkler drawings.

Why did it go on and on? Why didn't someone call a halt? Because all the added changes came piecemeal, scattered over time and among many different individuals. Those who implemented the accumulated changes were never aware of the original design decisions or their rationale. They just followed orders from memos and checkprints.

The chain reaction continued. An exterior elevation and cross section were affected, as was the window frame schedule. Ultimately the architectural and consultant plan changes had to be repeated through six floors. Six types of drawings at six levels--36 plan sheets--were affected.

It took awhile before the worst of the hidden costs appeared. Some changes had been made incorrectly, and some hadn't been made at all. This followed the rule that late project changes are the most rushed, most error prone, and least likely to receive thorough checking.

This example was extreme, but you can find some variant on it every day in every office. Still, what's the alternative? You can't ignore opportunities to make design improvements. And changes are often unavoidable for reasons no one could predict.

Here are some of the better techniques used to minimize nonessential designer- or client-initiated changes. They've all been used for years by A/E firms to keep unnecessary design changes to a minimum and/or to reduce the impact when changes are unavoidable.

To minimize nonessential A/E-initiated changes:

1) Run through a checklist of justifications before accepting any design change:

____ What's the reason for the change?

____ What's the benefit?

____ What's the direct cost?

____ What are the potential repercussions?

____ What architectural drawings might be affected?

____ What consultants' work and drawings might be affected?

____ What other action might provide the same anticipated benefit as this change?

2) Require a sketch of any proposed plan changes showing possible impact on surrounding parts of the building. Think of it as a miniature Environmental Impact Statement.

3) Use Value Analysis to uncover worthwhile changes that should be made early in the game. Value Analysis means taking a broad look at the project and asking what can be simplified, streamlined, consolidated, reduced, etc., to simplify, speed-up, and lower the cost of construction. The key questions to ask of every major component of a building during Value Analysis are: **What does this do? What would do it better at lower cost?**

To minimize client-initiated changes:

1) Include a detailed Scope-of-Work list and miniature mock-ups of working drawings during fee negotiation, and include them by reference as part of the contract. When changes occur in the scope of work, it has to mean changes in schedule and in scope of fee.

2) Create separate job numbers for separate billing of all work that pertains to client-initiated design changes. Send a memo of agreement before starting all such work, to assure that the scope of added work is understood and that it's agreed that it will be reimbursed.

3) Take the client on room-by-room tours, using 8-1/2" x 11" sheets with blowups of each room plan. Make sure furniture, fixtures, equipment, lighting and switching are shown and understood. Provide signature lines at the bottom of each sheet that say "Approved," and "Approved pending . . . ." When a client signs off a space as approved, any changes are then clearly billable.

4) Make sure interior and exterior components of the building are similarly understood and accepted by the client by means of drawing blow-ups, mock-ups, sample display boards and the like. Get agreement and sign-offs on fixtures, materials, etc.

5) As always, use a checklist. Create your own standard predesign and planning checklist to avoid oversights and unaccounted cost items that might lead to late-date design changes.

A GREAT DESIGN
AND DRAFTING
TOOL . . .
AND IT'S
FREE FOR THE
USING

There's an extraordinary, simple tool that speeds up design and drafting, reduces errors, simplifies construction and even helps improve building design.

There's just one catch; it doesn't cost a cent. So there's not much motivation to sell it. As a result, it's rarely taught in school and is used by only a handful of the most sophisticated design offices.

The simple tool in question: The modular grid; in particular, the 4-inch and 4-foot design/drafting/construction module.

4-inch/4-foot modules were publicized in the 50's and 60's as Modular Construction. Building product manufacturers were enlisted to make "modular" products. The modular drafting system was part of Time-Saver Standards and widely published in the professional journals.

Despite valiant efforts, modular planning never caught on as a national construction standard. The drafting standards were a little too complex and the realities of construction didn't fit the modular ideal. Designers and drafters spent too much time trying to force everything into the grid, whether it wanted to fit or not. Sometimes there was more waste and confusion at a "modular" construction site than at a traditional random-dimension cut-to-fit job.

But some firms made a simple version of modular drafting work extremely well. They found it to be a great all-around timesaver and adopted it as their permanent office standard.

Their key to success is in the word "simple." Rather than apply the many special rules of formal Modular Construction, they follow only two elementary stages in their work:

1) They lay out their buildings according to a modular grid that best suits the building type. It might be a grid based on 4-foot squares, or a 5 foot grid, 2 feet x 4 feet -- whatever works best for the building's dominant rooms and/or corridor arrangements. Sometimes it's determined by a pre-selected partition and integrated suspended ceiling/lighting system.

2) They keep all minor plan subdivisions of the grid as close as possible to 4 inches. That means door frames, opening sizes, windows, counters, trim, etc., are all generally aligned with a 4-inch grid.

If something doesn't fit into the planning module or the 4-inch sub-grid easily and naturally, they make an exception and don't worry about it.

The benefits:

--Plan dimensioning is far easier to compute and easy to check, since almost all broad-scope dimensions are a multiple of the larger module plus a multiple of 4 inches.

--The drawing lay-out process goes much faster because the modules reduce guesswork as to exactly where to position walls, doors, windows, and other major components. All the manual drafters work with grid sheet underlays on their drafting boards and CADD operators set their grid and dimension commands to default to 4 inches. Nobody has to think about using the 4-inch standard; it's just there.

--Far fewer plan dimensions are needed. If there's an overall reference grid, it's easy to visually count modules (such as 4-foot increments). Some offices, such as Taliesin Associated Architects, use a unit system grid and 1-2-3-. . . , A-B-C-. . . grid coordinate system instead of plan dimensions. It's like the coordinates structural engineers use when drawing column lines on modular framing plans.

--A total set of orthographic drawings are more readable and better coordinated because the module lines and modular coordinate system can be shown in elevations and sections as well as on plans. Details can also be keyed according to their modular coordinate numbers.

--Building elements that line up throughout a building are less likely to be drafted out of alignment or accidentally misdimensioned. Core and chase walls are often improperly positioned on different floors of a building or in different layers used by consulting engineers because of dimensioning errors, for example. A modular coordinate grid helps prevent this common error.

--Oddball unrealistic dimensions are avoided. Some drafters show 1/8" dimensions on plans and 1/32" on details. Usually it does no harm and just provides amusement for the contractors, but sometimes overtight dimensions lead to conflicts on the job and require subsequent change orders.

--A modular building will be highly integrated proportionally compared to any similar building that doesn't have an ordering system, especially if the module is applied in three dimensions--as say, a 4-inch vertical module in elevation combined with a 4-foot plan module.

--Construction layout is speeded up when the plan module is laid out in relief joints on a concrete slab. Construction workers don't have to measure wall lengths and intersections when the stop points are inscribed in the floor slab.

--Manual drafting goes faster in general because the grid line underlayment acts as a visual guide for drafting tool alignment and lettering. The grids also expedite high-speed semi-freehand drafting and the use of unattached long scales as straight edges instead of parallel bars and triangles.

DESIGN TIPS THAT HELP SPEED UP WORKING DRAWING PRODUCTION

From several top designers around the country:

--When programming a project, make as many rooms as possible the same size. If one space needs to be 11 x 12, another 10 x 12 and another 12 x 12, just make them all 12 x 12. That makes the spaces interchangeable for flexibility during design development. The more interchangeable spaces there are, the easier they are to draw and manipulate during design and working drawings. It simplifies things at every phase of the project from design to production to bidder takeoffs to construction.

--When rooms can't be the same size, at least make them multiples of the same size. That's what modular planning achieves, but you can go further and design conference spaces, for example as two or three times multiples of office spaces, and larger work spaces as multiples of conference spaces. In this example, corridor widths and utility spaces might be 50% subdivisions of office space sizes. Then when a space has to be relocated, it becomes a simple equivalent spatial trade, and not a major dislocation with unexpected repercussions throughout a whole section of the building.

111

HOW TO GAIN A 20% HEAD START ON EVERY SET OF WORKING DRAWINGS

You can be 20% finished with every set of working drawings the moment you start,

BUT:

--You must depart from a long-standing architectural tradition.

--You must listen and argue with those who will find all kinds of reasons not to do it.

--You must block the sabotage of those opposed to it.

--You must decide if the possibility of saving the office thousands of dollars each year is worth the effort just described.

If it still seems worth doing, read on:

"IT'S AN IDIOTIC WASTE OF TIME," says the architectural drafter about a tradition that costs most offices thousands of dollars a year.

The tradition: The drafter must take the designers' final presentation drawings and trace or otherwise redraw them from scratch in working drawing format. For years drafters have objected: *"Why can't they leave the rendering stuff off their drawings, render on copies or overlay sheets, and leave the originals for working drawings?"*

"IT WON'T SAVE TIME TO DO IT ANY OTHER WAY," says the designer, *"and it's a nuisance . . . and who's working for who?"*

"THERE'LL BE SO MANY CHANGES, THERE WON'T BE ANYTHING TO REUSE ANYHOW," says the project manager who has to keep the peace and explain why the designers can seemingly waste time and make pointless extra work for the drafters.

So it goes over the years in office after office: The drafters suggest changes to stop what they see as a big waste. The designers resist the changes and management sides with the designers.

But it doesn't always go that way.

The result, as originally reported years ago by the midwestern office of Hansen Lind Meyer:

WHEN YOU FIGURE OUT A WAY TO REUSE DESIGN DRAWINGS DIRECTLY IN PRODUCTION, YOU'LL BE 20% FINISHED WITH THE WORKING DRAWINGS ON THE DAY THEY'RE STARTED.

As many offices have since discovered, that 20% savings may be the only profit margin they have on a lot of their jobs. For a medium-size project with $20,000 for working drawings, 20% means a three- to four-week head start, and $4,000 extra in the bank at the end of it all.

To restate the problem: Designers draw plans, elevations, maybe some sections ... and they're often perfectly useful, accurate drawings of the building.

Then the designers RUIN them as far as working drawing production goes. They add furniture, "designer" titles, shadows, trees, and other fru-fru that make it impossible to reuse the drawings for production.

The tradition starts in design school. Rendered presentation drawings are the end goal of design studio experience. In most schools the design process consists only of the design presentation, the jury, and the grade. Period. Nothing follows. No permits. No budgets. No working drawings or specs. No bidding. No construction. Just rendered plans, elevations, and perspective.

Who can blame designers who enter the office and continue working as they were taught in school for four to six years? Why not do final design development drawings without thought to further applications? No one ever told them it should, or could, be otherwise.

The supposed options may not seem so advantageous either. Suppose the designer stops at the presumed critical moment before there are any special presentation notes or rendering components on the drawing. Then what? Make photocopies? Who gets the original? Well, the drafter does. But what happens when there are changes? Does the designer take the original back from the drafter? It's no wonder people get confused and wary.

There's ONE solution that makes it simple; but it raises resistance. It's spelled: O-V-E-R-L-A-Y.

"OH NO. WE DON'T DO OVERLAY DRAFTING."

This time around we're describing OVERLAY DESIGN, not overlay drafting, and designers have been doing overlay design for generations. It's traditional as can be to do design studies on bumwat design study overlays atop bumwat base sheets. It just hasn't been traditional to carry the process one step further as a bridge between design development and working drawings.

Here are the steps for success, to gain that 20% head start:

--**When doing late-stage design development drawings, draw in hard line as usual; then STOP** before adding notes, titles, furniture, trees, textures, etc. In other words, ONLY draw the solid lines of the building structure, walls, doors, and windows. That's it! Stop.

--**Treat those solid-line, no-frills, basic plans and elevations as base sheets** and do all the rendering, "designer" titles, furniture, and trees on overlay sheets. You can do it on tracing paper for small jobs; ink and polyester ("Mylar") for larger projects. You can even use transparent plastic as do the graphic artists.

--**Use reprographics to blend the base and overlay sheet images in final presentation drawings.** Base and overlay images can be blended on office copiers, in office diazo print machines, on vacuum frames, or on large-format equipment at repro shops. It's fast and inexpensive.

That leaves the plain, uncluttered, unrendered base plans, elevations, and sections for direct reuse as working drawing sheets.

Designers and clients benefit in two big ways:

1) **Overlay Design lets them try out far more variations and alternative design studies at every stage of design development.** This is because they don't have to spend time drawing the same base sheet information again and again. If the client wants to see another story, wing, or exterior treatment, just do a new layer for that work only.

2) **If major changes have to be made, it's easy to change base sheets** and salvage that which doesn't have to be changed without redrawing it.

If your designers never learned about Overlay Drafting, they'll wonder: "How on earth do you do these things?"

Make it short and easy; check the art store for books on reprographics for graphic artists. Graphic artists have done it all this way for years. Their simple guide books show how to draw background information on base sheets; add variable information on overlays; keep the two in registration; and make final composite prints for study or presentations. It's easy. It's fun. And it's possible for designers to get all manner of gorgeous presentation special effects with these techniques.

Design development and working drawings are successfully done this way by HUNDREDS of architectural firms across the U.S. and Canada. Some have been doing it consistently for over 20 years and they can't imagine why any firm would consider doing otherwise.

"DON'T TALK TO US ABOUT 'SYSTEMS,' WE DON'T DO REPETITIVE WORK"

Architect and systems pioneer Ned Abrams had an eye opening exercise for people who didn't think there was much repetition in their work. He would have A/E's sit down with a set of design development or preliminary working drawings and have them red-mark the repetitions.

They'd come back to Ned with a few red marks on their prints and say: *"See, I told you we don't do much repetitive work."* Then Ned would take their prints and go to work with red pencil and draw circle after circle after circle until they threw up their hands in surrender.

What Ned demonstrated was one of the most difficult aspects of working drawings: Repetition is so common and universal that it's almost invisible, like the air we breathe. It takes a special effort to open one's eyes and raise consciousness enough to see all the waste . . . and all the opportunities.

To see the repetitions clearly, you have to learn their differences, their types. It's like bird watching. Walk through the woods and you'll hear and see a few birds. Go with a committed "birder" and you'll see dozens and dozens of distinct species you never knew existed.

The main types of repetitive elements are: the Avoidables, Unavoidables, and the Alike but Different.

--The Avoidables are items like the same note or dimension repeated over a sheet of drawings that could be covered in a single general note: . . . *"unless noted otherwise."*

--The Unavoidables are items that have to appear again and again, like identical apartment units in different buildings of a complex. These have to be shown repeatedly but they're identical and can best be done by exact-size reprographic reprinting as with a vacuum frame.

--The Alike but Different items are those that are the same in most ways but vary in some aspect such as size. Or they're reversed, upside down, different on one side, etc.

--Then there are One-Time Repetitions and Multiple Repetitions.

Each type has to be identified for what it is, then handled in a different way.

Here are some further examples:

An "Alike but Different" "L" footing on two ends of a building cross section, for example. It's opposite at each end of the building, but it's still the same element. And the footing blown up in an architectural detail may look different because it's larger, but it remains the same item. And it may be shown in another detail with a different exterior sill and wall connection, but it's still the same element. These items that are reversed, repositioned, resized, and with different items attached to them are repetitions-- ideal subjects for standard symbol and detail libraries.

Some items aren't repetitive from job to job but only in the project. They're still subjects for copy and pasteup.

Some items are unique to a job but still may be repeated numerous times on a single drawing. Sometimes, as with window bays in elevation, and common room types in plan, the repetitions on a single large sheet might add up to days of manual drafting.

By the time you review all the types, and really LOOK at a set of working drawings with this viewpoint, you realize virtually every aspect of drafting is repetitive in some way or another.

Then you realize the simple choice: Keep doing the same work over and over year after year, or stop, sort them out and apply the systems principles which are fundamental to this book.

114

WHY WORKING DRAWING PROBLEMS OFTEN START IN THE DESIGN STUDIO

What happens when a designer lays out a building according to a ten-foot setback when the setback limit is really fifteen feet?

Or draws a five-foot wide corridor to scale and labels it as being six feet?

Or thinks a building shell can easily be changed from masonry infill to a metal curtain wall no matter what structural frame is used?

What commonly happens is that the problems are passed along to the drafting staff. Often as not, the hand-me-down problems remain unsolved or the solutions that are devised create other problems. Equally often both the old problems and the new ones are passed on further to the construction site.

Offices with the fewest ongoing design problems use a special sequence of design decision making. They know designers often leap into the fun of designing a building before they understand all the problems and conditions of the project. Naturally, the less designers understand about the project, the more changes there will be later. So these firms give their designers checklists that tell them what data has to be found and what decisions have to be made ahead of other decisions. Some top designers have made their own checklists along this line as personal reminders.

Checklists from the nation's best designers and project managers are based on the fact that there's a logical SEQUENCE OF DEPENDENCY in making any complex set of decisions. That is, some decisions **have** to depend on others. If they're made out of sequence, they'll be half-baked and may lead to other off-the-wall decisions--all of which have to be repaired later when it gets most expensive.

If designers don't use such sequential checklists, production managers benefit by doing it themselves. They check to see what decisions have been made, or not made, and whether they're based on logical sequence and coordination. They postpone moving deeply into those parts of working drawings where design decisions are still indefinite or questionable.

In one well-known New York office all design work is checked in much the same way working drawings are checked -- for completion and soundness of design decisions. Working drawings begin in earnest only when everything is checked and signed off. This process may delay the start of working drawings by as much as two weeks. But it can speed up their completion by a month or more depending on the size of project.

A production manager in Kansas City gets a similar result through verbal questioning during meetings: Has the designer been to the site yet? Are they working with the latest building code for the project city? Does the designer understand the span limits of the structural frame they've tentatively chosen? The questioning isn't interrogation, just gentle, helpful reminders.

A SEQUENCE OF DEPENDENCY checklist used by project managers before finalizing working drawing data follows below.

This list of considerations is also a list of the most common sources of changes during design development, working drawings, and during construction.

__ 1)   **Confirm final occupancy types** and actual final tenant occupant loads.

__ 2)   **Room sizes, corridors and exits match building occupant loads.**

__ 3)   **Room sizes match size required for equipment and storage** as well as occupant loads, as confirmed by the users of the rooms.

__ 4) **Building exterior boundary is per zoning,** setback limits, and building code.

__ 5) **Building parking is per zoning,** building, and neighborhood codes.

__ 6) **Building height in feet and in stories is per zoning** and building code.

__ 7) **Interior spaces are located per importance of their relationships** to one another--adjacency conflicts have been resolved.

__ 8) **Interior spaces are located relative to the exterior** as per importance of adjacency to exterior functions.

__ 9) **All handicap access and fire exit requirements are provided.**

__ 10) **A structural system and module** has been decided that matches:

__ Multiples of the spans of room sizes (where practical)

__ Building height and floor to floor height requirements

__ Fire code requirements

__ Site constraints

__ Exceptional mechanical service and load requirements

__ 11) **A plan unit module** has been decided that matches:

__ Logical increments of room sizes

__ The structural module

__ Partitioning, ceiling, and raised floor systems

__ 12) **An HVAC system** has been selected that is:

__ Centrally located

__ Provides uninterrupted main duct runs

__ Sized and planned for additions and easy maintenance

__ Exceeds minimum air change requirements

__ Considered relative to conditions of locale, climate, and specific site

__ 13) **Ceiling system and heights** have been decided that:

__ Work with the lighting system

__ Provide sufficient space for HVAC, lighting, plumbing, other services

__ Match the partition system

__ Work with above-floor construction that will support ceilings

__ 14) **Plumbing supply and drains** are as centralized as practical with clear access for maintenance.

__ 15) **Plumbing fixtures** are fully accessible to handicapped.

__ 16) **Electrical power and communications** systems are chosen and planned:

__ For ease of expansion and maintenance

__ For maximum safety and fire protection

__ 17) **Lighting** is chosen for:

__ Room functions per tenant approvals

__ Occupant health

__ 18) **Exterior materials** are compatible with:

__ Building construction budget

__ Considerations of locale, climate, and specific site

__ The structural system

__ Construction system

__ Maintenance considerations

__ 19) **Fenestration** is compatible with:

__ Building construction budget

__ HVAC system

__ Considerations of locale, climate, and specific site

__ Exterior shell materials

__ Maintenance considerations

__ 20) **Interior materials and finishes** are compatible with:

__ Fire code requirements

__ Occupancy needs per tenant approvals

__ Building construction budget

__ Maintenance considerations

Any of these that can't be checked off as coordinated and reasonably definite decisions are very likely to be revised during working drawings or later.

Feel free to edit the list as you see fit and pass it around for others to use. There are always a few prematurely decided or undecided items in every project that will cause troublesome changes later. This list will help you catch most of them.

WHAT DO THE DESIGNERS NEED FROM THE CLIENTS?

WHAT DO THE DRAFTERS NEED FROM THE DESIGNERS?

WHAT DO THE ENGINEERS NEED FROM THE ARCHITECTS?

Some checklists have to be written on a project-by-project basis, but at least you can provide handy starter lists, formats, and reminders for doing the lists.

For example, engineers always need additional data at crucial junctures from the architectural staff who are doing the working drawings. The architectural drafters may need to consult with the designer for that information, but the designer may not be able to say until he or she gets clarification from the client.

Not only is this merry-go-round time-consuming, it's fraught with potential errors in communication. There's every likelihood that some crucial data will fall through the cracks.

Thus the need for standard communications checklists; that is, simple lists of what each member of the team needs from the other members. For example:

--A few architectural firms have asked their structural, mechanical, and electrical engineers to write checklists of project information they need from the architects.

--Some have asked their production staff to make similar lists of essential data they usually need (and don't always get) from the designers.

--Some have also asked designers to note the decisions and data they typically need from the clients--particularly in the sequence they need it.

Such lists are made anyway in memos and checkprints done throughout any project, so they may as well be formalized and prewritten as standards for use in future projects.

Every office has its own Red Flags to emphasize in such lists. Some engineers **always** have to ask for exactly the same missing drawing data from firms they've worked with for years. And some clients are always late with certain kinds of data on job after job. The checklists capture these peculiarities and reduce the recurring time loss and frustration.

As such lists are developed, you'll find items that can be given multiple uses as reminders, supervisory instruction, and checkprint supplements, in predesign checklists, working drawing checklists, coordination checklists, and construction administration checklists. Thus one item may serve up to seven of the checklist functions and do so in four or more phases of work. That's 28 or more reuses of each piece of information on any one project!

118

HOW CAN YOU PREVENT EXCESSIVE LAST-MINUTE DESIGN CHANGES?

Midstream revisions are costly, demoralizing to staff, and they throw schedules to the winds.

Here are some good answers to the problem:

--A West Coast solar home designer programs all design decisions in advance. He leaves little opportunity for oversights and last-minute flashes. His central tool is an extensive "dream list" created during a client interview. He covers everything a client might possibly be concerned with--from ceiling heights to finishes to furnishings--before he draws a line. All items are systematically rechecked through design development. **By the time production drawings are under way, virtually everything has been decided and confirmed.**

--A large Midwest firm also does itemized, room-by-room walk-throughs with clients. Jerry Quebe, of Hansen Lind Meyer, says: *"The system catches those potential later embarrassments, such as finding out there's going to be a filing cabinet in front of a light switch."* When HLM completes detailed room layouts, they have the client formally check them off.

This checklist approach combined with systematic programming mean the design documents are basically complete outline working drawings at the close of design development.

That lets HLM gain direct reuse of design documents as production drawing. Working drawings are off to a flying start, with 20% of the work already completed and little chance of later, major changes of mind by the client.

--California architect Lee Ward has a special device that inhibits clients from making arbitrary late changes: A chart with time and costs recorded as originally estimated for design and production.

A copy of the diagram is part of the design service contract. When a client requests revisions after design approval, he or she has to review the job schedule diagram with the designer and participate in changing the time and cost budget. This makes the time and cost consequences of design revisions completely clear to the client.

A SPECIAL HELPER: THE ROOM-BY-ROOM DATA-CAPTURE FORM

Every room type is repetitive in most ways.

Every hotel sleeping room has essentially the same requirements as every other. As does every office conference room, medical exam room, mechanical room, etc.

So why should designers, drafters, and engineers have to look up the same requirements from scratch every time they start a new project? It's a heck of a research job too, since flooring will be in one or more handbooks and catalogs; walls in another; and fire regulations, electrical requirements, hardware, acoustics, etc. will all be scattered every which way in handbooks, Sweets, personal files, and building codes.

Here's a list of checklist items for capturing room-by-room research data for potential reuse on future projects. Use it to record project-specific information at first, then compile the information you collect as suggested options of the most likely wall finishes, door types, fire ratings, etc. to use in future jobs.

You have to collect this information for every project, anyway, so it may as well be compiled in a standard format and reused in future projects.

CHECKLIST OF ITEMS FOR ROOM-BY-ROOM DATA-CAPTURE FORM

___ **STANDARD ROOM TYPE**

___ **Subcategory**

___ **Project-Specific Information (project name, building type, etc.)**

___ **Project Room Name and/or Number**

ARCHITECTURAL

___ **FLOORING:**

___ Auxiliary Flooring

___ Subfloor

May also include:

___ Curbs

___ Recesses

___ Platforms

___ Waterproofing

___ Fireproofing

__ **BASE**

__ **WALLS:**

__ Wainscot

__ Special Walls

__ Glazed Walls

May also include:

__ Substructure

__ Backing for fixture supports

__ Waterproofing

__ Fireproofing

__ Openings/Passthroughs

__ Soundproofing

__ **CEILING:**

__ Auxiliary Ceiling

__ Soffit/Furred Ceiling

__ Acoustic Control

May also include:

__ Ceiling Substructure

__ Equipment Supports

__ Access

__ Fireproofing

__ **DOORS:**

__ Door Glazing

__ Standard Hardware

__ Special Door Hardware

May also include:

__ Fireproofing

__ Soundproofing

__ Vents or cuts

__ **Door Frames**

__ **Fire-rated Materials**

__ **Fire Protection:**

__ Sprinklers

__ Alarms

__ Emergency Signs & Lights

May also include:

__ Ceiling/Attic Fire Barriers

__ Chase/Duct Barriers

__ Floor Fire Barriers

__ Panic Hardware

__ Smoke Removal

__ **Acoustic Requirements:**

__ Allowable Sound Transmission

__ Soundproofing Treatment

__ **Built-ins, Fixtures & Equipment:**

    __ Room accessories

    __ Casework

    __ Cabinets

    __ Shelving

    __ Specialized storage

    __ Pin Boards

    __ Chalk Boards/Felt-Tip Boards

____ **Furnishings & Furniture**

ELECTRICAL

__ **POWER:**

    __ Electrical Power Outlet

    __ Electrical Equipment

    __ Electrical Supply

__ **LIGHTING:**

    __ Lighting Fixture

    __ Switching

    __ Illumination Required

__ **COMMUNICATIONS**

__ **COMPUTER**

__ **SECURITY**

PLUMBING

__ **Plumbing Fixtures:**

    __ Water Supply

    __ Drains

    __ Special Piping

HEATING VENTILATING AIR-CONDITIONING

__ **PRIMARY SYSTEM:**

    __ Fixtures

    __ Vents

    __ Supply

    __ Air Changes per Hour

    __ Temperature Controls

    __ Humidity Controls

CHAPTER NINE

QUALITY CONTROL AND FAILURE PREVENTION: EASIER THAN YOU THINK

THE MOST COMMON OMISSIONS IN WORKING DRAWINGS

This may surprise you, but working drawings that were done around 1900 have essentially the same problems as those done now. As do working drawings done in the 40's, the 50's, 60's, and 70's.

The oldest drawings often LOOK better. Ink and linen line quality was pleasing, ornament was abundant, and lettering more meticulous and less rushed. But, in general, the average drawings of yesteryear had about the same number and kinds of errors and omissions as any set of drawings you would find today.

A survey of current working drawing problems, included an analysis of over 200 sets of drawings representing a cross section of offices of every type and size. The problems of the divergent offices are very much the same, and are virtually unchanged since the early 60's.

Check your documents in an objective search for problems and see how your work compares with the averages that follow.

--40% of working drawing sheets have dimensioning gaps or errors, such as:

--Missing rough-opening size dimensions or notes (a big headache on the job).

--No vertical dimensions of sills and lintels.

--No location dimensions for floor drains in slabs (plumbing and architectural drawings often conflict visually and when neither has dimensions, the plumber has to guess).

--No cumulative dimensions to structural columns or grid for locating walls and partitions. (The building frame will be there when walls are built, so structural dimensions should govern but they're often ignored in architectural dimensioning)

--In 70% of plans for larger buildings, parallel strings of plan dimensions don't add up or don't match structural frame dimensions. This is a further result of failing to reference walls and partitions to structural grids or to use cumulative framing reference points.

--Over 40% of plan and elevation drawings include unrealistic fractions of inch dimensions such as 1/4", 1/8", and 1/16". Sometimes there are highly detailed studies of store front mullions so tightly dimensioned that only a machinist could make them.

--Over 55% of drawings lack clear reference points for height dimensions in exterior elevations. The datum start point is often unstated or unclear.

--In over 60% of sets of small-scale plans, it isn't clear if some dimensions are to partition wall finishes, wall framing, or centerlines.

--50% of exterior wall details lack specifics on flashing and other water-proofing. The words "caulking" and "sealants" in notes are used interchangeably as are notes of "waterproofing," "moisture barrier," and "vapor barrier."

--Construction joint locations, sizes, and types are missing, unclear, or contradictory in over 70% of the sets. The phrases "expansion joints," "movement joints," "contraction joints," and "construction joints" are used interchangeably. Joint notation, besides being inaccurate, is frequently overdone and either duplicates the specifications, contradicts them or manages to do both.

--40% of relevant details do not clearly show or note anchors, supports, backing, and fittings for mounted fixtures and equipment.

--Offsets in exterior walls that have openings or other special construction are missing in 20% of the exterior elevation drawing sheets.

--Reference notes that don't reference, such as "see Structural Drawings," "see Specifications," etc, appear throughout 80% of all working drawing sets. They're misleading nearly half the time since the reference doesn't exist, or is very difficult to find, or contradicts data on the reference drawing.

--Up to 40% of details in some sets of drawings are not referenced from small-scale drawings. The details are there but there's no way of knowing if they should be. These almost always require explanatory phone calls and memos during construction.

--Detail referencing is so bad it can take 10 to 30 minutes to find a detail in some large sets of documents. Obviously nobody will be willing to search for so long a time and if it happens once on a job, details won't be searched for again.

--Slightly over 33% of all details shown serve no purpose. They need not, or should not have been drawn at all.

In summary:

15% of all sets of working drawings from licensed professionals are inadequate for bidding and construction.

Between 30% to 40% of all working drawing sheets have many lapses in important data.

Only about 30% of working drawing sets can be considered fully professional.

If your office needs improvement, it may not take major housecleaning to do it. Several top production managers have told us they got great results from very simple means: Sit down with your drafters for an hour and review each item on this list. Most drafters have never had such a learning session. Just one meeting can cut most repetitions of the same problems in your office for most of the year to come.

WHAT'S MOST IMPORTANT? THE INSIDE SCOOP!

What do architects think is the most crucial in technical knowledge? Try WATERPROOFING!

Here's a list of the TOP NINE most important skills and competencies out of 400 listed in a survey sponsored by the California Board of Architectural Examiners.

1) Understanding techniques for making buildings water and moisture proof.

2) Understanding roof system slopes, applications, and flashing.

3) Knowledge of fire safety materials.

4) Knowledge of roof drainage and water disposal.

5) Preparing details for moisture and environmental control.

6) Knowledge of flashing, drainage, and weather stripping of wall openings.

7) Design and detailing of ramps and stairs.

8) Understand dampproofing and waterproofing subgrade walls.

9) Use of moisture barriers in concrete slabs on grade.

Seven of the top listed items deal with water. The criteria for ranking the most important aspects of practice was, according to the California State Board, those items: "... most critical to the public health, safety, and welfare and most frequently used in practice." Could be we're on to something.

FAILURES REMAIN EPIDEMIC -- CAUSES REMAIN FEW AND SIMPLE

Urgent reminder: A rash of building failures and lawsuits this past year remind us once again that the leading factors in most building failures are:

1) Last-minute changes and substitutions of products or an abrupt change in some aspect of design.

2) Over-reliance on building codes and handbook minimums. Minimum standards are simply not acceptable and not adequate when it comes to preventing major failures.

3) Lack of final checking such as when the last set of working drawings is red-marked, yet the final prints go out the door without a look to see if the final corrections were made as specified or not. Or a contractor is told to change or correct some work and it's assumed it will be done without a final confirmation.

4) Miscommunications and lack of communication. Prior to the Challenger shuttle failure, for example, there were memos of warning -- they were not read at the right time by the right people.

These are the leading general causes of failures as reported time and again by insurers and forensic investigators of building failures.

They also happen to be the leading general causes of all kinds of failures: Business, manufacturing, medical . . . every trade and profession has its failures and most of their causes boil down to variations of the above.

Last-minute changes, for example, induce two kinds of potential failure problems.

First, late changes are usually made under pressure and rushed. The original decisions may have been carefully thought out months earlier, but nobody remembers the sequence of reasoning behind them. The change may be a quick-fix remedy or cost reduction offered by a shady contractor or supplier. This is a very common source of roofing failures, for example, when bids come in too high and in the search to reduce costs, somebody offers a terrific last-minute bargain on a new kind of roofing . . . A year later it's wetter inside the building than outside and the contractor is long-gone.

Second, all the repercussions aren't fully considered. Every change requires additional changes to accommodate it and they may or may not be considered. As when a poured concrete frame is substituted for steel but all the details for infill brick panels remain the same. The brick fails when it's squeezed between its own moisture swelling and the shrinkage of the concrete frame.

It's not widely appreciated, but building codes and "handbook engineering" are a major cause of failure. For example, long-span roofs, shallow dome roofs, trusses and space frames will be loaded with piles of snow this winter on one side of the roofs and not the other. The code dead load allowances used in calculations will have assumed a totally unrealistic even distribution of snow. And even those load allowances can be different for two adjacent communities. The only bright side is that such failures occur during overnight blizzards when people aren't in the buildings.

Codes, handbooks, and rules of thumb have to be understood as being only minimal preliminary working assumptions. If minimum standards are strictly adhered to throughout a design -- as they often are -- failure is a virtual certainty. This particularly pertains to issues of life safety -- exit corridors and stairs, fire sprinklers, etc. Minimum code standards are compromises; they don't allow for the vagaries of building maintenance and "unexpected" circumstances. Minimums aren't adequate and people may die if those minimums are taken to be "maximums" by your design and drafting staff members.

Just a few examples of major historic failures due in part or whole to over-reliance on code standards: The Hancock Tower -- 10,344 curtain wall panes of glass replaced -- plus other repairs requiring four years and an extra $84 million. The MGM Grand Hotel Fire, 85 deaths and 500 injuries due to inadequate but "acceptable" fire protection. Hundreds more such events that you don't hear about occur every year, but they're just as tragic for the individuals involved.

Project managers who let drawings go out the door without a final quality control check say they have a reason. There's never enough time for a final-final check and they have to rely on "self checking."

You couldn't find a more certain prescription for errors in any realm than "self checking." Especially at the last minute when people are most tired, distracted, rushed, and pretty much fed up with the whole project.

A tip for those facing fee-services negotiations with clients: Let clients know that the offices that charge the least are the ones who are able to spend the least time on cross coordination and final quality control checking. That's been noted by liability insurers time after time.

Finally, it seems in every case of a major failure there was ample advance warning that things were going haywire:

--Conflicts and antagonism between designer and production staff. Production people say they're being required to "do the impossible" to make an unrealistic design workable. The designer says the drafting people are incompetent and too lazy to do their job. After awhile the communication is no communication and everyone assumes the other will take the blame when things go wrong in bidding and/or during construction.

--Architectural and engineering staff get into conflict and ignore one another's memos and check-sets. It often happens along the same lines as the design-production conflicts.

--A major problem is allowed to remain unresolved throughout the project and major last-minute changes will be required to remedy it. Along the way there's no resolution, suggestions are ignored, and all that's left is the secret anticipation that somebody will "get theirs" when it all comes out into the open.

--In these cases, memos and meetings are NOT used to communicate useful information but as self defense devices to get notes "on the record" to avoid or assign blame when disaster strikes.

To avoid these conflict problems that lead to building failures is to acknowledge that the people involved know there are problems and need to get them into the open. Secretaries know, the lowest level drafters will know, and the key contentious people involved in the situation will know. To get them to talk means to raise issues at every warning sign.

Several office principals and project managers we know start every in-office meeting with variations of:

"OK, what are the problems we should be dealing with?"

"What might be going wrong here that we're not looking at?"

"Things are too quiet and seem to be going well; that may be a sign that something's wrong."

"Do any of you know about any possible ways we might improve on what we're doing?"

And one of the most effective ways to force everyone to at least say something:

"I want every one of you to name at least one potential problem in this situation that we might not be considering."

Then when somebody has the guts to hint about a real problem that's simmering, these managers have enough sense not to recoil at bad news and hang the messenger. They listen, then investigate, and then do some troubleshooting.

127

HOW BAD IS IT TO HAVE FIVE COORDINATION ERRORS PER DRAWING?

The average working drawing sheet has about five coordination errors. That's what government personnel find when they do in-house checking of A/E construction documents.

People respond in one of two ways to that piece of information:

The Hawk: *"That's five too many! "*

The Dove: *"Five isn't so bad when you consider the thousands of items in drawings."*

There's truth in each response. Five IS too many. One is too many, for that matter. But many of the lapses **are** trivial, they just **look** dumb (like North arrows that point South).

What slip by the most and cause the most trouble are obscure technical items. As cited by former Navy A/E coordinator William Nigro:

--Architectural/structural coordination blank spots, such as positioning of rainwater leaders when they come to the foundations. Do they go through, under, or over? Some sets of drawings show all three.

--Architectural drawings that don't show enough space for ductwork in the ceilings and HVAC drawings that don't care.

--Mechanical drawings that refer to non-existent equipment supports supposedly shown in architectural or structural drawings.

--Plumbing drawings with chases and drains sized differently from those shown in the architectural floor plans and/or allowed for in structural drawings.

--Electrical drawings that show light fixtures that fail to match the architectural reflected ceiling plans. Incredibly, sometimes they're totally mismatched in location, types, sizes, and quantity.

--Electrical drawings that show mismatched horsepower ratings and voltages, and voltages shown differently on different electrical sheets.

--Disturbingly common and troublesome architectural drawing errors:

___ Segmented plan match-up lines that don't match

___ Incorrect scales

___ Specifications and drawings that refer to nonexistent data elsewhere in the documents

___ No ceiling, wall, or floor backup and support detail for fixtures, fittings, railings, corner guards, equipment mounts, etc.

BUILDING FAILURES AND THE MISSING LINK BETWEEN WORKING DRAWINGS AND CONSTRUCTION

Time for a pop quiz:

Question: What are the three worst and most common points of construction failure, claims, and lawsuits?

Answer:

1) Roofing -- 40% to 50% of all claims.

2) Exterior walls and fenestration (especially involving brick masonry) about 30% of claims.

3) Site drainage and below-grade waterproofing -- 8% to 12%.

Question: What is the most common project management point of failure?

Answer: Rushed acceptance of last-minute changes and substitutions, especially during construction. Last-minute changes have been rated by insurers as the project administration cause of half of all failures.

Question: How can you avoid the great majority of failures, claims, and lawsuits?

Answer: See the previous two answers. Establish stringent office policy against rushed consideration of last-minute changes and substitutions, **especially** in roofing, exterior walls, fenestration, and sitework. Enforce this and you eliminate most of your potential construction failures.

Here's another answer; one that enforces the previous one but isn't obvious or well known:

ALWAYS hold a preconstruction planning meeting before each of the major Red Flag phases of construction.

A preconstruction planning meeting must involve everyone whose work has to be coordinated in a particular phase: the prime contractor's site supervisor, the subcontractors who will do the work or whose work is affected, and consulting engineers, such as the civil engineer, to review the planning of sitework grading.

Preconstruction meetings get right to the heart of the worst and most common construction problems. For example:

--Even the best construction documents contain lots of opportunities for misunderstandings. The preconstruction meeting is a rehearsal that finds problems on paper instead of on the jobsite. That cuts back on the rush and pressure that force people into ill-considered decisions.

--Construction details are often misunderstood or not even seen. The meeting clarifies their meaning and uncovers possible discrepancies between the details as drawn and the realities of the job.

--Specifications are not carefully read and always have to be clarified on the job. The preconstruction meeting brings out the primary conditions required for acceptance of the work. If there are contradictions between specs, detail drawings, and/or shop drawings, they're worked out on the spot or noted for correction.

--Engineers' intent isn't always completely understood by architect or contractor. Yet it may be the architect's representative who watches over the civil engineer's grades and slopes for drainage, for example. Better to clear it up in an office meeting than on the phone in the midst of site grading and excavation.

--Subcontractors have unanticipated conflicts in work. Although the prime contractor is supposed to coordinate all the subs, some aspects of coordination invariably fall between the cracks.

The architect and engineer(s) at the preconstruction meeting can raise these issues directly with the subs just to make sure they're being handled. For example, is all the trenching and fill coordinated among the crews doing drainage, utilities, supply lines, and paving? Is there anything special the contractor who handles the roof deck and/or skylight curbs should know about preparation for roofing application? Have there been any changes in the roof deck and insulation since the roofing was originally detailed and specified? Among the various trades involved with the brick walls, who is responsible for placing window flashing?

Even if these issues have been worked out previously, the preconstruction meeting is insurance that everyone's memories are intact. And it tells the prime design firm what to double-check for compliance during the construction process.

Helpful tips from offices that have used preconstruction meetings for many years:

--To make sure preconstruction meetings happen, they have to be part of the specifications. Then the specifications have to be enforced or that portion of them will be nullified.

--Schedule the meetings with reasonable leeway before each Red Flag phase of construction. Don't let meeting times sneak up on you or the others; they will be inconvenient and it will be too tempting to skip them. Make them a part of the construction calendar.

Those who use preconstruction meetings rank them as among the best of their production, problem prevention, and quality control tools. It's the kind of action that we like to boost the most: Huge timesaving benefits with relatively trivial investment.

A MINOR CHANGE EQUALS A $60 MILLION LOSS

The marble cladding of the 1,100' tall Amoco Tower in Chicago failed and was stripped and resurfaced at an estimated cost of $60 to $80 million.

The existing marble cladding is between 1.25 and 1.5" thick. The new, stronger, stone will be 2" thick. (The reported source of failure is loss of strength due to thermal cycling.)

This type of disaster was foreseen two years ago by Lee Nelson, FAIA, chief of preservation assistance for the National Park Service. The July, 1987 issue of ARCHITECTURE magazine reported:

"Of particular concern to Nelson is the current resurgence of stone panel building facades. He believes that many stone panels specified today are just too thin. He fears that water will find its way through them to corrode and rust the metal pins that attach the panels to the building frame, causing panels to fall off and shatter.

"Additionally, thermal damage may plague these stone facades. A common stone spandrel panel might be six feet long, two feet wide, and only an inch thick. Nelson compares such a unit to the tall, narrow chimneys on 18th-century buildings which expand and contract daily because of cyclical heat from the sun. Over time, the chimneys developed permanent curves due to this thermal stress. Nelson is convinced that the thin stone panels will respond as the chimneys did.

" 'Architects need to find a better way to attach these stone panels, and for that matter architectural concrete, to allow for the curvature that will result,' he says. 'Or else the industry should return to thicker sections so that there is more mass to resist thermal stresses and more material to be anchored.' "

130

LOTS OF NEW FAILURES, STILL THE SAME OLD CAUSES

Several years ago I discovered a fresh air intake installed beneath a loading dock where trucks back up. I thought at the time: "This has to be one of the strangest freaks in the world of architecture."

A little later I found another one. And another.

The production manager in one of the nation's top hospital design firms told me they had done the same thing at a San Francisco Bay Area hospital. As did a manager in one of Chicago's largest architectural firms. They had designed an air intake above the alley parking garage entrance to their headquarters building.

And then a news story from LA: UCLA Medical Center found a fresh air intake at its loading dock that pipes the truck diesel fumes right to the nurseries for newborns. Infants were relocated after high levels of carbon monoxide were discovered.

Then there are the fresh air intakes and exhausts placed at opposite ends of high ceilings. The "fresh" air enters and leaves nine feet above the floor, never reaching the lungs of the occupants below. And there are untold numbers of "fresh-air" intakes on roofs downwind from exhaust vents and chimneys.

What has been the ruling force here? By what logic are fresh air intakes made ineffectual or poisonous?

One rule above all: Appearance. Some designers don't know about such things as air supply to buildings and are surprised and offended when engineers want to plop ugly metal grilles on the facades.

But why is it that production people and engineers don't notice such things?

We've asked some of the culprits, with guarantees of anonymity: "Why didn't you notice this, or if you did, why didn't you do something about it?" The answers are telling:

--*"This office is an emotional environment. The designers yell the loudest and they get their way. Sometimes we raise a technical objection and they say: 'That's what you're hired for. So fix it.' And if there is no solution, we just have to let it slide and hope somebody will see it and work it out during construction."*

--*"They hide the mechanical gear on the rear corner roof of the building and you know it's going to cause problems, but if it means changing their design, they won't listen."*

--*"I tell them what will happen and if they reject what I say, I write it in a memo and file the original. If somebody gets hurt, at least I'm in the clear."*

--*"I used to believe in collaboration among design professionals. But in the trendy firms I worked for, that's bull . . .!"*

OK, that's one problem, "designer dominance" over technical realities.

But what about glaring errors on the technical side?

--Five buildings were closed last month at the Cal State Long Beach campus because the load-bearing stud walls were undersized for the building height--16' high walls framed with 2x4 studs instead of 2x6's or 2x8's. (This followed a roof collapse early this year.)

--Brick, stone veneer, and tile continue to fall off of hundreds of relatively new buildings because of inadequate allowance or detailing for thermal expansion.

--Built-up roofing failures remain the primary source of claims, mainly because of inadequate drainage, faulty installation, and last-minute substitutions and deviations from the specs.

DETAIL DRAWINGS TELL A STORY; BUT IS IT THE ONE YOU WANT?

On July 17, 1981, two skybridges in the Kansas City Hyatt hotel collapsed on a lobby filled with party-goers. 113 died and 180 were injured; many crippled for life.

The collapse started with a vaguely defined detail drawing of a suspension rod and connective nut. The detail wasn't buildable in any obvious way and a buildable but structurally unsound substitute detail was used instead.

A similar detail failure led to a massive ceiling collapse at the Jersey City Transit Terminal in 1983. "Only" two people were killed. Had it been rush hour when the terminal was usually crowded, hundreds of people would probably have been crushed.

That failure started with a construction detail that "implied" an insert in a concrete roof slab to support hangers for the suspended plaster ceiling. A series of errors and misunderstandings followed which culminated in a workman stepping into an inadequately supported ceiling plenum. He fell to his death in a pile of rubble that buried another person.

The Las Vegas MGM Grand Hotel fire started because of detailing that allowed dissimilar metals to come into contact: copper tubing next to aluminum electrical conduit in a soffit above a pastry display case. Electrolytic corrosion led to an electrical short, and eventual combustion. The blaze killed 85, injured 500, and left over $1 billion in claims to settle.

The mullion detail of the Boston Hancock Tower, built in 1972, led to a building that couldn't be occupied for four years, nearly doubled the cost from $96 million to $180 million, and required replacement of 10,344 panes of glass. (Since the parties involved have refused to share technical data on what they know about the causes of the failure, their fellow professionals are left to draw their own conclusions. Underdesigned mullions along with under-engineered glazing are widely considered to be the likeliest source factors.)

These few famous failures are dwarfed by the total number of major and minor lapses plague buildings everywhere. But all have a lot in common, in particular they all share an acute procedural problem: There is no standard SYSTEM for making and checking details.

Think about that for a minute: There are no profession-wide standards for designing drafting, and checking construction details.

There are national standards for contracts, bid forms, change orders, and every other major document in the project manual. Specifications format and standards have been standardized through the Construction Specifications Institute. But details? Virtually nothing!

Under the circumstances it would seem strange if detail problems WEREN'T rampant and universal.

EASIER WAYS TO QUALITY CONTROL

Your quality control system can have the major features of the most effective systems in the country. It doesn't take much; just a few key steps take care of most of your potential errors and liability problems. Adapt the following action list to your needs:

1) Establish a Red Flag system that alerts staff to the most common problem spots in design, working drawings, bidding, and construction.

The main Red Flag hot spots during planning, detailing, and construction are:

--**Roofing: materials, application, slopes, and drainage** (the source of nearly 50% of all claims)

--**Below-grade waterproofing**

--**Exterior walls: fenestration, joints, and caulking** (the source of 30% of all claims)

--**Site grading and drainage**

--**HVAC:** jumbled, unworkable systems due to lack of preplanning and coordination

--**Structural: Detail errors,** especially those not checked during fabrication and erection

And special warning about a Red Flag which doesn't lead to many lawsuits, but is a constant complaint among clients:

--**Poorly considered materials and fixtures,** especially those that interact electrochemically, such as dissimilar metals and those not suited to long-term maintenance, wear, and abrasion

2) Educate the staff with jobsite feedback.

--**Circulate or display prints of drawings** that illustrate standards of office drafting quality, clarity, and completeness.

--**Circulate samples of past errors.** Assuming this can be done without unduly embarrassing any current staff member, every office has plenty of raw material that can illustrate mishaps in design and drafting.

3) Systematize the checking process. The traditional sit-down session with red pencil and a set of check prints just isn't adequate any more.

--**Use transparencies and layered checking on a light table to compare work of different disciplines.**

--**Use a "checking checklist."**

--**Do follow-up checks** to make sure the items that have been marked for change and correction are actually completed on the original drawings as directed. They often are not.

ABOUT WATERPROOFING

Here's a checklist for dealing with the most common sources of water entry; mainly how to deal with them in terms of design and detailing:

--**Design a generous number and size of weep holes** to drain the water that inevitably enters any masonry wall or curtain wall. Such drains, even if properly designed, sometimes aren't executed. And if properly built, they may become clogged with construction debris. A minimum standard: 1/2" weep holes in cavity walls at 24" o.c. at each story, and at flashing at lintels and headers. 32" o.c. is OK for concrete block.

--**Don't let the water that normally enters an exterior wall reach steel framing.** Rusted framing is a frequent cause of long vertical masonry cracks at the corners of steel framed masonry buildings.

--**Whenever possible, keep steel reinforcing three or more inches below the surface of masonry and concrete.** Water commonly enters hairline cracks, or just seeps through the surface of reinforced materials, and oxidizes rebars if they're within a couple of inches of the surface. Then the reinforcing rusts, expands, and spalls the surfaces.

--**Don't let water corrode and swell metal anchors, stirrups, straps, etc.** Don't place steel anchors, angles, etc. where they'll be exposed to weather or moisture because whatever protective coating they have will wear off eventually. When the metal expands it will open up adjacent construction to more water entry, and so on. Very often, exposed metals that should receive added protective coatings at the jobsite don't get it.

--**Never allow dissimilar metals to be in contact.** Such combinations in the presence of moisture are virtually certain to cause corrosive electrolytic action.

--**Design mortar joints to match the weather.** Concave tooling reduces water entry and should be used in wet and windy climates. Rough and flat tooled or raked joints are more vulnerable to water entry and should be avoided where there's much weather exposure.

--**Allow for capillary action and building air-pressure suction of air and water.** Water can be pulled out of the ground or sucked in from the outside not because of wind, but because the exterior air pressure is higher than interior. Block the water AND provide pathways to get it out when it does get in.

--**Design flashing at lintels and flashing that abuts expansion joints to be "dammed" or turned upwards at the ends 6" to 9".** Otherwise the flashing can channel water inside internal joints and cavities instead of outwards to drain the wall.

--**Don't cut thru-wall flashing off at the wall surface.** Thru-wall flashing is often stopped short of the exterior face of a building for reasons of appearance. This invites water re-entry beneath the flashing. Internal flashing should project outward from the building face by 3/4" and turn down at 45 degrees to form an effective drip edge.

--**Be generous with internal concealed wall flashing to redirect incoming water back out again.** In walls that face heavy weather but lack external protection, such as overhangs, use internal flashing at all sills, heads, wall bases, and lintels.

UPDATED ADVICE ON VAPOR BARRIERS AND SLABS

What's commonly used may not be right.

Douglas Sordyl, PE, Assistant Director of the Structural Engineering Dept. at Giffels Associates in Southfield, Michigan writes that the typical recommendation of "6 mil polyethylene, under the slab, atop a 4" fine gravel base" is: *"not the current practice recommended by the American Concrete Institute and knowledgeable floor specifiers."*

His complete statement to us:

"We have found that a 3" layer is difficult to compact over a vapor barrier and that it is easily disturbed during placement of steel and concrete. Our policy is to use a minimum of 6" of granular base fill over the vapor barrier. In addition to being easily compacted, this also permits a neat division of work. The site grading contractor can rough grade to within 6" of the slab bottom, and the floor contractor is then responsible for placing the vapor barrier, the base course material, the reinforcing steel (if any), and the concrete floor slab.

"Mr. Robert F. Ytterberg (an expert on floor construction) . . . recommends 'a minimum of 50 mil thick impermeable vapor barrier covered with about 6" of crushed stone and topped with a 1/2" thick layer of sand.'

"Our firm designs slabs on grade for primarily industrial and manufacturing type applications. We avoid the use of a vapor barrier since the slab surface is free to release water vapor which may migrate from below. In commercial applications where a floor covering, household good, or equip- *ment must be protected from damage by moist floor conditions, we typically utilize a vapor barrier where the possibility of vapor migration in the soil is possible."*

The real debate has been whether vapor barriers should be used at all, and it's clear to most in the industry that, yes, whenever water vapor can cause problems, it should be used.

Then the question seems to remain, should it be on the grade or on the crushed stone? And why is there such disagreement?

H. Maynard Blumer, a consulting architect who does forensic consulting and troubleshooting of construction problems, writes in the February, 1990 Construction Specifier:

"Many in the concrete industry think vapor barriers under concrete slabs-on-grade cause concrete to crack. Conversely, I and many others believe that a vapor barrier below a concrete slab-on-grade is usually necessary for a variety of reasons.

"Vapor barriers perform six functions, any one of which may be sufficient justification for their use."

He then lists:

1) Improved concrete curing.

2) Prevention of shrinkage cracks.

3) Prevention of slab curling.

4) Protection of floor finishes.

5) Blocking radon gas.

6) Preservation of termiticide treatment until it's covered by concrete.

Mr. Blumer adds this interesting tip:

"I normally use 6 mil clear. I use clear rather than black because I believe it is treated better by persons walking on it when they can see what's under it."

THE EXTRAORDINARY DANGERS OF ROOF PONDING

Ponds on built-up roofs are so common that they often don't even arouse comment. How can anything that common be very important? What's important are the following:

--**The roof structural system may be underdesigned and gradually failing.**

--**Roof slopes and drains aren't working** because they're underdesigned or inadequately constructed (or drains may be clogged from a combination of underdesign and lack of maintenance).

--**Roofing membranes are being gradually penetrated by water.** It's not visible yet, but membranes are cracking and separating, and waterproofing properties of the system are being destroyed.

--**Insulation under the roof may be under compression which will increase ponding,** and/or insulation will absorb moisture through pin-hole leaks and lose its insulative value.

--**The building owner may have installed added roof-mounted equipment.** Designed as an afterthought, such equipment usually means poorly designed curbs, inadequate flashing, leaks at penetrations, and more roof deck deflection.

--**The roof will have to be replaced years sooner than the roofing warranty would indicate,** and the replacement may involve structural changes and rerouting of drains.

--**The roof may abruptly collapse,** destroy tenant's property, and injure or kill occupants.

Ponding is so common on low-sloping built-up roofs that it indicates the typical design standards of building codes and handbooks are way off the mark.

And they are.

Building codes that were faulted for requiring 1/8" slopes or less per linear foot, which were not adequate, now just say the slope should be "adequate" or "sufficient." Some code writers realize that even doubling the old minimums would not be adequate or practical in some cases, so they've stopped setting a specific standard. (In general 1/4" to 1/2" slope is adequate if drains are ample, plentiful, and properly positioned.)

Rainfall charts commonly used to establish the number and size of drains are either misleading, misunderstood, or misused. Someone may prudently use the tables and find the minimum size and number of drains supposedly required. That will not allow for the more important factors of roof structure, positioning of drains, drain clogs, bad design of drain connections and manifolds further down the line, and massive sudden downpours. If the drains are designed next to columns, for example, roof spans will deflect and the drains will end up at high points of the roof.

Recommended:

--**Resize the drains by at least one size upwards from handbook requirements.** A 6" drain will handle three times the gallons per minute (540 gpm) that a 4" drain will carry (180 gpm), and the difference in overall construction cost will be zilch.

--**Put the drains at the centers of spans,** not at the wall lines or at columns. If a drain has to be at a bearing wall or column, increase the height of structural support of the opposite side by about 1-1/2" for every 25 foot of span.

--**Always provide large overflow scuppers at parapets.** Small scuppers can quickly clog up with accumulated roof debris. A reasonable minimum for overflow scuppers is 8" to 10" high by 12" wide, with the bottom pan no more than 4" above the roof at every exterior wall parapet.

FIRE - THE BEAST

In the movie "Backdraft" about Chicago fire fighters, fire is called the "beast." It's depicted as a wild, unpredictable creature that stalks, hides, breathes, consumes and fights for its life like a monstrous predator.

The fire fighters in the film are like knights, heroic but small and weak compared to their adversary. Some die, many are injured; most end up winners. Winners who risk their lives mainly making up for the incompetence of building owners, builders, and architects.

The day I write this, our town had major brush fires on three sides. We lucked out because of the winds. A few miles away, more than two thousand homes and apartment units burned to the ground in a matter of hours.

When viewing the remnants of the holocaust from a hilltop, there are vivid lessons displayed by the homes burned and the homes NOT burned. The difference is usually the roofing: fire-resistant roofs versus wood shakes or shingles. Sometimes the difference is exterior construction -- cement stucco or masonry versus wood siding. Sometimes there were sprinklers. Sometimes, of course, the difference had nothing to do with construction, just some heroic soul with a garden hose keeping his or his neighbor's house wet until the storm passed.

Most of you aren't concerned with brush fires or forest fires. It's just fire, in all its forms that's the enemy and there's ample evidence that the enemy isn't taken seriously by most designers.

In general, the design professions are not aggressively designing for fire safety. Old lessons continue to be ignored in every city in the country.

Sprinklers, which are proven beyond question to be a major saver of life and property, are usually not installed wherever clients and architects have a choice.

We've even heard architects argue against their use. Here are the facts:

--Fire deaths have declined by nearly half in the past ten years due primarily to increased use of sprinklers and smoke alarms. That's 50,000 lives saved in a ten year period.

--The deadliest and most costly fires year after year are in buildings that lacked sprinklers or, for whatever reason, where sprinklers were inoperative.

--Only one or two sprinkler heads manage to control half of all fires that start in sprinklered buildings.

So why wait for the code book to say what simple professionalism would require: namely, sprinkler every new building and, except for smaller homes, retrofit the old ones.

I recently heard an architect argue that talk of sprinklers was silly. They're expensive, he said, and they wouldn't have helped in a fire like the one in the Berkeley-Oakland hills anyway. He had never heard of exterior eaves sprinklers which decidedly would have helped. Too expensive? Many of these homes were worth a half-million dollars and many were worth much more.

What will work or not work in the California hills is not the issue. Millions of dollars of construction -- new construction -- goes up in flames annually and thousands of people are injured and killed. Most of it is preventable. Preventable by design professionals following the simplest of fire safety standards: sprinklers, alarms, stringent observation of jobsite fire-safety rules, and more than minimal fire exits and corridors. These measures must be a part of every office manual if the design professions are to continue to think of themselves as professional.

THE TIME TO WORRY ABOUT EARTHQUAKES IS NOW -- NO MATTER WHERE YOU LIVE

Architects tend to think of seismic design as the engineer's turf, but it isn't so.

Here's a succinct summary of the true architect's role as stated by seismic authority Stanley Scott.

"Many architects do not understand how much initial decisions on a project's structural concepts can influence ultimate seismic resistance, for better or worse. The opportunity to affect a project's quality or cost is greatest in the earliest phases of the design process, after which it drops precipitously.

"Architects need to take this into account, otherwise early decisions in the design process, made independently of structural and seismic considerations and without conferring with the structural engineer, may commit a project to a building configuration or design concept that makes it hard to achieve good lateral-force resistance. Accordingly, from the outset, close collaboration is needed between the architect and the structural engineer -- as well as the mechanical and electrical engineers.

". . . It should be understood that simply complying with the code does not assure adequate seismic design."*

It could hardly be said better: The architect chooses the siting, the construction system, structure, configuration, materials, and appurtenances and the earliest decisions in these realms ALL have seismic significance. Any architect who says earthquake resistance is strictly up to the engineer has dropped out of the design process.

Any designer who believes his or her buildings aren't subject to quakes because they're not in a zone of frequent seismic activity is living in dreamland. Texas, Missouri, Illinois, Tennessee, South Carolina, Massachusetts, and New York are particularly vulnerable.

*The author, Stanley Scott, is Chair of the California Seismic Safety Commission Architect's Role in Earthquake Hazard Mitigation Committee. Published in the October, 1989 issue of **The Examiner** from the California State Board of Architectural Examiners.

DESIGN AND DETAILING OF MOVEMENT JOINTS

Aside from roofs, the focal point of most building failures is their joints, particularly movement joints.

THE JOINTS AND FLASHING DESIGN AND DETAILING CHECKLIST

CSI Reference Numbers: Waterproofing (07100) Dampproofing (07150) Flashing (07600) Sealants (07900)

Movement joints that don't move and sealed joints that don't seal are not only common, they may be in the majority.

Most technical sources cite lapses in workmanship and maintenance as the two dominant sources of trouble. But many of the workmanship and maintenance problems are built in by design and detailing. Joints are often under-designed in distribution and size. Aspects of sealant application are often not designed at all and are left to specifications and/or contractor know-how.

This checklist deals with considerations of wall expansion/contraction joints and separation joints.

The central material component in joint design is the sealant. There's confusion in many people's minds as to the difference between a "sealant" and a "caulk."

"Caulks" are oil- and resin-based sealants widely used up through the 1950's in bearing-wall construction. They were displaced in more demanding applications by materials such as polysulfide, polyurethane, silicone, and acrylics.

Caulks and the newer products have the same purpose -- the sealing of joints -- so it makes sense to refer to both types generically as "sealants." That is now becoming standard practice.

The best of materials and design cannot compensate for workers who don't clean the surfaces to be adhered, or don't include proper backing, or allow mortar or debris to make movement joints immovable.

Even if the original material decisions, details, and workmanship are correct, there's still a likelihood of trouble. All sealants have a limited life expectancy and that life is cut short without proper maintenance. (Sealant life is estimated to range from two to eleven years on metal, three to ten years on wood, four to nine years on masonry. If nobody is watching over the joints, a building may experience joint failures and leaks within only a couple of years after construction.)

Allowing for full awareness of the potential for jobsite and maintenance problems, problem prevention starts with design and detailing considerations such as the following:

WALL JOINTS

__ **Continuous vertical expansion joints** are required near corners, offsets, L turns, and where the building changes in height or volume. 1/2" vertical control joints are common, typically spaced at 25' intervals.

__ **Masonry wall** vertical expansion joints must be continuous and uninterrupted from grade through parapets or fascias.

__ **Stucco** over metal reinforcement may require shrinkage relief joints as close as 18' intervals with total areas between joints restricted to as little as 150 sq. ft.

__ **Stucco** also requires movement relief and crack control joints at wall turns, intersections, and at wall penetrations of elements such as door or window frames at openings. (Relief joints at openings may be no more than trowel cuts made in the fresh base coat.)

__ **Plywood siding** panels require 1/8" to 1/4" gaps to allow for swelling expansion.

139

__ **Tile** veneer may require horizontal separation joints at 10' and vertical joints at 13' intervals or less.

__ **Metal curtain** wall panels expand and contract to an extreme (up to 1" for a 50-degree change in temperature over a 100' length.) You can start with a design assumption of 1/2" joints between each panel but the subsequent design, detailing, and fabrication documents must be engineered and checked with great care.

__ **Joints** between large precast wall panels may enlarge or shrink 1/4" or more in a thermal expansion/contraction cycle. They require 1/2" wide joints for panels up to 15' in width or height and larger panels require 5/8" to 3/4" wide joints.

JOINT SIZES FOR SEALANTS (07900)

__ **The smallest any sealant joints can be** is 1/4" x 1/4" and that size is hard to seal properly.

__ **1/4" to 1/2" wide** joints should have depth equal to the width.

__ **1/2" to 1" wide** joints should be 1/2" deep.

__ **1" and wider** joints should, in general, be from 1/2" to 3/4" deep.

__ **Extra wide** joints are especially subject to external damage especially by people just picking away at the sealant, so provide protection.

THE EXTERIOR MASONRY WALL DESIGN AND DETAILING CHECKLIST

Whether you build with masonry, metal curtain walls, window walls or ordinary fenestration, the primary source of trouble boils down to one thing: movement.

Designers and drafters don't usually design and detail for movement. They think of a building as a solid monolith. Their details are drawings of static assemblies, not moving parts.

In a related vein, most detailing is expressly for construction -- how to put things together. The bigger issue is how to hold things together over time, how to prevent failures. That's a perspective that isn't widely taught in the schools or the offices.

Detailing to prevent failure in any part of the building shell mainly means detailing for movement. Sources of movement include:

--Different rates of expansion and contraction between different adjacent or attached materials.

--Corrosive or expansive chemical reactions. (Such reactions require the presence of water.)

--Wind and water, heat and cold.

The checklist that follows names essential points in designing and detailing for motion in masonry walls. We'll provide a similar list in the next section on design and detailing considerations for curtain walls, window walls, and fenestration.

THRU-BUILDING CONSTRUCTION JOINTS

__ **Buildings experience great stresses from soil movement,** thermal expansion and contraction, frame shrinkage and creep, and wind loads. Thru-building isolation joints to absorb these forces should occur at the juncture points of large T-, L-, and U-shaped buildings. The rule of thumb for long buildings is to include such joints roughly every 100/150 linear feet. This is an item to be engineered more precisely early in the design process. Thru-building joints are also desirable at stress points such as large elevator shafts and wells that penetrate the structure.

__ **Include construction and expansion joint considerations early in the design development stage** of the project. Otherwise such joints tend to become an afterthought and may not mesh with other components that have already been decided. Then they tend to be minimal in number and unevenly distributed.

__ **Don't rely on rules of thumb or design tables** except for rough early estimates of the spacing of joints. Actual conditions often differ radically from what those approximations can handle. Get detailed engineering estimates as early as possible in the design development stage and follow up with refinements throughout the later phases of the project.

WALL EXPANSION/CONTRACTION JOINTS

__ **Wall expansion joints must be continuous and uninterrupted** from grade through parapets or fascias.

__ **Continuous vertical expansion joints are required** near corners, offsets, L turns, and where the building changes in height or volume. 1/2" vertical control joints are common, typically spaced at 25' intervals.

__ **Provide extra vertical joints in parapets at hallway points between the wall joints** in addition to the continuous vertical expansion joints that penetrate parapets. If for any reason a grade-to-parapet joint isn't provided near a building corner to offset, the parapet should at least have its own relief.

__ **Expansion joints are often ruined** by deliberate or accidental blockage with mortar or by badly executed caulking and caulking backing. Emphasize correct procedures in exact, insistent details and specifications, thorough construction inspection, and cautionary maintenance instructions to the client.

MOVEMENT GAPS BETWEEN CONSTRUCTION, FRAMING AND SLABS

__ **Provide a gap between the face of roof spandrels or slabs and the inside face of the exterior wall.** The general minimum is 1". Roof slabs are prone to extreme thermal expansion.

__ **Provide generous continuous gaps between roof terrace quarry tiles** or other paving that is likely to expand and push against parapet construction.

__ **Provide gaps of 1" (or more) around steel or concrete columns** enclosed with masonry.

__ **Use bond-breaks of felt, flashing, or other durable insert material** to separate masonry expansion from masonry-supporting slabs or shelf angles that are anchored to concrete.

__ **Include horizontal relief caulking beneath compressive elements** where extreme compression might be imposed on brick due to concrete frame shrinkage.

__ **Provide deeply embedded flexible ties and anchors** that allow for differential movement between exterior masonry and interior masonry or framing.

SAFETY RULES OF THUMB FOR STAIRS

Here's how people are crippled for life or killed by design errors:

--A young man finishes dressing for a dinner appointment, rushes out of his motel room and heads for the stairs. He's rushed, and as he starts down the flight of stairs he grabs for the the railing. The extended bar of the handrail slides up the sleeve of his sports coat and sends him hurtling head over heels down the stairs.

--A couple enters a newly-remodeled restaurant. They're busy talking to each other and as they step into the restaurant from bar lounge, they don't notice that there is a six-inch step between the rooms. They stumble and the lady is thrown head first onto the marble tile floor.

--A student is walking up a set of open riser stairs in a new auditorium building. A friend calls to him from below and as the student turns around he jams one foot into the open riser. With one foot jammed while turning to look over his shoulder, he falls in a twisted backwards roll and lands with permanent spinal damage.

The two dominant safety hazards in stairs are just plain awkward tread to riser ratios or constructed inconsistencies of tread or riser sizes within a single run. Thousands of people are hurt or killed annually by these problems alone.

The next most common cause of falls is the single riser between rooms or, especially, single steps on a ramp or plaza.

Don't the building codes cover these safety hazards? Some codes do, some don't.

Are designers held responsible for the injuries even when the code is obeyed? You bet they are.

Here's our Safety Checklist for design and detailing of stairs, ramps, and handrails. It deals with the primary sources of injury and death:

__ Never allow single steps anywhere except at unavoidable and clearly indicated points such as at street curbs. Especially avoid single or double steps at:

    __ Ramps
    __ Walkways
    __ Plazas/terraces
    __ Lobbies/entrances
    __ Entry or exit platform landings
    __ Divisions between rooms or spaces

__ If a change in levels is unavoidable, provide clear demarcation with:

    __ Handrails
    __ Warning signs
    __ Bright lighting
    __ Stripes and/or changes in floor texture, material, and/or color

__ Restrict tread and riser ratios to the range of 4" to 7" risers and 11" to 14" treads. (Low risers such as 2" to 3-1/2" high cause as many falls as overly high risers.)

__ Don't allow stair construction to vary from riser to riser or tread to tread more than 1/8" in any run of stairs.

__ Don't allow open risers of a size that can catch and grip people's feet when they're ascending a stair.

__ Don't allow hard, smooth materials, especially those likely to be polished, for stairs, landings, or flooring adjacent to stairs.

__ Require safety nosing and non-slip surfaces.

__ Use extra care in detailing and specifying nosings and non-slip surfaces at:

    __ Exterior stairs

__ Interior steps or stairs near
    entrances that might be exposed
    to tracked water from the outside.

__ Provide very positive drainage slopes at
    all exterior stairs.

__ Avoid tubular or rod handrails, if
    possible, otherwise they:

    __ Must return to the wall or floor.
    __ Be of such size or shape that they
        cannot catch people's sleeves.
    __ Must have smooth, non-sharp
        connectors at balustrades or to
        wall that will not cut or abrade
        people's hands.

__ Provide handrails at all interior or
    exterior stairs.

    __ This especially applies where there
        is a short run of only a few steps

THE LEGAL SYSTEM IS COVERING UP VITAL INFORMATION

Before leaving the topic of building failures, a final word of warning:

Architects and engineers are forced into a totally immoral stance by the legal profession. The legal profession operates according to an adversary system. The system works reasonably well as a means of fact-finding and sorting out conflicting testimony.

But the adversary system is a disaster for revealing the truth of building failures or for helping the profession prevent these failures in the future. It's a disaster on two fronts:

1) **A/E's are suppose to deny everything.** Windows crack, pavers heave, roofs leak, roofs collapse . . . instead of getting together and solving the problems, the A/E's lawyers say: *"Stonewall it! Say nothing! If we lose, we'll appeal."*

2) When the A/E's, contractors, and materials manufacturers have to settle, their **attorneys throw in a final barrier to knowledge -- the vow of silence.** *"Seal the documents! Make everyone swear they'll keep their mouths shut."*

That leaves the rest of the profession free to repeat the same mistakes, destroy property, and risk people's lives through ignorance of what other professionals already know. One could hardly find more immoral or unprofessional behavior.

Lawyers turn the issue of a building failure into an issue of who is at fault. That's not the important issue; in those situations everyone is "at fault" to some degree. What's important is: **What went wrong and how do we get the word out so that others don't repeat the same mistakes?**

143

Until the lawmakers and lawyers clean up their act, bail out of the system as much as possible.

Establish thorough-going quality control, insist on mediation and arbitration of disputes in all contract relations, and get those you contract with to do the same with their contracts.

Let's just say out loud what most people already know: Lawmakers and the lawyers, particularly trial lawyers, have created a monstrosity.

Courts are clogged for years, people cannot get fast and fair hearings, and the most effort is directed at hiding the truth instead of discovering it.

For these reasons, the design professions must step out of the legal system as it exists. Until it's clean, insist that all contracts go to mediation and arbitration. Encourage alternative conflict disputes in third party suits.

And let your professional society know that you and everyone else in the profession is hurt by the seeming indifference to the need to share data on failures. Eventually they'll get the message and this scandal will come to an end.

CHAPTER TEN

CHECKING THE DRAWINGS: THERE ARE BETTER WAYS

OFFICES ARE NOT CHECKING THEIR DRAWINGS BEFORE THEY GO TO BID

THE WORLD'S GREATEST METHOD FOR CHECKING WORKING DRAWINGS

OTHER REFORMS IN CHECKING DRAWINGS

REFORMING SHOP DRAWING CHECKING, COORDINATION, & CHANGES

VALUE ENGINEERING FOR WORKING DRAWING REFORMS

PROBLEMS AND SOLUTIONS IN CHECKING DRAWINGS

A CHECKLIST OF THE MOST COMMON DRAWING COORDINATION PROBLEMS

OFFICES ARE NOT CHECKING THEIR DRAWINGS BEFORE THEY GO TO BID

We asked a room full of architectural employees how many of their offices did a **final** check of working drawings before they went to bid.

Three out of twenty raised their hand!

"They just get checked as we go along," said one drafter, and as with many firms, *"We don't usually have time to do any more than that."*

"Our senior people do self-checking," said a project architect. *Others agreed that that was fairly common practice. In other words, nobody checked many of the drawings except those doing them. You could hardly ask for a more hazardous situation."*

This partly explains the ever-higher rate of errors and coordination problems that plague today's A/E working drawings.

There are more claims for change orders and extras, too. **Contractors say that when they see a set of drawings that are incomplete and uncoordinated--obviously not checked properly--they know they can bid low and pick up their profit in extras.**

"We can pick up $500 to $1,000 in extras on even simple omissions or conflicts in the plans. If we find a couple of dozen of them in a small job, we're sitting pretty," says a Los Angeles housing contractor. "On top of that, every time an architect writes a change order, it screws something else up and leads to another change order."

A case in point - the design and drafting errors in drawings for the new Stamford, Connecticut train station. The station's problems have led to three years of repair work, adding $11 million dollars to a $15 million dollar project. As reported in *Engineering News Record:*

> *". . . the Justice Dept. alleges that critical portions of the station, including a tunnel lobby roof, were designed by an inexperienced employee who was not a registered engineer and that the firm failed to supervise his work. The complaint charges that a staff member appears either as the designer or checker on 52 of 71 pages of design drawings even though he designed only two small minor areas of the station. In a deposition read in court by the U.S. Attorney, he claims his initials were forged.*

> *" . . . depositions read indicate that the design was a team effort with a changing cast. No quality control system was in place."*

These are classic symptoms of a fragmented design, production, and checking process. Further testimony is yet to come, but it's already clear that a lack of coordinated overall checking is a major factor in this case.

William Nigro, formerly of the Navy's Civil Engineering Corps, comments:

> *"While administering hundreds of construction contracts that were designed by civilian A/E firms for the U.S. Navy, I found that about 50% of the thousands of change orders that were negotiated were due to coordination errors that could have been avoided."*

So better check *your* checking processes. Are drawings getting a final quality control check? Or are they just going out the door after only the interim progress checks that occur in the major phases of any job? Remember the quality control rule known to industrial quality control inspectors:

It's the last part of any production job that's likely to have the highest error rate.

That's because that's when people are most rushed, most distracted, most tired, and most eager to move on to something else.

146

THE WORLD'S GREATEST METHOD FOR CHECKING WORKING DRAWINGS

When you do a normal set of check prints, you note your changes with a color code -- red, green, etc. -- and the drafters yellow mark or check mark the changes on the check print as they make changes on their originals drawings or CADD files.

Then, if time permits, you double check to see that changes have been made as instructed, or more likely, you rely on the next check print phase to pick up any lapses.

So far, so good. But during the next check print phase, **you have to review and mark the check set all over again;** what's done, not done, and what needs changes.

Nothing unusual about that; in fact it's almost universal practice.

A MUCH BETTER WAY TO DO IT.

A lot of the rechecking work is a repeat of what was done before. Marking the same items as completed that were marked before, redoing red-marked items. In the redundancy, some changes that were marked before but were not done properly may slip by.

Avoid the repetitive remarking and possible lapses by doing the check print marking on a thin transparency sheet that's taped or stapled to each check print.

Your green, red, yellow marks go onto the overlay, not the check print itself.

In the next phase of checking, you unfasten the transparencies from the old check prints and attach them to the new check prints. Then, behold the timesavings!

--Most of the marks you made on work that's OK will be there and you don't have to redo them. (In three or four check print phases on a typical large job, hundreds of the same items get remarked in every phase. This method avoids all that remarking.)

--If you asked for a change in red and it wasn't made, it's clear to see right there on the check print under the marked transparency.

--If you marked an item as OK but it got mysteriously changed along the way, that shows up immediately.

Then you proceed with your new marks, using new shades of color markers to identify the checking phase. When the next check print phase comes up, you take those same transparencies and lay them atop the newest check prints.

The timesavings accrue with each check print phase, and the errors and oversights decline. Just the opposite of the pattern that's typical with the traditional check print system.

ANOTHER CHECKING TIMESAVER

Job tracings and checkprints are always "out" and have to be tracked down by drafting staff. And they're usually not kept in order, so even when they're in the stack, they're hard to find.

There's a slick solution that solves both problems at once. It's the "file plate." This is a conspicuously large -- but not full-size -- sheet of thin cardboard that's inserted at the place a tracing or checkprint is removed. Whoever takes a tracing or print writes in sheet number, date, and initials as the last entry on the cardboard. When the tracing or print is returned, the borrower locates the correct file plate, sets the sheet in its correct place, crosses off the notation and puts the file plate away. Most staffers adjust to this procedure without much arm twisting, since it's a convenience that benefits everyone.

OTHER REFORMS IN CHECKING DRAWINGS

Whether you're a pencil pusher or have the finest CADD system, **check prints are most likely to be your primary tool for job coordination.**

Check prints are the primary means of communication, instruction, supervision, and quality control. But they don't always do the job they're intended to do.

Here are some reasons:

Supervisors send out a final red-marked set and expect that all the items marked will be done as instructed. This is least likely to be true in the final phase of a project. The final phase of work is the most rushed and distracted, and it's when the drafters or CADD operators miss the most red-marks, or check off red mark items but fail to make the changes.

It's the same in industry and construction -- the final phase of all work is where you'll find the most errors and at the same time the inspectors become least attentive; the worst of all worlds.

You've probably run into this when punch-listing a construction job. You may tell the contractor and the contractor may tell the sub, and you assume a problem is being taken care of. But it ain't necessarily so. **Everything has to be double checked.**

The project manager may be the least effective person to do a detailed, final quality control check. People who have been on a job for months (sometimes years) become blind to problems in the drawings; problems that would jump right out at someone else.

That's why the best quality-control firms hire picky outsiders, such as retired contractors or construction administrators, to do separate quality control checks of all production drawings.

Typical checking is done at about 25% completion and about 90-95% completion.

Side-by-side checking of drawings that have to be coordinated is still the dominant method of checking and it's the worst way to do it.

Multi-story plans, elevations and cross-sections, architectural and engineering drawings SHOULD ALL BE PRINTED AS TRANSPARENCY CHECK PRINTS AND OVERLAYED ON A LIGHT TABLE FOR ONE-ON-ONE COORDINATION COMPARISONS.

As common-sensical as this is and as mandatory as it is in the best offices, only a minority of design firms actually do this. **One reason why:**

No light tables because they're "too expensive" or "take up too much room." So get a sheet of 1/4" Plexiglass, set it on a couple of wood blocks and put a couple of cabinet fluorescent lights from the hardware store underneath it. It'll cost less than $100 and you can take it out and put it away as needed within a couple of minutes.

REFORMING SHOP DRAWING CHECKING, COORDINATION, & CHANGES

The shop drawing process is so loaded with ambiguities and potential liability problems, many A/E's now question whether they should review shop drawings at all.

For example, when the design firm sends back, without comment, a shop drawing that shows a change from the original design, the contractor is likely to assume the design change has been approved.

But it isn't an approval; **changes are allowable only through Change Orders.**.

A detail may "imply" a connection, for example, but not be explicit as to how it's to be made. A fabricator or contractor works out a solution, the shop drawing is sent to the design firm, and then it's returned without question or disapproval.

Then what? Is that an approval? There's been no change in the detail, since the original design was ambiguous to begin with. There's no rejection of the detail. There's no stated approval either.

So what will happen if the detail fails? The design firm will say, correctly, they did NOT approve it. The contractor will say the design firm didn't find a flaw in the shop drawing.

Likely outcome: The court will probably find that the design firm was purposely not taking responsibility for the original detail, letting others do it without formal approval, and then holding them responsible for the detail failure. At best for the design firm, the court may require the designer and the contractor to split the damages.

There are four rules to follow to bring clarity to the situation:

1) **In the Contract, spell out the exact role of the design firm in reviewing shop drawings,** and make sure it's understood. Make special note that a deviation from the documents is not approved without a written Change Order.

2) **Establish an office rule** that whenever there is a proposed change from the original design or specification, such a change can be OK'd ONLY after a systematic review in accordance with clear-cut analytic procedures. This applies whether the change is proposed by the owner, design firm, or contractor.

3) **If the contractor does not understand a construction detail or specification** or if there's an apparent flaw in the documents, the contractor must ask for a document to clarify the situation.

4) **Shop drawings are not to be used as a means of proposing a change** or substitution without a clear notice on the drawing that a change from the original documents is, or may be, involved.

A RESTATEMENT AND CLARIFICATION ABOUT THE NATURE OF SHOP DRAWINGS . . . make sure you and your staff members are in full understanding and agreement on the following two paragraphs. Then make them an explicit part of your office manual and your standard contract terms:

Shop drawings are communications between the contractor and various suppliers, fabricators, and subcontractors. The design professional's role is to review the drawings to answer questions that arise about design intent and to alert the contractor to errors or misinterpretations.

*Even if an approved shop drawing has deviations from the original design and documents, it in itself is not a Change Order and it is not, in itself, an approval of the change. **Changes can only be approved by Change Order.***

VALUE ENGINEERING FOR WORKING DRAWING REFORMS

Value Engineering or Value Analysis is frequently used these days when a project is in trouble. The bids come in way over budget, or there are delays and the client wants a quick fix. Then everyone involved sits down for a little systematic analysis and brainstorming. VE has been used for over twenty years and it's going strong for one good reason: IT WORKS.

Here's how it works and how some offices have applied it to their production process BEFORE a crisis hits or before they built all their old problems into a new CADD system.

The steps:

1)   List each aspect of construction document production, including design development documents and phases.

2)   Define the primary function of each phase, step, and document. Think about it: Can you name the primary purposes of the main elements of your site plans, or exterior elevations?

"Why do . . . ?"

3)   Define the auxiliary functions of each phase, step, document, and component. What else does this do? What good is it? Why do large-scale and small-scale drawings of the same toilet rooms if they both show all the same data? Why these reflected ceiling plans when there's virtually nothing to show? Why arrowheads? Why feet and inch marks?

4)   Define the time and cost spent for each activity. Take a few representative past projects, and average out what time was spent on plans, elevations, door schedules, etc. (Don't have a way of getting that information? GOOD.

That tells you something important and you can introduce time and cost measurements for all aspects of production before completing the Value Engineering process)

5)   Question each activity:

--What part of the activity serves the defined primary function?

--What part serves the auxiliary functions?

--Are the functions necessary?

--Can the functions be provided by something or someone else?

--Where does this part of the work come from? Who does it?

--What else would serve the purpose?

--What would that something else cost?

Value Engineering requires more than just an office brainstorm session. It's a multistep process, requiring several sessions, to untangle all the issues.

Also, you can ask the questions but there won't always be immediate answers. The process is ANALYTICAL not creative. It provokes the thought processes that bring up answers later on. That's not good enough for some people. They want answers, not the thought process to get the answers.

The point is first to identify the problems. See if what you thought were problems really are or not. Then, with a clean list of possible problem spots, go after the solutions.

As you'll see, many parts of the production process are traditional, not functional. Ask: "Why is this done this way?" You get: "We've always done it that way." That's a sure troublespot.

And some solutions may surprise you. Sometimes it boils down to doing the wrong projects, or having the wrong people on the job.

PROBLEMS AND SOLUTIONS IN CHECKING DRAWINGS

Problem: Inspection and checking are always worst on the last batch of work to go out. Boredom or last-minute rushing leads to cursory checking at the time inspection should be most thorough.

Solution: This is when it especially pays to have someone who's not connected with the project do a separate quality control check. Months of familiarity with a job creates blind-spots. Someone from within or outside the office who is not involved in the project will always find more items and more crucial items to correct than project veterans.

Problem: Color codes in drawing checking aren't consistent or don't go as far as they should to catch commonly repeated errors, error-prone drafters, and client initiated problems.

Solution: Use the most common checking color code (red for changes/question, yellow for items that have been corrected, and green for OK items) and elaborate on it to get added value from the checking process. For example, try using blue notes for corrections and omissions, particularly items that should have been done after the previous checking process. Then, on reviewing the job check prints, it will be clear which drafters have the worst error and production rates and you'll know who needs the most help to upgrade their work.

Problem: Sometimes clients make excessive last minute changes and sometimes it's the design firm that's overdoing the changes. How can you keep track of who is doing what and how much it costs?

Solution: You can keep track of who is overdoing the revisions by keeping a record of the changes through further elaboration on color coding during checking. Choose a special color, such as orange, to mark and draw client-initiated revisions. If client-initiated changes are numerous, retain checkprints in the archives along with original final drawings as part of your legal records. The color coded prints will help resolve payment or legal disputes over added job costs.

Problem: Drafting staff members have to hunt all over the office for misplaced tracings, plots, and checkprint sheets if sheets aren't kept in order and if there's no checkout control system.

Solution: Use a large binder clip to fasten checkprint sheets together and provide 8-1/2" x 11" cardboard "file out" cards to insert wherever a checkprint sheet is removed by drafter. Require drafters to note on the file out cards the date a checkprint sheet is removed and who has it.

Also, insist that tracings be filed in sequence, not on the top of the pile, whenever originals are returned to the files. Provide either large file out cards to show who has missing sheets, or provide a checkout list on a clipboard by the flat files for staff members to write names, dates and drawing numbers to record who has what sheets.

Problem: Flat drawer files for prints, drawings, and plots discourage sequential filing. They're bulky, inefficient, and they seem to invite clutter.

Solution: Spring-loaded vertical files. Drawings are stored in large folders which are held upright in top-loaded vertical file cabinets. These are the most space-efficient and convenient filing cabinets we've seen in use anywhere. Look at "VIP Plan Hold" or comparable units.

A CHECKLIST OF THE MOST COMMON DRAWING COORDINATION PROBLEMS

The checklist that follows was a Toronto architect's revision of a very useful checklist created by William Nigro which he called the "Redi-Check system.

This simple list has had a noticeable impact on the office's error reduction program and it certainly looks sound to us. Review it, add any special notes you consider important, and put it to work.

COMMON ARCHITECTURAL DRAWING ERRORS AND COORDINATION PROBLEMS:

__ Existing work, work to remain or to remove, and new work are not clearly identified and differentiated

__ Exterior elevations don't match doors, windows, roof lines, and expansion joints on the plans

__ Building cross sections lag behind and aren't coordinated with plans and elevations

__ Rough openings for doors and windows are too large or too small (especially in masonry)

__ Expansion joints aren't continuous throughout the building

__ Room wall/floor/ceiling construction doesn't match the finish schedule

__ Door, window, and frame schedules don't reflect changes in doors and windows on the plans

SPECIFICATIONS COORDINATION:

__ Items that are to be bid.

__ Point by point, do they agree with the drawings? How easy are they to locate and coordinate with the drawings?

__ Construction phases.

__ Are the phases clearly defined?

__ Items that are included in the finish schedule.

__ Are finish materials specified?

__ Quantities, sizes, spacings, thicknesses, gauges, etc. should not be in the specs.

__ Are such data referenced to drawings?

__ Are the data in the drawings as referenced?

__ Items referenced to other sections of the specifications.

__ Are they elsewhere in the specs as referenced?

CIVIL ENGINEERING COORDINATION:

__ Check site plans for interferences of underground utilities.

__ Do they allow for:
   __ Power
   __ Telephone
   __ Water
   __ Sewer
   __ Gas
   __ Storm drainage
   __ Fuel lines
   __ Grease traps
   __ Fuel tanks

__ Check site plans for interferences between new drives, sidewalks, and other new sitework.

___ Do they allow for:
  ___ Telephone poles
  ___ Pole guy wires
  ___ Street signs
  ___ Drainage inlets
  ___ Valve boxes
  ___ Manholes

___ Check civil earthwork grading and excavation plans.

___ Are they coordinated with architectural and landscape planning for:
  ___ Clearing
  ___ Grading
  ___ Sodding, grass, and mulch
  ___ Other landscaping

___ Compare civil drawings with fire hydrant and streetlight pole locations.

___ Are they coordinated with other drawings:
  ___ Architectural sitework
  ___ Electrical
  ___ Plumbing

___ Check profile sheets showing underground utilities.

___ Are conflicts avoided in elevation as well as plan?

___ Check plan and profile sheets of drainage systems and manholes.

___ Do scaled dimensions match with written dimensions?
  ___ In plans
  ___ In profiles

___ Check final finish grade and pavement elevations relative to manholes and valve boxes.

___ Will tops of manholes/utility boxes be flush with finish grade, pavement, walks, streets:
  ___ Sewer
  ___ Power
  ___ Telephone
  ___ Drains

___ Check that all existing and final grades are noted at every point of change.

___ Are dimensions reasonable, without widely varying differences?

___ Do they allow for run-off drainage without danger of ponding?

STRUCTURAL DRAWINGS COORDINATION:

___ Overlay and compare:
  ___ Column lines on structural and architectural.
  ___ Column locations on structural and architectural.
  ___ Perimeter slab on structural and architectural.
  ___ Depressed and raised slabs on structural and architectural.
  Slab elevations on structural and architectural.

STRUCTURAL DRAWINGS:

___ Foundation piers are identified.

___ Foundation beams are identified.

___ Roof framing plan column lines and columns, and foundation plan column lines and columns.

___ Perimeter roof line matches Architectural roof plan.

___ All columns and beams are identified and listed in column and beam schedules.

___ Column lengths are all shown in column schedule.

___ All sections are properly labeled.

___ All expansion joint locations match Architectural.

___ Dimensions agree with Architectural.

___ Drawing notes agree with Specifications.

153

MECHANICAL AND PLUMBING COORDINATION:

___ All new electrical, gas, water, sewer, etc. lines connect to existing.

___ All plumbing fixture locations coordinated with architectural.

___ All plumbing fixtures coordinated with fixture schedule and/or specs.

___ Storm drain system against architectural roof plan.

___ Pipes are sized and all drains are connected and do not interfere with foundations.

___ Wall chases are provided to conceal vertical piping.

___ Sanitary drain system pipes are sized and all fixtures are connected.

___ HVAC floor plans match architect's.

___ Sprinkler heads are shown in all rooms.

___ All sections are identical to architectural/structural.

___ Adequate ceiling height exists at worst case duct intersections.

___ All structural supports required for mechanical equipment are indicated on structural drawings.

___ Dampers are indicated at smoke and fire walls.

___ Diffusers are coordinated with architectural reflected ceiling plan.

___ All roof penetrations (ducts, fans, etc.) are indicated on roof plans.

___ All ductwork is sized.

___ Air conditioning units, heaters, and exhaust fans match roof plans and mechanical schedules.

___ Mechanical equipment will fit in spaces allocated.

ELECTRICAL COORDINATION:

___ All plans match the architectural.

___ All light fixtures match the architectural reflected ceiling plan.

___ All major pieces of equipment have electrical connections.

___ All panel boards are properly located and are shown on the electrical riser diagram.

___ There is sufficient space for all electrical panels to fit.

___ Electrical panels are not recessed in fire walls.

___ Electrical equipment locations are coordinated with site paving and grading.

___ Motorized equipment:

___ Equipment shown is coordinated with Electrical Drawings.

___ Horsepower ratings are verified.

___ Voltage requirements verified.

CEILING PLAN COORDINATION:

___ Verify reflected ceiling plan against architectural floor plan to ensure no variance with rooms.

___ Ceiling materials match finish schedule.

___ Light fixture layout matches Electrical.

___ Ceiling diffusers/registers match Mechanical, including all soffits and vent locations.

CHAPTER ELEVEN

SHARPER, CLEARER DRAWINGS WITH LAYERING AND SCREENING

USE OVERLAY PRINTING AND SCREENING
TO GET THE CLEAREST OUTPUT FROM CADD OR
HAND DRAFTING

THE OLD STANDBY
OVERLAY DRAFTING TO CONTROL AND
COORDINATE CONSULTANTS DRAWINGS

SCREENED SITE PLANS

SCREENED IMAGES FOR EXTERIOR ELEVATIONS

SCREENED IMAGES FOR ROOF PLANS

USE OVERLAY PRINTING AND SCREENING

TO GET THE CLEAREST OUTPUT FROM CADD OR HAND DRAFTING

In the rush to CADD and plotted output, many offices have forgotten the primary advantages of overlay drafting: clear differentiation of base and overlay information in the final prints.

Screened prints ("shadow prints") get lots of extra mileage. Screening breaks up the image of a drawing into small dots or lines, so linework fades as background reference. Data you want to emphasize are printed in

solid line for contrast against the subdued line background.

Architectural data should usually be subdued as shadow print background when combined with consultant drawings such as HVAC, electrical, and plumbing. That clearly separates the contracts and avoids confusion between walls and ducts, dimension lines and pipes, and especially, between architectural work and the other disciplines.

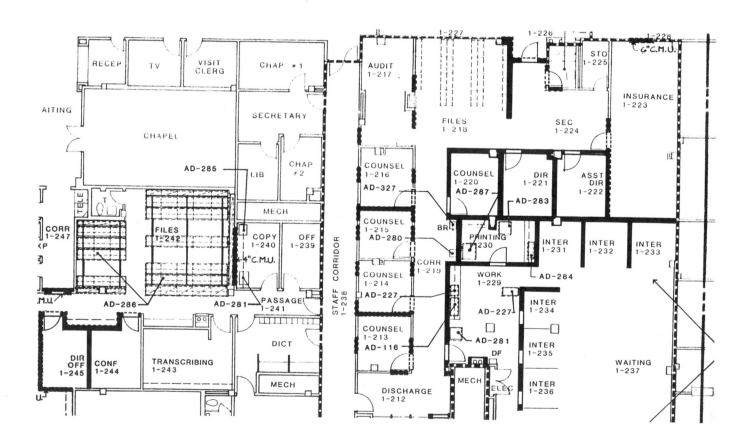

156

Drawing Type	Screened Background	Solid Line for Contrast
Renovation, rehab	Existing construction	Demolition, new work, additions
Foundation plan	Floor plan above foundation	Foundation, basement, crawlspace
Roof plan	Floor plan below roof	Roof surface, drains, vents, etc.
Consultants' drawings	Architectural floor plan background	Separate building trades: electrical, plumbing, structural, etc.
Consultants' reference	Architectural background with furnishings, equipment	Electrical, communications
Plans, elevations, sections	Base construction	Alternative bid construction or additions
Site plan	Surveyor's map, existing contours, foliage	New contours, demolition, construction, site work
Consultants' site plans	Contours, new construction	Site electrical, mechanical, irrigation, landscaping details
Punch list reference	Construction as designed	Work requiring completion or correction
Change orders	Construction as designed	Revisions with cloud bubble or with revision symbol tags keyed to a revision schedule
As-builts	Construction as designed	Construction as changed

THE OLD STANDBY

OVERLAY DRAFTING
TO CONTROL AND
COORDINATE
CONSULTANTS DRAWINGS

You'll have 100% coordination of plans as long as the architect prepares the base sheets and the consultants all do their work on layers coordinated with those base sheets.

This is a natural for CADD and has been one of the success stories

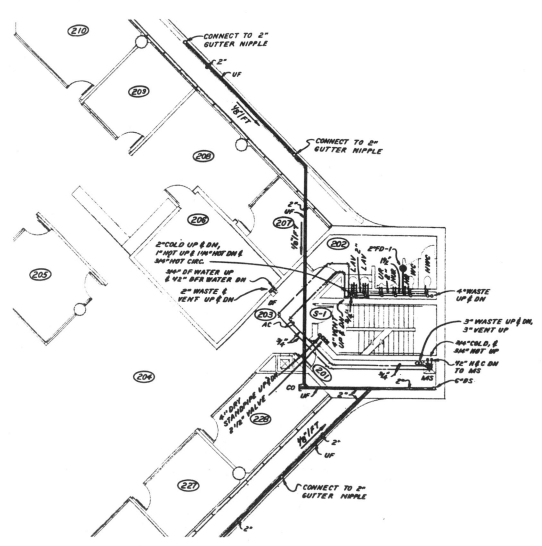

SCREENED
SITE PLANS

Too often the surveyor's drawing is re-drawn by the architect. Avoid this waste by using the survey as a base sheet and printing it as screened background in combination with new sitework.

Below is a portion of a site survey before screening followed by the same survey as a subdued screened print combined with the solid line image showing new construction.

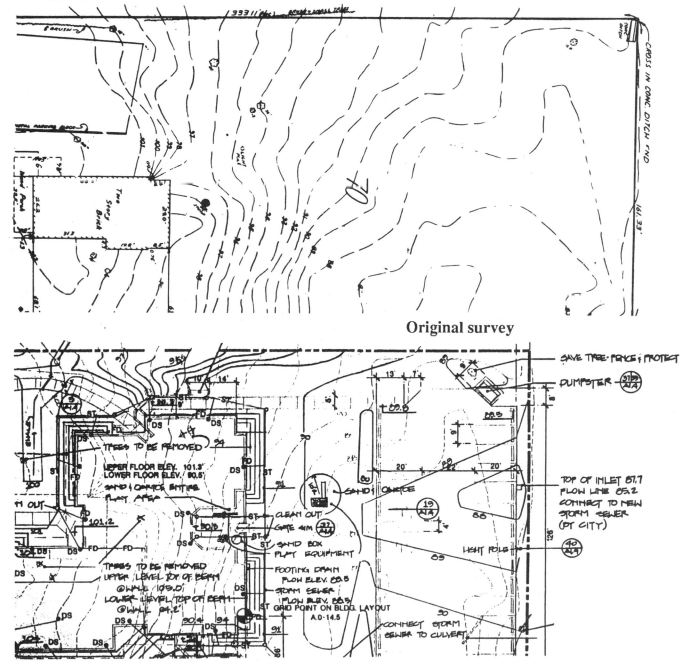

Original survey

Survey as screened background image.

159

SCREENED IMAGES FOR EXTERIOR ELEVATIONS

Screening and subdued background images are widely used in plans, but rarely in exterior or interior elevations.

They're most useful for showing existing conditions in contrast to new construction -- existing screened, new in solid line.

Or screening can be used show alternate bids -- base bid in screened line, and alternates in solid.

Screening is also used to show optional additions, alterations and retrofit work.

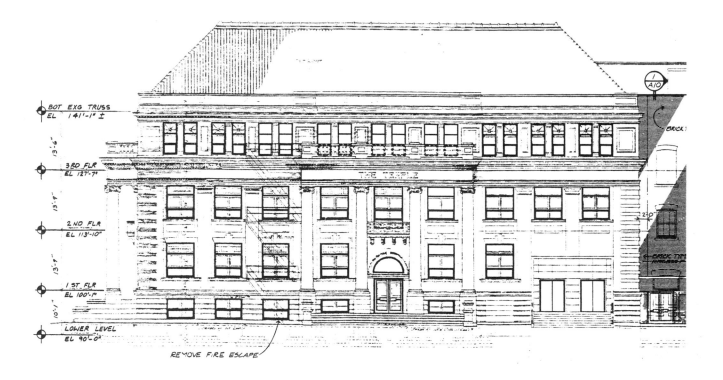

Screened exterior elevation showing new work in solid line contrast to existing work.

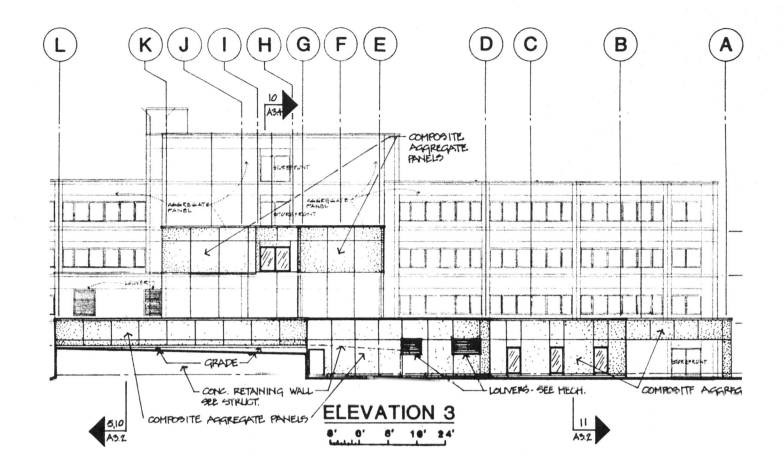

Circle labels (left to right): L K J I H G F E D C B A

10
A3.4

COMPOSITE
AGGREGATE
PANELS

AGGREGATE
PANEL

STOREFRONT

AGGREGATE
PANEL

STOREFRONT

LOUVER

STOREFRONT

GRADE

CONC. RETAINING WALL
SEE STRUCT.

COMPOSITE AGGREGATE PANELS

LOUVERS · SEE MECH.

COMPOSITE AGGREG

5,10
A3.2

11
A3.2

ELEVATION 3

8' 0' 8' 16' 24'

Screened exterior elevation showing building
additions in solid line.

161

SCREENED IMAGES FOR ROOF PLANS

There's no better way to show the relationship of roof construction to the floor below it than to combine the images -- screened floor plan and solid line roof plan.

This helps assure that roof drains aren't in the wrong place and that skylights are centered as intended over the rooms below.

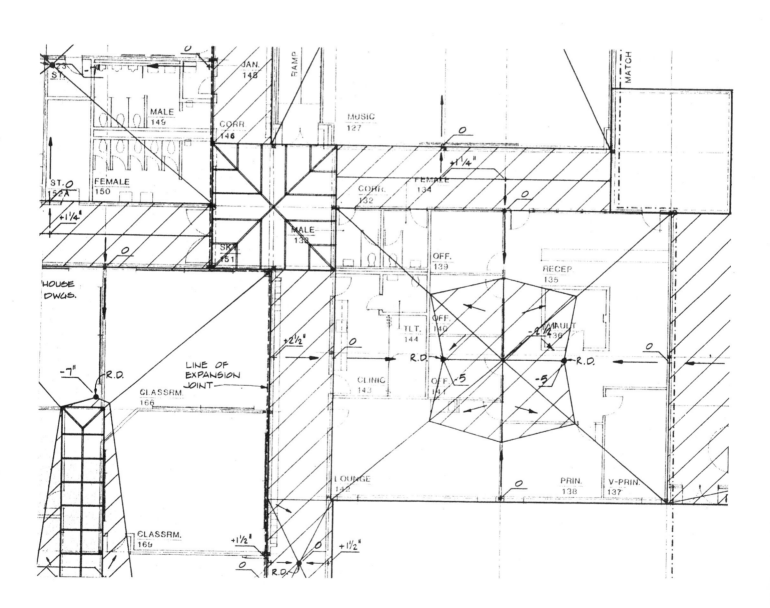

CHAPTER TWELVE

KEYNOTING: THE MALIGNED AND ABUSED TIMESAVER

KEYNOTING --
A MULTI-PURPOSE
TIMESAVER IN NINE
EASY STEPS

KEYNOTING is a problem solver that creates its own peculiar problems.

The trouble is that it's so plain and simple. It doesn't require any special tools or products, so no companies can profit by advertising it or by teaching people how to use it.

The rules for using it correctly are so obvious that it throws some people. The system has thus been discredited in some offices because of gross misuse of the system.

And although it saves an enormous amount of time, the individual segments of timesaving are so small that it hardly seems worth the bother.

"Keynoting" means listing notes by number on a drawing rather than repeatedly writing them out. It's common practice in industrial engineering. An item is drawn and then surrounded by numbers with arrows pointing to various parts. The numbers are keyed to identifying notes that are typed at one side of the sheet.

The advantages build as you follow from one logical step to the next:

1) When starting, write your notes on check prints as your normally would and write a number in front of it. Whereas you might write a single note several times on different parts of a drawing, just write it once with its keynote number. Wherever you would write the note again on a drawing, instead write in the keynote number with a leader arrow to the noted item. Later the typist or word processing operator will print out the notes you've written as a single keynote list to be attached at the right hand side of the original drawing sheet. The

drafters will add in keynote numbers and leader arrows in the drawing. You may notice some notes are repeated many times on a drawing but you (and the drafters) only do it once. A minor timesaver, but the start point for many others.

2) When there's a change in a note, it only has to be made once, in the keynote legend, and not over and over again in the field of the drawing.

3) Notes are typed or computer printed instead of hand-lettered. That saves time in the long run no matter how fast your drafters are at lettering, and it makes for much neater looking and more legible drawings.

4) Once a set of notes are created for any type of drawing on a job, the same list can be used with minor additions on all other similar drawings. The notes for one level floor plan or the exterior elevations of one side of a building are usually identical to the others, so they're done as a keynote list only once, not over and over.

5) And after a set of keynotes are written for a project drawing, chances are that they can be reused with minor editing for the same drawings on future jobs. The roof notes, site notes, cross section and wall-section notes, door and window detail notes, etc., will be essentially the same for all similar building types.

6) Once you've created a set of keynotes for a project, then the list of notes can become the start of your office checklist of contents for any drawing. Use the list later when planning a set of working drawings and sketching mini-mockups of drawings.

7) Use the same standardized list of keynotes as a list of working drawing contents to guide the drafters. Once a standard list of keynotes for a drawing type is created, the process turns around. Instead of having to write a list of notes to match the items on a drawing, the keynotes will come first as a checklist of what to put on those drawings.

8) Since notes become reusable from drawing to drawing and job to job, they can be improved upon with experience. Staff members can give more attention to the content and wording of notation than has been practical before, and that reduces errors.

9) You can add a CSI number at the end of most keynotes and link up the note with the relevant specifications section. (If you have a standard detail system filed by CSI numbers, the same CSI number can also identify the file number of your relevant standard detail.)

A/E's who have followed all nine steps say the seemingly minor timesaver becomes worth thousands of dollars each year.

SOME USEFUL NOTES ABOUT KEYNOTES

There are three main kinds of notation: Identification Notes, Reference Notes, and Assembly Notes. An Identification Note names the generic object or material being described in the note. That will usually be a single word or phrase such as CONCRETE SLAB or ANCHOR BOLT. Then follow with more particulars such as a reference note such as SEE DETAIL 12-3. Or the note may require assembly information such as sizes, gages, spacings, etc., and some sequence of action

SPECIAL APPLICATIONS FOR KEYNOTING

--**Photodrawings.** Photodrafting is most often used for rehabs, adaptive reuse, etc. Photos of the existing structure are printed, and all the various demolitions, repairs and additions are noted. There's tremendous repetition in this kind of notation. A photo detail sheet for one restoration job had notes such as "remove loose paint," "remove curb," etc., lettered 20 to 30 times each. Overall there were relatively few different notes, but there were many places they had to be applied. That's the kind of situation that's made for keynoting.

--**Wall materials indications.** Some firms use keynotes to avoid drawing intricate texture and materials indications on floor plans. Each length of wall is identified with a small circled number or letter. The circle has a tail that cuts through the wall. The circled key refers to a small wall section detail and/or note that identifies wall construction and materials. Sometimes finish is identified at the same time. When wall construction is the same through most of a building, that is noted, and only the exceptions are identified by key symbol.

--**Repeat dimensions of small components on detail drawings.** A sheet of millwork details, for example, may have a limited number of special shapes that are repeated many times on the sheet. Instead of dimensioning the repeat sections over and over, each has a key number or letter that refers to a separate schedule. The schedule shows a blowup of each piece with complete dimensions, hatching, and notation.

--**Keynoting unclutters working drawing sheets and keeps them readable even when you go to "half-size" printing.** The rule, of course, is to make your notes large enough to remain clear after the sheets are reduced in size.

--**A simpler version of keynoting for smaller jobs:** typewritten keynotes on applique

are placed directly beside exterior elevations and building cross sections. Identification arrows are drawn from notation to the noted part(s) of the drawing, without using intervening key reference numbers. Small building floor plans, site plans, foundation and framing plans have all been successfully notated this way.

--**General notes.** Most firms use general construction notes that repeat from job to job but they don't usually show exactly where the general notes apply on the drawings. Some code the General Notes and add code numbers and arrows on the drawings where they particularly apply. One office has wall, floor, and ceiling General Notes which they use again and again and have coded by abbreviation "W-1, W-2," etc. for wall notes, "F-1," etc. for floor notes.

--**Detail keys augment keynote lists.** As a further improvement to keynoting, some firms place their detail and section key symbols at the keynote legend. That way a single keynote reference number in the field of the drawing does double duty and identifies the detail and note simultaneously. Thus one simple number and arrow does the same job as a full note and a separate detail key used to do. Ultimately, details, specification sections and notation are linked by the same identification numbers.

THE FINAL TOUCH -- INTEGRATE KEYNOTES WITH CSI NUMBERS

As cited in item 9 on the first page, you can add relevant five-digit CSI numbers after keynotes and link those notes directly with your specifications and standard detail file numbers.

As you do so, you'll cluster similar notes under the same CSI division numbers. Then all sitework notes on a sheet will start with a 2., concrete notes with a 3., masonry notes with 4. etc.

Here's how a string of such keynotes might look:

3.0   CONCRETE (03000)

    3.1   Concrete Slab (03300)
    3.3   Construction Joint (03251)
    3.5   Anchor Bolt (03252)
    3.7   Concrete Curb (03336)

4.0   MASONRY (04000)

    4.1   Brick Planter Wall (04210)
    4.3   Weep Holes (04215)
    Etc.

Observe several important points about this list:

The complete five-digit CSI numbers are NOT used as the keynotes. Keynotes are kept to simple two or three digit numbers used in the field of the drawing and as the identifying list of numbers on the left of the notes

The second digit of the keynote code numbers start as odd numbers. That makes it easy to slip in additional numbers later near related notes by using even numbers such as 1.2, 1.4, etc.

Elaborations on notes, such as sizes, spacings, references to other parts of the documents, and assembly instructions can be added after the "title" or "Identification Note" or they can be included with keynote code numbers in the drawing. Or the keynote can be more elaborate and look like this:

3.0   CONCRETE (03000)

    3.1   Concrete Slab (03300)
        6" slab over 6" crushed rock
        (see specifications)

    3.3   Construction Joint (03251)
        1/2" joint (see specifications)

    3.5   Anchor Bolt (03252)
        1/2" diam. x 6" @ 36" o.c.

    3.7   Concrete Curb (03336)
        4" x 4" curb

Don't start adding CSI reference numbers at the end of identification notes until after you've started standardizing your notes and reused them for awhile. At that point it becomes practical to add in the helpful extra data but it's too much to do at the outset when it would just distract from doing project work.

Contractors respond very well to this kind of documentation, by the way. One wrote to congratulate an architect for using integrated keynote format, saying it was "extremely helpful to us and the subcontractors and I wish all architects could be persuaded to use such rational systems in their working drawings."

IF YOU DON'T WANT TO ALIENATE THE CONTRACTORS, FOLLOW THESE RULES:

1) Don't put keynotes and a drawing on separate sheets. It's been done, and it destroys any and all convenience for users of the system

2) If any note is highly particularized and crucial, go ahead and include it in the field of the drawing where you normally would. There's no reason to fear mixing keynotes and non-keynotes in the same drawings.

SIMPLE COMMON NOTES IN A SIMPLE KEYNOTE LIST

   Notes that are most common to a particular type of details, such as footings, are preprinted on the sheets as simple "1" "2" "3" "4" notes. The more variable notes are included with the detail itself.

   Clearly more of notes could be keynoted, and probably would be as this office developed this system further.

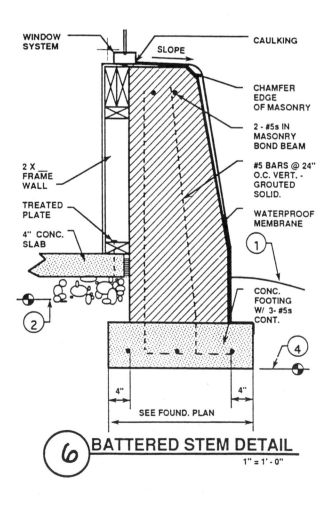

BATTERED STEM DETAIL

1" = 1'- 0"

NOTE KEY

1   FINISH GRADE.

2   COMPACTED ROUGH GRADE.

3   12" MIN. BELOW UNDISTURBED SOIL OR ENGINEER CERTIFIED COMPACTED SOIL.

4   18" MIN. BELOW UNDISTURBED SOIL OR ENGINEER CERTIFIED COMPACTED SOIL.

BASIC KEYNOTING WITH NOTES CLUSTERED BY CSI CATEGORY

A simple, easy to use system. Notes have been standardized and are computer printed, then edited for the specific job.

The keynote legend would be improved if it had a complete 5-digit keynote number after each note for convenient, direct reference to specifications.

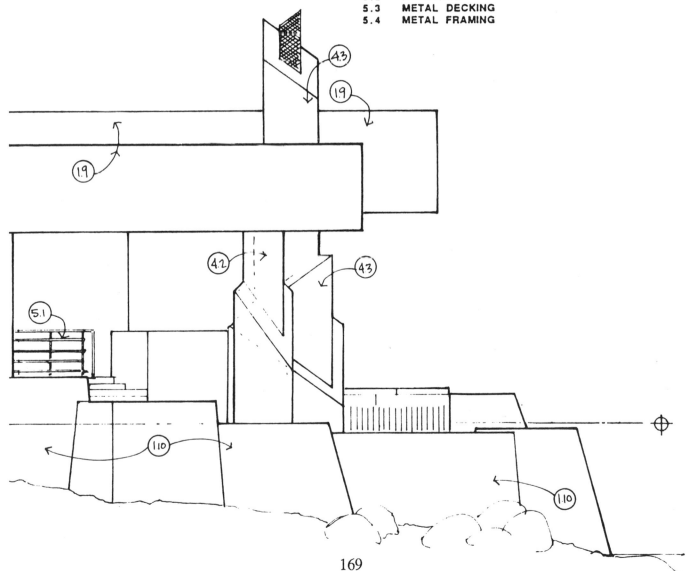

169

THE RECOMMENDED SYSTEM

The best of all worlds:

Simple, easy to use and easy to remember three-digit keynote numbers are used in the field of he drawing

The reference numbers are listed in the keynote legend which shows each note followed by the complete five-digit CSI reference number.

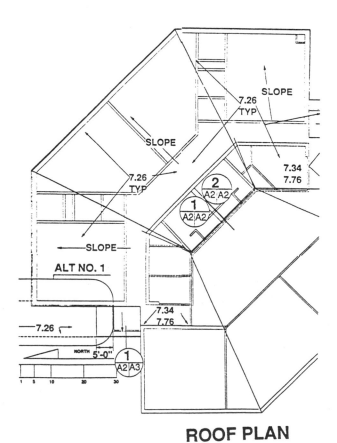

ROOF PLAN

KEY NO.	KEYNOTE
7.26	STANDING SEAM MET ROOF - 07410
7.32	CLOSURE/DRIP FLASHING - 07410
7.33	MET SOFFIT, COLOR TO MATCH ROOF - 07410
7.34	REUSE EXIST RETAINING BAR & FLASHING. RE-SET BAR IF REQ'D 07410
**	AL FLASHING & TRIM - 07600
7.36	AL SILL FLASHING - 07600
**	SEALANTS - 07900
7.50	SILICONE CAULK W/BACKUP - 07900
7.58	SET THRESHOLD IN BED OF SEALANT 07900
7.70	CAULK & BACKUP, BOTH SIDES - 07900
7.76	RE-CAULK RETAINING BAR & FLASHING - 07900
**	MET DRS & FRAMES - 08110
8.10	HOLLOW MET DR - 08110
8.16	HC MET TRANSOM PANEL - 08110
8.26	HOLLOW MET DR FRAME GROUTED SOLID - 08110
8.30	MET DR FRAME ANCHOR - 08110
**	WOOD & PLASTIC DRS - 08211
8.34	SOLID CORE WOOD DOOR - 08211
**	SPECIAL DRS - 08305
8.40	MET ACCESS DR - 08305
8.44	MAS ANCHORS FOR MET ACCESS DR FRAME 08305
**	ENTRANCES & STOREFRONTS - 08410

170

"SYSTEMS NOTES"

In this example, the office includes standard floor, wall, roof, construction system notation. The office adds a notation reference number where applicable in the field of the drawing.

The reference system is alphabetical, "F" for "Floors," etc. It could be improved upon by adding CSI reference numbers after the notes in the legend.

SYSTEMS NOTES

FLOORS:

F1 - HARDENED, SEALED, STEEL TROWELED REINFORCED CONCRETE SLAB.
F2 - HARDENED, SEALED, STEEL TROWELED CONCRETE SLAB ON STRUCTURAL SYSTEM.
F3 - 4" ARCHITECTURAL CONCRETE SLAB ON 2" RIGID INSULATION (R-11) ON STRUCTURAL CONCRETE ON PRECAST CONCRETE ON DOUBLE TEES, (3-HOUR RATED FLOOR SYSTEM).
F4 - EXPOSED PRECAST CONCRETE.

WALLS:

W1 - POURED-IN-PLACE CONCRETE.
W2 - PRECAST CONCRETE.
W3 - CONCRETE BLOCK.
W4 - THRU-WALL BRICK.
W5 - BRICK VENEER.
W6 - 1x8 CHANNEL LAP ROUGH SAWN WOOD SIDING ON #15 BUILDING FELT ON 2-LAYERS 5/8" EXTERIOR TYPE "X" GYPSUM SHEATHING ON METAL STUDS WITH R-19 BATT INSULATION.
W7 - CEMENTITIOUS SPRAY-ON ON 1/4" CEMENT ASBESTOS BOARD ON 5/8" EXTERIOR TYPE "X" GYPSUM SHEATHING ON 6" METAL STUDS WITH R-19 BATT INSULATION.
W8 - BRONZE ANODIZED ALUMINUM STOREFRONT SYSTEM.
W9 - CEMENTITIOUS SPRAY-ON.
W10 - 5/8" TYPE "X" EXT. GYPSUM SHEATHING ON STRUCTURAL STEEL STUDS WITH R-19 BATT INSULATION.
W11 - 5/8" TYPE "X" GYPSUM BOARD ON METAL STUDS.
W12 - 5/8" WATER-RESISTANT TYPE "X" GYPSUM BOARD ON METAL STUDS.
W13 - 5/8" TYPE "X" GYPSUM BOARD ON METAL FURRING.
W14 - 5/8" TYPE "X" GYPSUM BOARD MASTIC APPLIED.
W15 - 2-LAYERS 5/8" TYPE "X" GYPSUM BOARD ON 2½" C-H METAL STUDS WITH 1" GYPSUM BOARD SHAFT LINER (2-HOUR SHAFT).
W16 - 2-LAYERS 5/8" TYPE "X" GYPSUM BOARD ON METAL STUD.

ROOFS:

R1 - ELASTIC SHEET ROOFING SYSTEM ON 2" MINIMUM RIGID INSULATION ON METAL DECK ON STRUCTURE WITH BATT INSULATION PIN ATTACHED, R-38.
R2 - ELASTIC SHEET ROOFING SYSTEM ON 2" MINIMUM RIGID INSULATION ON PRECAST CONCRETE DOUBLE TEE WITH R-38 BATT INSULATION PIN ATTACHED.
R3 - PREFINISHED METAL ROOF ON #15 BUILDING FELT ON ½" EXTERIOR PLYWOOD ON 2x FURRING 24" O.C. ON #15 BUILDING FELT ON 5/8" EXT. PLYWOOD ON STRUCTURE WITH R-38 BATT INSULATION.
R4 - PREFINISHED METAL ROOF ON #15 BUILDING FELT ON 5/8" EXTERIOR PLYWOOD ON METAL STUD JOISTS.
R5 - 4" MINIMUM ARCHITECTURAL CONCRETE ON 1/8" PROTECTION BOARD ON WATERPROOFING MEMBRANE ON STRUCTURAL TOPPING ON PRECAST CONCRETE DOUBLE TEES.
R6 - 1/8" PROTECTION BOARD ON WATERPROOFING MEMBRANE ON STRUCTURAL TOPPING ON PRECAST CONCRETE DOUBLE TEES.

CEILINGS:

C1 - EXPOSED STRUCTURE.
C2 - 5/8" TYPE "X" GYPSUM BOARD ON 7/8" METAL FURRING CHANNELS 24" O.C. ON 11/2" COLD ROLLED CHANNELS 48" O.C. OR WITHOUT CHANNELS DIRECTLY TO STRUCTURE.
C3 - 5/8" WATER-RESISTANT TYPE "X" GYPSUM BOARD ON 7/8 FURRING CHANNELS 48" O C. METAL SUSP. SYSTEM, SEE C2.
C4 - 5/8" TYPE "X" GYPSUM BOARD ON 1/2" RESILIENT METAL CHANNELS ON 5/8" GYPSUM BOARD ON METAL SUSP. SYSTEM, SEE C2.
C5 - 2'x2' NON-RATED SUSPENDED ACOUSTICAL CEILING SYSTEM WITH RECESSED GRID SUSPENDED FROM STRUCTURE.
C6 - 2'x4' NON-RATED MYLAR FACED SUSPENDED ACOUSTICAL CEILING SYSTEM SUSPENDED FROM STRUCTURE.
C7 - 1/2" EXTERIOR GYPSUM CEILING BOARD ON METAL STUDS OR METAL SUSPENSION SYSTEM, SEE C2.

COLUMNS:

CL1 - PRECAST CONCRETE EXPOSED.
CL2 - BRICK VENEER ON STRUCTURAL STEEL COLUMN.
CL3 - 5/8" TYPE "X" GYPSUM BOARD ON METAL FURRING ON STRUCTURAL STEEL COLUMN WITH 3-HOUR SPRAY-ON FIREPROOFING.
CL4 - 5/8" TYPE "X" GYPSUM BOARD MASTIC APPLIED TO PRECAST CONCRETE.
CL5 - BRONZE ANODIZED ALUMINUM COVER ON METAL FURRING ON STRUCTURAL COLUMN WITH 3-HOUR FIREPROOFING.

GLAZING:

G1 - 1/4" WIRE GLASS.
G2 - 1/4" HEAT TREATED OPAQUE SPANDREL GLASS.
G3 - 1" CLEAR TEMPERED INSULATED GLASS.
G4 - 1" TINTED INSULATED GLASS.
G5 - 1/4" TINTED TEMPERED GLASS.

ANOTHER VARIATION ON STANDARDIZED SYSTEMS NOTES

Wall, Roof, Ceiling, etc. assemblies are identified on the drawings as "Type 1," Type 2," etc.

Then the types are described in great detail, virtually as a checklist of elements in the note legend. It could be improved by adding CSI reference numbers with the notes.

ROOF ASSEMBLY

TYPE 1   Along East Elevation (1 hr. rated)
Prefinished metal roofing over
15# building paper over
Plywood decking over
1/2" gypsum sheathing extending
5' 0" up roof from face of exterior wall on
2x12 rafters with
R-30 batt insulation and with
5/8" gypsum board interior finish (1 hr. rated
     under protected roof area)

TYPE 2   Prefinished metal roofing over
15# building paper over
Plywood decking over
1x4 furring at 2' 0" o.c. on
2x12 rafters with
R-30 batt insulation and with
5/8" gypsum board interior finish

SOFFIT

TYPE 1   South Elevation (1 hr. rated)
1/2" plywood over
1/2" 1 hr. rated gypsum board nailed at 6" o.c. to
2x12 roof rafter tails

TYPE 2   East, North, and West Elevations (non-fire rated)
1/2" plywood nailed to
2" roof rafter tails
Provide continuous vent

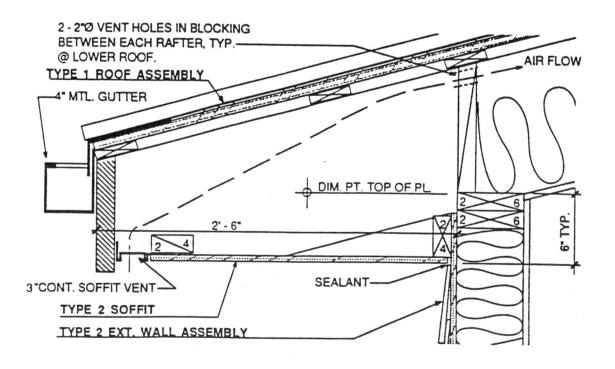

2 - 2"Ø VENT HOLES IN BLOCKING BETWEEN EACH RAFTER, TYP. @ LOWER ROOF.
TYPE 1 ROOF ASSEMBLY
4" MTL. GUTTER
AIR FLOW
DIM. PT. TOP OF PL.
2' - 6"
3" CONT. SOFFIT VENT
TYPE 2 SOFFIT
TYPE 2 EXT. WALL ASSEMBLY
SEALANT
6" TYP.

AN INGENIOUS VARIATION ON KEYNOTING:

COMBINING DETAIL KEYS WITH THE NOTES

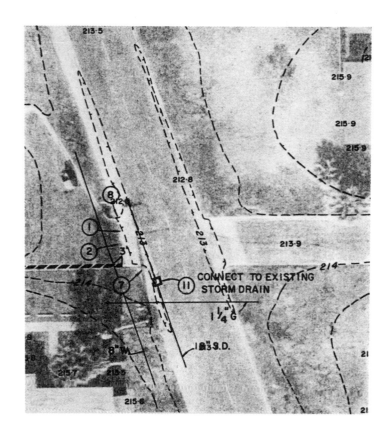

This civil engineering firm used an aerial photograph as a base sheet, combined it with an overlay plan to show sitework and keyed the sitework notes shown as in the illustration

The notes in the keynote legend are preceded by detail keys. This way each detail key only needs to be created once, as does each note, instead of being repeated on the drawing sheet.

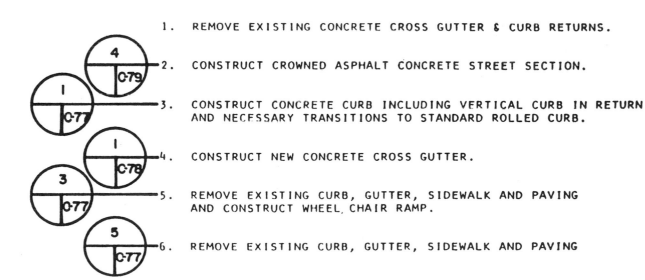

1. REMOVE EXISTING CONCRETE CROSS GUTTER & CURB RETURNS.

2. CONSTRUCT CROWNED ASPHALT CONCRETE STREET SECTION.

3. CONSTRUCT CONCRETE CURB INCLUDING VERTICAL CURB IN RETURN AND NECESSARY TRANSITIONS TO STANDARD ROLLED CURB.

4. CONSTRUCT NEW CONCRETE CROSS GUTTER.

5. REMOVE EXISTING CURB, GUTTER, SIDEWALK AND PAVING AND CONSTRUCT WHEEL CHAIR RAMP.

6. REMOVE EXISTING CURB, GUTTER, SIDEWALK AND PAVING

KEYNOTES WITH EXTENDED CSI NUMBERS

The example below is from the ConDoc system. This system uses extended five- digit numbers as the keynote numbers themselves.

Some users, especially contractors, complain that this increases their work rather than decreases it. It's hard to remember the full five-digit number when looking from drawing to keynote legend and back again.

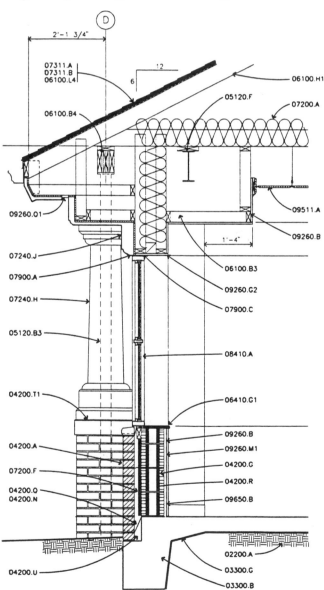

DIV. 2 SITEWORK

02200.A	COMPACTED SUBGRADE
02520.N2	1/2" PRE-MOLDED E.J. FILLER
02712.B	FOUNDATION DRAIN

DIV. 3 CONCRETE

03300.B	FOUNDATION — SEE STRUCTURAL
03300.C	3" CONCRETE SLAB
03300.G	MOISTURE BARRIER

DIV. 4 MASONRY

04200.A	FACE BRICK
04200.B	SOLDIER COURSE
04200.G	8" CONCRETE BLOCK
04200.J	BOND BEAM — SEE STRUCTURAL
04200.K2	CMU RETAINING WALL — SEE STRUCT.
04200.M	MASONRY REINFORCING
04200.N	WEEP HOLES
04200.Q	MEMBRANE FLASHING
04200.R	BLOCK INSULATION
04200.T1	CAST STONE CAP
04200.T2	CAST STONE LINTEL
04200.U	NON-SHRINK GROUT

DIV. 5 METALS

05120.B3	PIPE COLUMN — SEE STRUCTURAL
05120.F	FRAMING — SEE STRUCTURAL
05210.A	STEEL JOISTS — SEE STRUCTURAL
05300.B	CORRUGATED METAL DECK

DIV. 6 WOOD AND PLASTICS

06100.B3	2 x 4 FRAMING @ 24" O.C.
06100.B4	FRAMING — SEE STRUCTURAL
06100.B5	2 x 8 FRAMING @ 24" O.C.
06100.F1	WOOD BLOCKING AS REQUIRED
06100.H1	WOOD ROOF TRUSSES
06100.L4	5/8" T&G STRAND BOARD DECKING
06410.G1	CULTURED MARBLE SILL

DIV. 7 THERMAL & MOISTURE PROTECTION

07120.A	BITUMINOUS WATERPROOFING
07120.B	PROTECTION BOARD
07200.A	R-30 BATT INSULATION
07200.F	1" BOARD INSULATION (R-8)
07200.G	CHICKEN WIRE
07240.B1	DRYVIT ON 1" INSULATION BOARD
07240.H	DRYVIT COLUMN — SEE DETAILS
07240.J	DRYVIT FASCIA — SEE DETAILS
07311.A	ASPHALT SHINGLES
07311.B	15# FELT
07311.C	METAL DRIP EDGE
07600.A1	GUTTER
07900.A	SILICONE SEALANT
07900.C	CAULK

DIV. 8 DOORS AND WINDOWS

08110.A	HOLLOW METAL FRAME
08410.A	ALUMINUM FRAMING — SEE SCHEDULE
08710.A	METAL THRESHOLD

DIV. 9 FINISHES

09260.B	5/8" GYP. BD.
09260.G2	METAL CORNER BEAD (TYP)
09260.M1	7/8" FURRING CHANNEL
09260.Q1	3" VENTED CHANNEL SCREED
09511.A	ACOUSTICAL LAY-IN CEILING
09650.B	4" RUBBER BASE

CHAPTER THIRTEEN

FAST TRACK CONSTRUCTION DETAILS

STANDARD ARGUMENTS AGAINST STANDARD
DETAIL LIBRARIES

THE BASIC STEPS AND RULES OF STANDARD DETAILING

THINK "TRACER SHEETS" ALONG WITH STANDARD
DETAIL SHEETS

ELIMINATING REDRAFTING OF DETAILS BY FREEHANDING

SLOW TIMES -- NO REASON NOT TO KEEP BUSY

CHECKLIST: HOW TO SET UP A STANDARD DETAIL LIBRARY

STANDARD DETAILS REQUIRE A STANDARD FORMAT

STANDARD DETAIL COMPONENTS CAN BE
REASSEMBLED IN INFINITE VARIETY

STANDARD WALL SECTION DRAWINGS AND OTHER
COMPOSITES PAY OFF

"CLUSTERED" STANDARD DETAILS

STANDARD
ARGUMENTS
AGAINST
STANDARD
DETAIL LIBRARIES

"None of our buildings are alike; it isn't worth the time to reuse a few details from time to time."

Those who use standard detail libraries report a consistent 60% to 80% reuse factor on all their projects. That is, no matter how diverse their building types and designs, **as much as 80% of all details used often come directly from the detail library.** That's because nobody invents whole new roof drains, sliding doors, grab bars, suspended ceilings, curbs, etc.; they come out of boxes and are combined with equally standard roof, floor, and wall construction.

Imagine what 60% to 80% reuse means financially. A typical detail will take two to four hours to design, sketch, draft, check, correct, and recheck. If your average drafting time costs $12 direct hourly pay plus 30% benefits, for a total of $15.60 per hour, then a medium-size project with **40 detail drawings at 3 hours per detail** costs $1,872. **If you get just 70% reuse of details on the same project, the savings is $1,310. Multiply that by 12 projects a year for an annual** potential savings of $15,724. **After a few years of that, you're talking substantial money.**

"We don't have time to sit down and draw all those details."

You'll be drawing all those details on future jobs anyway, so why not start by making the next detail you draw for the standard file as well as the job. Then the next . . . and the next. Eventually you'll have a complete file.

"But most details are unique; they won't be reused again."

Maybe. **Actually most users are pleasantly surprised to find a much higher reuse factor than they expect.** Besides, those details which don't end up in a totally reusable Standard Detail File can go into an auxiliary Reference Detail File. You may only do a residential swimming pool detail once in a blue moon but you can be sure there will be a need for related information some day, so why not keep it for possible future reference?

"It's too much trouble to draw them on small sheets, file them, find them . . . ."

Details drawn on small sheets (8-1/2 x 11) **are faster to draft than those drawn across back-stretching full-size drawings.** And with the commonly used enlargement reduction copiers and large-format copiers, it's never been easier, faster, or cheaper to handle copying, paste-up, and reproduction. Not only are half of most details on a job reused on a project, the process of pasting up new and old details is markedly faster than drafting them all in the traditional manner.

"We have CADD, and standard details are still hard to manage."

Computers are good for storing standard details, but it's time-consuming and wasteful to plot them out on full-size sheets. If you're using a pen plotter, it's faster to print details on small sheets with high resolution dot matrix or laser printers and have them cut out and pasted up manually. That makes the inevitable changes easier to manage, too.

"You can't possibly make enough details for all the different combinations of construction--like wood windows in wood frame walls, in brick walls, block walls . . . "

A few hundred standard detail components can give you a library of tens of thousands of possible detail combinations. For example, just thirty drawings--including ten window sections, ten sill types, and ten different wall types--will give you the raw material for one thousand possible detail drawings. You can recombine them quickly by tracing, by combining transparency overlays, by paste-up, or on CADD.

"If you make a mistake on a standard detail, it'll be repeated over and over."

A Standard Detail Library gives you your best chance of truly reducing recurring errors and omissions in details. You include a detail history form on each original standard detail sheet so that jobsite experience and suggested improvements can be noted on the drawing for later reference. That way the details get better and better. Errors are corrected instead of repeated and the people who make the errors learn about them. That often doesn't happen in the typical situation where there's virtually no feedback from the construction site to the designers and drafters.

"Offices start a standard detail library and after awhile it gathers dust."

Any new improvement requires changes of habit and the requirement to think and question what one is doing. That invokes resistance. The only way a new Standard Detail Library will be initiated, maintained, and used to its full potential is if someone who really cares is in charge of it. The higher that person is in the office hierarchy, the better. But if it can't be an associate or partner, it should be someone who reports directly to an office principal and who will have top management support. Otherwise, it's true, it will be "too much trouble." The system won't be used, and the office will continue to throw thousands of dollars out the window every year.

THE BASIC STEPS AND RULES OF STANDARD DETAILING

1) Establish a standard detail format. We recommend 8-1/2" x 11" sheets with a "cut out detail module" on the top half for the detail drawing which is 6" wide x 5-3/4" high. The 6" x 5-3/4" detail module will fit reasonably neatly within most standard working drawing sheet sizes. And establish standard positioning of notation (a right-hand column is usually preferable).

A standard format for all details will allow them to fit together neatly when composed on larger sheets; otherwise assembled details of different sizes will become a patchwork.

2) Establish standard scales for basic detail types. Younger drafters tend to draw details too large or too small. Here are the best recommended standards:

-- **3" = 1'-0"** is usually the best scale for doors and window details, wall fixtures and connectors and the more elaborate components of walls, floors, and ceilings.

-- **1-1/2" = 1'-0"** is for very simple construction components in roofs, walls, floors, etc.

-- **1" = 1'-0"** or **3/4" = 1'-0"** is suited only for the very simplest light framing, landscaping, and cabinet details.

-- **Half-size** and **full-size** details are only for shop drawings or to show the smallest construction components.

3) Create a file number system and index of details, preferably coordinated with the CSI specification division numbers.

4) File reproducible master copies of all details by file number in file folders or in plastic sleeves in three-ring binders -- in CSI number sequence. Keep originals stored separately as backups in case any reproducible copies disappear or are damaged.

5) Make a separate office "Catalog" of photocopies of the filed standard details for easy reference. These can be filed alphabetically for ease of use -- Stair details under "**S**", window details under "**W**", etc. That makes it easy for someone to find the types of details he or she might need. Each catalog copy will show the file number which shows where to get the master copy.

THINK "TRACER SHEETS" ALONG WITH STANDARD DETAIL SHEETS

"Tracer sheets" are standard format sheets that are laid on drafting boards or kept in CADD files as template guides for drafting new work. They save time by showing the sizes and positions of typical drawing components so drafters don't have to look them up and then scale them out on the drawing sheets. Scaling, marking, and checking measurements are the most time-consuming parts of drafting; these tracer sheets "pre-measure" the most vital dimensions for the drafters.

Tracer sheets used for wood frame or masonry construction include:

--Foundation plans showing the typical corner start point, foundation widths, pier and joist sizes and spacings, slab reinforcing patterns, vent spacings, etc.

--Exterior elevations with typical footing depths, floor-to-header and floor-to-floor heights, and **examples** of typical roof pitches.

--Partial longitudinal and lateral cross sections with complete one- and two-story framing sections with parapet and overhang options.

These tracer sheets have variable data on them that's not practical to show on reusable standard detail and section sheets. They're for reference and can show lots of options on the same sheet (such as different roof slopes) -- all purely as guides and reference. While no match in timesavings compared to fully reusable standards sheets, they expedite drafting enormously and save 20% to 30% of drafting time on many common types of drawings.

ELIMINATING REDRAFTING OF DETAILS BY "FREEHANDING"

Robert Stauder at Hellmuth Obata Kassabaum in St. Louis noticed a strange phenomenon. He observed that senior drafters were sketching construction details to scale, usually complete with notes and dimensions, and then handing them to junior drafters to copy and redraw all over again. Then the details went back to the senior people for checking, then back to the drafters . . . and maybe back and forth one more time after that. This was standard operating procedure and nobody thought much about it.

Robert thought about it and asked: "Why not have the senior people work out their details freehand at extra large scale and then we can paste them up and photo-reduce them to look like finish drafting."

They did. It worked. It eliminated a major interim step and bottleneck in the production process. Stauder observed that overall drafting savings amounted to about 20% on most projects, and total time savings on one university building was about 35%. The word then got around to hundreds of other offices, many of which have been doing freehand drafting ever since.

There were arguments at the time, like: *"How are we going to train our junior drafters how to do details?"* The reply was: **Drafters have to learn materials and construction** in order to do good detailing and they don't learn that by copying other people's drawings. Those who were willing to learn materials and construction were encouraged to do so; those who did not were allowed to focus mainly on non-detail drafting such as plans and elevations.

Two related examples:

A San Diego architect lost two drafters and found out it was no loss. His habit had been to do careful freehand drawings of his residential designs and hand them over to the drafters to "finish up." Both drafters were out with the flu one week and the architect had to get a set of drawings out to bid on his own. He had to complete a set using mainly freehand design development drawings and quickly realized that they were fine for the purpose. The redrafting he had always had done had served no particular constructive purpose.

Similarly, a large remodeling project was in crisis and had to be finished off in a hurry. Up to that point, there were only flimsy, yellow tracing paper sketches and studies. After some analysis, sorting, and beefing up the linework of some of the sketches, the project manager was able to paste up a set of drawings with the help of two people in just four days. (Normally several drafters would have required two to three weeks to redraw all the material in traditional finish form.)

Reminder, for successful freehand drafting: Use no-print blue tracing paper or underlay with 1/8" and 1" grid line, and long scales for quick-draw of long lines.

SLOW TIMES -- NO REASON NOT TO KEEP BUSY

When work is slow, it's a great opportunity to set up new productivity systems.

For example:

Standard details are great, but sometimes they're better done on large sheets. For years A/E's have noticed that some common kinds of working drawing sheets show the same basic construction again and again. They exploit that situation by drawing the sheets once instead of separately for project after project.

Typical full detail sheets are:

--Civil engineering construction: Curbs, paving, catch basins, manholes, trenches, erosion control, etc. Several dozen typical sitework details are common to these sheets.

--Landscaping: Tree and shrub planting and protectors, retaining walls, fencing, yard walls, signs, site furniture, etc.

--Exterior appurtenances (maybe not worth a full sheet in itself, but it may be a standard half sheet to which other drawings are added): Wall flag mounts, plaques, wall louvers, etc.

--Interior details: Fire rated wall construction, standard casework and cabinets, especially as used on often repeated building types like medical facilities, labs, and schools. Standard suspended ceiling details usually go together and are often combined with standard furring, soffit, and fireproofing details.

--Door types and frames are highly standardized according to building types -- housing, hospitals, schools, etc. -- and make another excellent partial sheet to go with standard forms for door and hardware schedules

--Electrical and HVAC drawings have their own standard details, schedules, stock riser diagrams, etc.

Standard Interior Elevation sheets are among the greatest timesavers of all. Floor plans show the locations of stock fixtures and equipment such as drinking fountains, fire hose cabinets, shelves, etc.; the standardized interior elevations show fixture heights and appearance.

A single standard fixtures and equipment schedule can take the place of dozens of interior elevation drawings for clinics, hospitals, office interiors, housing, schools, kitchens, recreation rooms, libraries, jails, etc. That especially goes for one of the most pointless drawings found in most sets of working drawings -- interior elevations of rest rooms.

Unless there's something very special being shown, most restroom fixtures and handicap aids are best shown by a **standard** fixture schedule rather than customized interior elevation drawings.

General Information sheets are almost universal now, but some firms still don't use them or at least they need to edit and update their General Information. Besides the common symbols, materials indications, abbreviations, and standard code provisions, these sheets should include explanations and samples of special systems such as Keynoting, the use of the CSI number references with construction notes to coordinate with specs, **and explanations** of timesavers such as partial hatching, wall material keys, short-form schedule formats, etc.

Typical standard detail sheet drawing sizes are 24" x 36" for medium to large size buildings. Except for highly varied civil and landscaping work, it's hard to fill a 30" x 42" sheet.

CHECKLIST:
HOW TO SET UP
A STANDARD
DETAIL LIBRARY

This is a model action checklist for setting up a Standard Detail Library. Modify the checklist as you see fit.

Most users of standard details typically adapt 80% of details for any particular project from their library. That's considerable time and money savings and it's a reasonable expectation for any office no matter how varied or original your work may be.

ALL details you do have potential multiple uses, either as standards for reuse or as reference details, so don't let them go to waste.

You can start a standard detail library with the very next detail drawn in your office. It'll be drawn anyway so you may as well do it in such a way as to gain multiple benefits. It doesn't require any significant investment in time and money to set up a system; it just takes communication and monitoring to make sure it's done right. Use this checklist accordingly.

PLANNING AND SCHEDULING
A STANDARD DETAIL SYSTEM

__ Select from the actions listed in the pages that follow and add others of your choosing.

__ Assign tasks.

__ Establish dates for starts, progress reviews, meetings, and completions.

__ Create a file or binder to store data related to planning and managing a detail system.

__ Decide who should start or be in charge of the detail system.

__ Meet with office staff members who will participate in creating or maintaining the standard detail system to discuss the action steps in this checklist.

__ If your office size justifies it, establish a task force to research, plan, and implement the system.

__ Start a list of top priority problems and concerns in detailing.

__ Collect comments and suggestions from supervisors.

__ Review past sets of working drawings to identify the types of details that are most readily reusable.

__ Locate any office detail files that might have been used in the past, or files currently being used by individual personnel.

__ List possible sources of existing details that might be usable as standards or as reference details.

__ Consider retaining a qualified general contractor to critique some typical samples of your working drawings with you and to help establish rules for effective detailing.

__ Decide how to integrate details with CADD and/or computer data base management.

__ Decide whether you want details to be notated individually or if detail types or clusters should be keynoted on final drawing sheets.

__ Establish a date for deciding who will ultimately be in charge of the system.

__ Create a calendar time line and delegated assignment list for implementing the major phases of researching, planning, and implementing the new or improved standard detail system.

__ Notify staff of three requirements in detailing:

__ All details created for new projects will henceforth be filed for long-term

181

reference or for the standard detail system.

__ Since details will be used for long-term reference and possible reuse, they will have to follow consistent formats and drafting standards, as described later in this checklist.

__ Since details used as standards will often go through several generations of photo-reproduction and possibly be enlarged or reduced in size, they must be done in ways that ensure top-quality reproduction at every step. Extra sharp, clear drafting will be a must.

Create a master list of action STEPS for getting the system going. Here's a suggested model list:

__ Establish office drafting and graphics standards for good readability and reproducibility of details.

__ Establish an office detail sheet format so that all details on file are consistent in appearance.

__ Convey the best rules on the steps for creating original details: drawing sequence, notation, scales, dimensioning, leader lines, and simplification techniques.

__ Review the options of detail files you might use and your choices in filing and retrieval systems.

    __ Filing of Master Details for reuse.
    __ Filing of reference or "design" details.
    __ Filing of original construction details.
    __ Computer filing and retrieval.
    __ A detail reference catalog.

__ Create file folders and/or three-ring binders, related materials, and procedures for using the details.
__ Establish a jobsite feedback system for upgrading and improving the details.

__ Select reprographic methods for printing final detail sheets.

Establish and enforce consistent drafting standards:

__ Small lettering is a problem on any job. All hand notation lettering should be at least 1/8" high. Computer-printing or clear typing is preferable for notes and can be a little smaller than 1/8".

__ Light linework tends to fade away, so lines should be consistently black, more differentiated by line width rather than "darkness" or "lightness."

__ Small symbols tend to clog up, so symbols should be large and open. That includes arrowheads, circles, triangles, etc.

__ Crosshatch patterns tend to run together when reproduced, so line patterns should be spaced at least 1/16" apart.

__ Poche made by using grey tone drawing or graphite dust does not reproduce well, so use dot or Zipatone-type patterns to achieve comparable results.

__ Numbers and lettering should not touch linework or they will tend to flow together in reproduction and lose clarity.

Establish a consistent standard detail sheet format:

__ Set a size for the detail "cutout window." (I recommend 6" wide x 5-3/4" high as a size and shape that accommodates the largest number of details of different types, scales, and sizes, and still fits evenly within most standard working drawing sheet sizes.)

__ Set standard margin limits for maximum detail profiles: left and right, top and bottom.

__ Set standard sizes and positions for:

    __ Notation string or keynote code string.
        __ Dimensions.
        __ Detail key.
        __ Title.
        __ Scale.
        __ Detail File Number.

__ Create a form for detail reference information and a Detail History Log to be included on standard detail sheets.

TRAINING STAFF IN
THE BEST DETAILING PROCEDURES

Create office rules and recommendations for clear and consistent detail layout:

__ Draw in sequence from the most general to the most particular.

__ Draw in "layered" phases so that a detail is substantially visually complete at every stage.

__ Keep the exterior face of construction facing to the left and the interior face to the right. (This is a general rule that works well for consistency and readability most of the time. Abandon the rule whenever it fails to support the objective of maximum clarity and readability.)

Keep most notation as a list column on the right-hand side of a detail window. Place other notes as appropriate for clarity and to avoid crowding the information.

Create office rules FOR notation:

___ Most notes should be simple names of materials or parts. If more information is required, add the data as assembly or reference notes after the material or part names.

__ Notes should provide information in a consistent sequence:

    __ Size of the material or part, where size is relevant and not duplicated by a dimension.
    __ The name of the material or part--generic, not specific names are usually preferred. Don't use product brand names, workmanship standards, or code and reference standards unless your drawings and specifications are one and the same.

    __ Noted position or spacing of parts unless they are dimensioned.

__ Provide an office nomenclature and abbreviations list, so all drafters will use the same terminology. Such terminology should be consistent with specifications.

__ Arrow leader lines from notes should follow a consistent office standard. Recommended: a short straight line starting horizontally from the note and breaking at an angle to lead to the designated material or part.

Create office rules and recommendations regarding proper scales for detail drawings:

__ In general stick to two scales: 1-1/2" = 1'-0" and 3" = 1'-0". The scales used throughout this book are the ones most widely used for these details.

__ Use 3/4" = 1'-0" only for the very simplest light framing, landscaping, and cabinet details.

__ 1-1/2" = 1'-0" is used to show simple construction components in roofs, walls, floors, etc.

__ 3" = 1'-0" is mainly used for doors and windows, wall fixtures and connectors, and more elaborate components of the walls, floors, ceilings, etc. When in doubt, this larger scale is preferable.

__ Half- and full-size details are only for shop drawings or to show the smallest of construction components. Extra-large details take up space and time out of proportion to their usefulness and are rarely justified.

Create office rules regarding dimensioning:

__ Avoid fractions in dimensions as much as possible. The smallest practical fraction in most dimensioning is 1/4".

__ Dimension lines should connect only to lead lines extended from the faces of materials,

183

and should not connect directly to material profile lines themselves.

__ Don't duplicate the same dimension on two sides of a detail. Such redundancy is confusing.

__ Use consistent and simple dimension connection symbols--arrows, slashes, or dots.

ESTABLISHING A FILING AND RETRIEVAL SYSTEM

__ Create two primary files for office-wide use: a Master Detail File for details that are potentially reusable, and a Reference Detail File for all other details. These should be top-quality reproducible copies of details, but not the originals themselves.

__ Keep ORIGINAL Master and Reference original drawings in separate, secure, files controlled by the person in charge of the detail system. (Original details may get lost or misfiled if staff members have easy access to them.)

__ The most popular file product is the "Pendaflex" type of hanging file folder with subsection folders within the hangers. Many useful color code and tab identification systems are available. Sideways slider file drawers are most convenient for ready access.

__ Establish a file number system, preferably based on CSI Masterformat. Every Master and Reference Detail should have a file number corresponding to the 16-division CSI system.

The file numbers we favor are based on CSI Masterformat MP-2-1, February 1979.

We've ignored subsequent CSI number changes because such changes are arbitrary and only confuse the system if they're changed periodically.

The detail file numbers work like this:

__ The first number identifies the detail's CSI-coordinated division number. All sitework details start with 02, for example; concrete details with 03; masonry with 04; etc. (These divisions are set no matter which variation of CSI specification numbers you use.)

__ The secondary numbers name subcategories of the CSI division, as per the CSI Masterformat. Thus 03300 identifies a detail as Concrete, and in particular as Cast-in-Place Concrete.

__ Details should also have suffix numbers so that individual details may be differentiated from their type. Thus 03300-1 would be the first of the Cast-in-Place Concrete details, 03300-3 might be another similar detail, and so on.

__ Provide gaps in the expandable suffix number sequences so you can add in new details later as they're developed. Thus your Cast-in-Place Concrete details might be numbered 03300-1, 03300-3, 03300-5, etc. Or to allow even more space for future details, the numbers might start as: 03300-1, 03300-11, 03300-21, 03300-31, etc.

__ Make the detail file number a conspicuous part of each detail drawing as part of the title.

__ Consistently position the file numbers in a place where they won't be accidentally cut off in cut-and-paste operations.

__ The person in charge of the detail system should be responsible for assigning file numbers.

__ Post or distribute an index of your details and their file numbers to guide those trying to find details in your files.

__ Consider keeping a "catalog" of details on file as a further aid to retrieving details. A catalog is a 3-ring binder with good-quality photocopies of all details stored in the Master File.

__ If making a catalog, decide whether to keep one or two copies in a central location or whether to give every staff member his or her own copy.

__ Although details should be filed according to the CSI 16 Divisions, consider listing or cataloging them for retrieval in a way that will feel natural to the average staff member.

That might be:

__ By alphabet: "Concrete" details under "C," "Windows" under "W," "Stairs" under "S."

__ By the major phases of construction:
    __ Sitework--Excavation and Grading, Drainage, Site Furniture
    __ Foundations
    __ Framing--Concrete, Steel, Wood
    __ Interior Substructure--Floors, Walls, and Stairs
    __ Roofing
    __ Exterior Enclosure
    __ Windows and Glazing
    __ Suspended Ceilings
    __ Interior Partitions
    __ Interior Finishes
    __ Flooring
    __ Doors
    __ Finish Carpentry
    __ Interior Finishes
    __ Fixtures and Furnishings
    __ Appurtenances
    __ Finish Sitework--Paving and Finish Landscaping

__ By your standard sequence of working drawing sheets.

ESTABLISHING EFFICIENT PROCEDURES FOR YOUR SYSTEM

NOTE: MOST OF THESE PROCEDURES APPLY WHETHER YOU'RE USING CADD, SYSTEMS DRAFTING, OR MANUAL DRAFTING.

Create a procedural checklist so staff members can fully understand the correct way to use the standard and reference detail files.

INCLUDE THESE INSTRUCTIONS IN YOUR CHECKLIST:

__ Plan and sketch each working drawing detail sheet in miniature on 8-1/2" x 11" grid paper.

__ Identify the detail types that will go on each sheet.

__ Consult the Standard Detail File Index or Catalog and find details available that fit the situation. Write down their file numbers.

__ Make photocopies of selected details from the Catalog or from the Master Detail File to use as checkprints.

__ Make a full-size mockup of the final sheet with pasted up photocopies. This will often help clarify the relationship of construction details to one another.

__ Mark the photocopies to show changes to be made, and sketch additional new details required for the sheet.

WHEN DRAWING NEW DETAILS:

__ Use the office Reference Detail Library for examples that relate to the new detail.

__ Create new details with an eye toward potential reusability in the office's Standard Detail File.

__ If the new detail might be more readily reusable in a somewhat uncompleted state, make an exact-size Master copy at that stage of completion and file it for later review for use in your Standard Detail File.

__ Follow office drafting and format standards to assure consistency in appearance.

__ File revised details as alternatives to the original standard details. File number and index them accordingly.

AS NOTED BEFORE, MOST OF THESE PROCEDURES APPLY WHETHER YOU'RE

USING CADD, SYSTEMS DRAFTING, OR
MANUAL DRAFTING.

EXCEPTIONS ARE NOTED BELOW.

WHEN READY TO COMPLETE A DETAIL SHEET:

__ Make exact-size duplicates of chosen
    Master Details in the Standard Detail File.

__ Make duplicates of any new details you've
    created to go on the same sheet.

__ Revise and finish the copy of the Master
    Detail as required.

__ Assemble the detail copies on a carrying
    sheet, according to the office's Systems
    Drafting or reprographics system.

__ Make a check-print of the paste-up and file
    the original.

__ If it's time to do a phase submittal requiring
    multiple printing of drawings, make a sepia
    paper reproducible of the paste-up sheet
    and use that as the "second original."

__ Mark revisions on checkprints in the normal
    fashion.

__ Save revisions, and redo the original paste-up
    sheet periodically, to keep it up to date.

__ At the close of the job, make a reproducible
    of the detail paste-up sheet as the final
    project "original."

CADD PROCEDURES DIFFER IN THE FOLLOWING WAYS:

__ The detail index will be available on screen,
    as well as in printout form, and retrieval
    procedures may be restricted by rules that
    apply to your particular CADD system.

__ Your sketched revisions may go to a
    computer operator for finish computer
    drafting. In that case, it's most important
    that you provide exact dimensions.

__ New details should be put in a "holding"
    file, so that whoever is in charge of the
    detail system can review them later and
    decide if they should become part of the
    Standard Detail Library or be filed as
    Reference Details.

__ Don't create arbitrary new file names on the
    CADD system. All filing and retrieval
    must follow a predetermined standard
    system of names and file numbers.

__ Circulate a memo about the detail system,
    the different types of detail files and their
    uses, and the procedures for using the
    system.

__ Add all explanatory notes and checklists
    for using the detail system to your
    Office Manual or CADD Operations
    Manual.

ESTABLISHING A SYSTEM FOR ONGOING DETAIL REVIEW AND IMPROVEMENT

__ Include a Detail History Form on each
    standard detail sheet, so that job use can be
    noted every time a Master is taken from the
    file to be copied for use on a job.

__ Establish semiannual or quarterly dates to
    review Detail History Forms on all details
    in the Master Detail File, to see how
    frequently or infrequently the details are
    being used. Make additional variations of
    the most popular details, and move unused
    ones to the Reference Detail File.

__ Require jobsite representatives to take
    photos and send back film and notes
    regarding all detail-related problems.

__ Frequently review jobsite problem memos
    and photos, to identify detail problems and
    make appropriate corrections and revisions.

__ Each detail printed in the working drawings
    should include its detail file number. That
    will expedite the search for details that
    have caused problems or that can be
    improved from jobsite feedback.

__ Staff members who conduct post-occupancy surveys should also be required to note detail-related problems and send memos and photos back to the detail system manager.

__ Copy all relevant construction problem photos as Jobsite Feedback Memos with explanatory notes about the problem, the cost of the problem, and tips on how to avoid it in the future.

__ Distribute Jobsite Feedback Memos to all staff members, and keep a permanent, 3-ring binder file on hand for staff study.

__ Maintain a file of articles and reports on roofing, waterproofing, fenestration, materials failures, etc., as a supplemental reference source for detailers.

STANDARD DETAILS REQUIRE A STANDARD FORMAT

A standard detail format is necessary for details to fit neatly on large assembly sheets, and for consistency in positioning notes, titles, and dimension lines.

We strongly recommend the format shown on the next page.

Below is the detail drawing space. The recommended module, and the module used for the Guidelines Architect's Detail Library, is 6" wide x 5 - 3/4" high. If you prefer another module, you can still use this format sheet for detail modules of 6 - 1/2" x 5 - 3/4" high maximum, and 5 - 1/2" x 5 - 1/4" high minimum.

(Space for the detail name, scale and file number.)

DETAIL FILE NUMBER: 03305-41

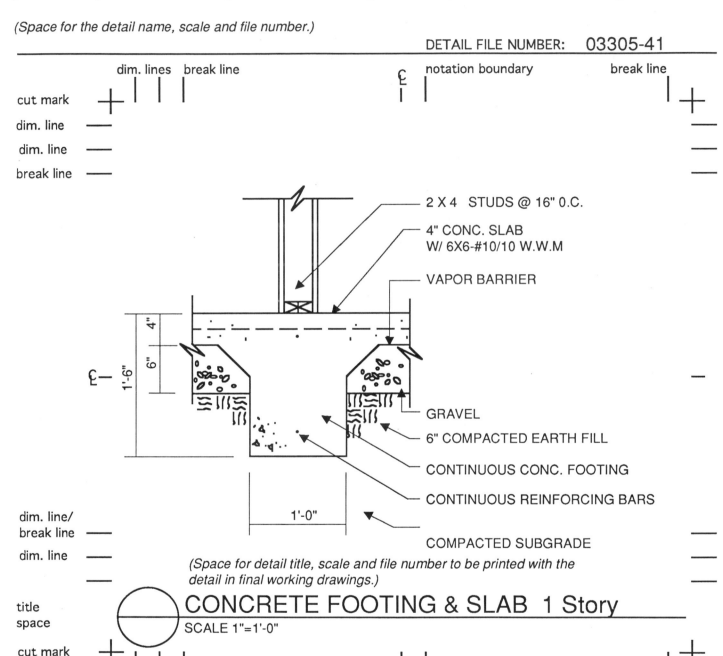

DETAIL INFORMATION
References, jobsite feedback, job history

Marks show various cut lines, notation, dimension, and face of construction lines. Exterior face of construction is normally at the left, notation blocks normally on the right-hand side. These suggested spacings aid visual consistency throughout your detail system but should be ignored if using them interferes with clarity and ease of detail drawing.

STANDARD DETAIL COMPONENTS CAN BE REASSEMBLED IN INFINITE VARIETY

Window types can be filed separately from wall types, then brought together as needed for a particular job detail.

08511 STEEL AWNING WINDOWS

Steel Awning Window -- Single Glazed

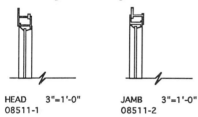

HEAD   3"=1'-0"
08511-1

JAMB   3"=1'-0"
08511-2

SILL   3"=1'-0"
08511-3

Steel Awning Window -- Double Glazed

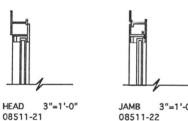

HEAD   3"=1'-0"
08511-21

JAMB   3"=1'-0"
08511-22

SILL   3"=1'-0"
08511-23

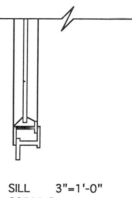

SILL   3"=1'-0"
08511-3

COMPOSITE MASTER DETAIL   DETAIL FILE NUMBER: 08511-3-6

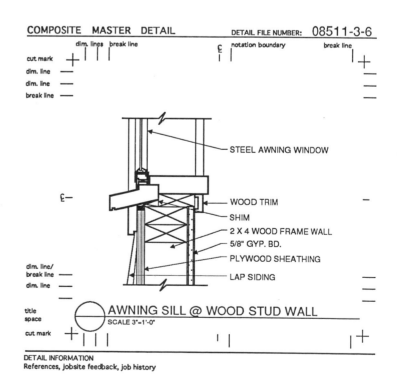

AWNING SILL @ WOOD STUD WALL
SCALE 3"=1'-0"

STEEL AWNING WINDOW
WOOD TRIM
SHIM
2 X 4 WOOD FRAME WALL
5/8" GYP. BD.
PLYWOOD SHEATHING
LAP SIDING

cut mark
dim. line
dim. line
break line

dim. lines   break line   notation boundary   break line

dim. line/ break line
dim. line

title space
cut mark

DETAIL INFORMATION
References, jobsite feedback, job history

08511 STEEL AWNING WINDOWS

NOTATION CHECKLIST, SAMPLE NOTES

WALL CONSTRUCTION
SHIM SPACE
DRIP
CAP/WEATHERSTRIPPING/FLASHING
CAULKING/GROUT
FINISH HEAD/SILL/JAMB
WINDOW TYPE, MATERIAL & FINISH
HARDWARE/OPERATOR
VENT/WEEP HOLE/WIND GUARD
GLAZING:
SINGLE/DOUBLE/REMOVABLE
SCREEN/SCREEN FRAME
CASING/TRIM/ADJACENT FINISH
ROUGH OPENING/FINISH OPENING

STEEL WINDOW SYSTEM
METAL SILL BELOW
THERMOPANE INSULATING GLASS
SINGLE PANE GLASS
HEAD, SEALANT BOTH SIDES
SILL W/SEALANT EACH SIDE
WOOD NAILER
TREATED BLOCKING
1 X   WOOD TRIM
METAL EDGE BEAD
STUD
FLASHING
SEALANT

The drawings below show the window type from the preceding page combined with wood framing for head, jamb and sill sections.

Once the generic components are combined to make a complete composite detail, they can then be filed for reuse in the office's customized standard detail library.

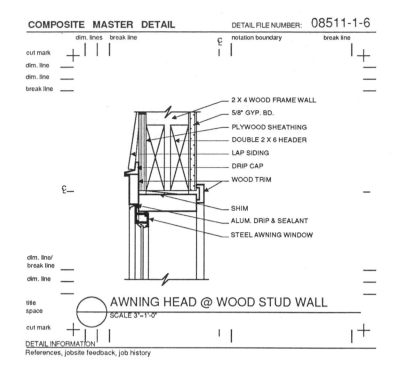

COMPOSITE MASTER DETAIL

DETAIL FILE NUMBER: 08511-1-6

dim. lines break line notation boundary break line

cut mark

dim. line

dim. line

break line

- 2 X 4 WOOD FRAME WALL
- 5/8" GYP. BD.
- PLYWOOD SHEATHING
- DOUBLE 2 X 6 HEADER
- LAP SIDING
- DRIP CAP
- WOOD TRIM
- SHIM
- ALUM. DRIP & SEALANT
- STEEL AWNING WINDOW

dim. line/ break line

dim. line

title space

AWNING HEAD @ WOOD STUD WALL

SCALE 3"=1'-0"

cut mark

DETAIL INFORMATION

References, jobsite feedback, job history

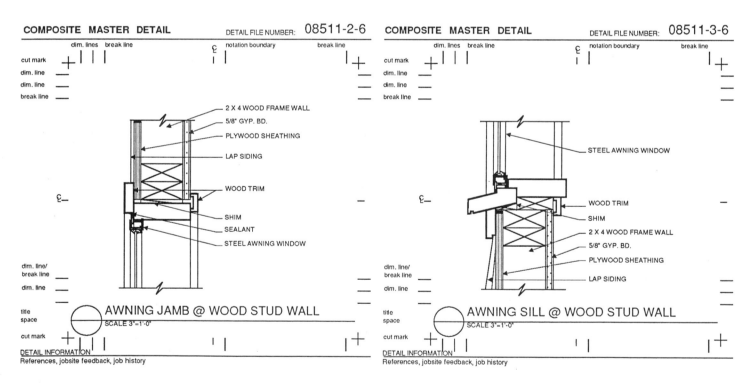

COMPOSITE MASTER DETAIL

DETAIL FILE NUMBER: 08511-2-6

dim. lines break line notation boundary break line

cut mark

dim. line

dim. line

break line

- 2 X 4 WOOD FRAME WALL
- 5/8" GYP. BD.
- PLYWOOD SHEATHING
- LAP SIDING
- WOOD TRIM
- SHIM
- SEALANT
- STEEL AWNING WINDOW

dim. line/ break line

dim. line

title space

AWNING JAMB @ WOOD STUD WALL

SCALE 3"=1'-0"

cut mark

DETAIL INFORMATION

References, jobsite feedback, job history

COMPOSITE MASTER DETAIL

DETAIL FILE NUMBER: 08511-3-6

dim. lines break line notation boundary break line

cut mark

dim. line

dim. line

break line

- STEEL AWNING WINDOW
- WOOD TRIM
- SHIM
- 2 X 4 WOOD FRAME WALL
- 5/8" GYP. BD.
- PLYWOOD SHEATHING
- LAP SIDING

dim. line/ break line

dim. line

title space

AWNING SILL @ WOOD STUD WALL

SCALE 3"=1'-0"

cut mark

DETAIL INFORMATION

References, jobsite feedback, job history

More examples of recombinations of generic components into final details for a job. Completed project details are then added to the office's standard detail library.

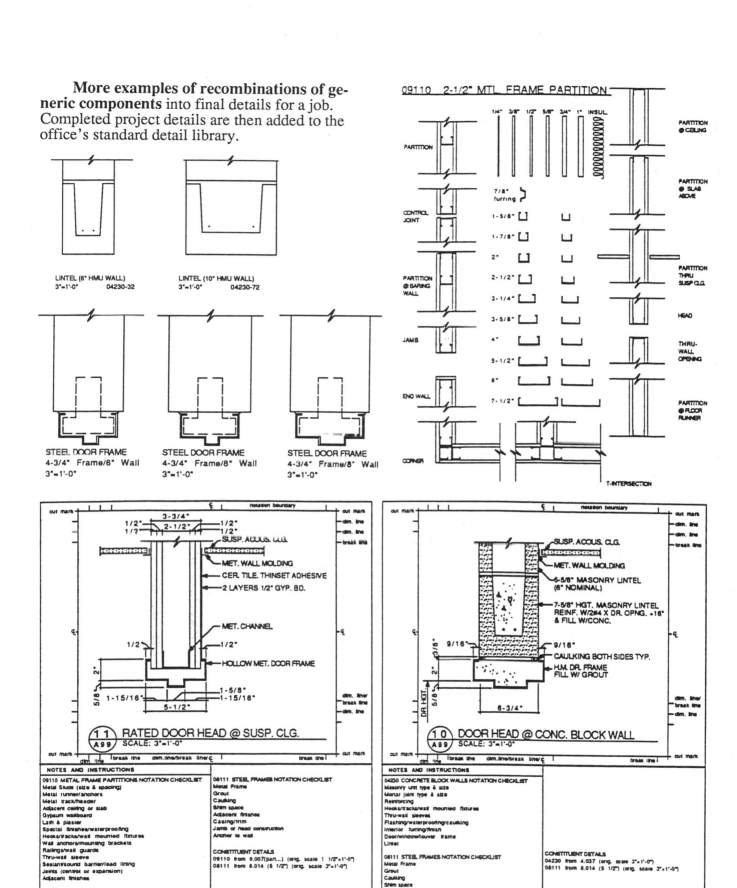

LINTEL (6" HMU WALL)
3"=1'-0" 04230-32

LINTEL (10" HMU WALL)
3"=1'-0" 04230-72

STEEL DOOR FRAME
4-3/4" Frame/6" Wall
3"=1'-0"

STEEL DOOR FRAME
4-3/4" Frame/8" Wall
3"=1'-0"

STEEL DOOR FRAME
4-3/4" Frame/8" Wall
3"=1'-0"

09110 2-1/2" MTL. FRAME PARTITION

1/4" 3/8" 1/2" 5/8" 3/4" 1" INSUL

PARTITION

CONTROL JOINT

PARTITION @ BEARING WALL

JAMB

END WALL

CORNER

7/8" furring
1-5/8"
1-7/8"
2"
2-1/2"
3-1/4"
3-5/8"
4"
5-1/2"
6"
7-1/2"

PARTITION @ CEILING
PARTITION @ SLAB ABOVE
PARTITION THRU SUSP CLG
HEAD
THRU-WALL OPENING
PARTITION @ FLOOR RUNNER

T-INTERSECTION

Detail 11 / A99 — RATED DOOR HEAD @ SUSP. CLG. — SCALE: 3"=1'-0"

SUSP. ACOUS. CLG.
MET. WALL MOLDING
CER. TILE. THINSET ADHESIVE
2 LAYERS 1/2" GYP. BD.
MET. CHANNEL
HOLLOW MET. DOOR FRAME
3-3/4"
2-1/2"
1/2" 1/2"
1/2" 1/2"
1/2" 1/2"
2"
5/8"
1-15/16"
1-15/16"
1-5/8"
5-1/2"

Detail 10 / A99 — DOOR HEAD @ CONC. BLOCK WALL — SCALE: 3"=1'-0"

SUSP. ACOUS. CLG.
MET. WALL MOLDING
5-5/8" MASONRY LINTEL (6" NOMINAL)
7-5/8" HGT. MASONRY LINTEL REINF. W/2#4 X DR. OPNG. +16" & FILL W/CONC.
CAULKING BOTH SIDES TYP.
H.M. DR. FRAME FILL W/ GROUT
3/8"
9/16"
9/16"
2"
5/8"
DR. HGT.
6-3/4"

NOTES AND INSTRUCTIONS

09110 METAL FRAME PARTITIONS NOTATION CHECKLIST
Metal Studs (size & spacing)
Metal runner/anchors
Metal track/header
Adjacent ceiling or slab
Gypsum wallboard
Lath & plaster
Special finishes/waterproofing
Hooks/tracks/wall mounted fixtures
Wall anchors/mounting brackets
Railings/wall guards
Thru-wall sleeve
Sealant/sound barrier/lead lining
Joints (control or expansion)
Adjacent finishes

08111 STEEL FRAMES NOTATION CHECKLIST
Metal Frame
Grout
Caulking
Shim space
Adjacent finishes
Casing/trim
Jamb or head construction
Anchor to wall

CONSTITUENT DETAILS
09110 from 9.007(part...) (orig. scale 1 1/2"=1'-0")
08111 from 8.014 (5 1/2") (orig. scale 3"=1'-0")

NOTES AND INSTRUCTIONS

04230 CONCRETE BLOCK WALLS NOTATION CHECKLIST
Masonry unit type & size
Mortar joint type & size
Reinforcing
Hooks/tracks/wall mounted fixtures
Thru-wall sleeves
Flashing/waterproofing/caulking
Interior furring/finish
Door/window/louver frame
Lintel

08111 STEEL FRAMES NOTATION CHECKLIST
Metal Frame
Grout
Caulking
Shim space
Adjacent finishes
Casing/trim
Jamb or head construction
Anchor to wall

CONSTITUENT DETAILS
04230 from 4.037 (orig. scale 3"=1'-0")
08111 from 8.014 (5 1/2") (orig. scale 3"=1'-0")

STANDARD WALL SECTION DRAWINGS AND OTHER COMPOSITES PAY OFF

Many offices include standard roofing, parapets, etc. in their detail libraries, but how about large combinations of details such as those encompassed in wall sections? Are they too individual to standardize?

Wall sections may be too variable to use directly in full standard sheets, but they make excellent unfinished starter drawings or tracer sheets. They show brick and block courses, glass block, standard header heights, etc. Designers and drafters draw their customized formal drawings right over standard construction without having to measure and compute the sizes of common construction components.

"CLUSTERED" STANDARD DETAILS

Many groups of details will be repeated from job to job, so they can be created and used as clusters rather than individual detail drawings. The wall sections on the next two pages are examples of this type of clustered standard detail sheet.

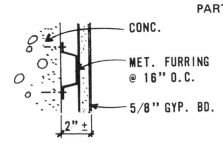

CONC.

MET. FURRING
@ 16" O.C.

5/8" GYP. BD.

2" ±

A TYPICAL FURRING

PLAN INDICATION

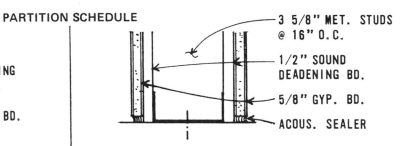

3 5/8" MET. STUDS
@ 16" O.C.

1/2" SOUND
DEADENING BD.

5/8" GYP. BD.

ACOUS. SEALER

E SOUND RETARDANT

PLAN INDICATION

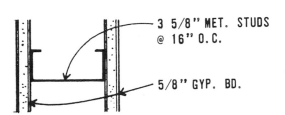

3 5/8" MET. STUDS
@ 16" O.C.

5/8" GYP. BD.

B TYPICAL

PLAN INDICATION

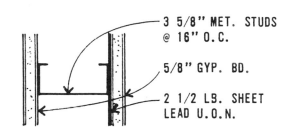

3 5/8" MET. STUDS
@ 16" O.C.

5/8" GYP. BD.

2 1/2 LB. SHEET
LEAD U.O.N.

F LEAD LINED

PLAN INDICATION

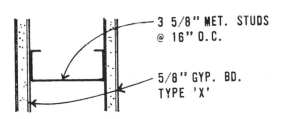

3 5/8" MET. STUDS
@ 16" O.C.

5/8" GYP. BD.
TYPE 'X'

C ONE HOUR

PLAN INDICATION

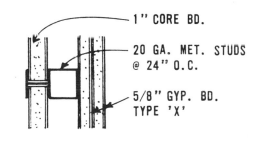

1" CORE BD.

20 GA. MET. STUDS
@ 24" O.C.

5/8" GYP. BD.
TYPE 'X'

G [G] TWO HOUR SHAFT

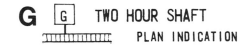 PLAN INDICATION

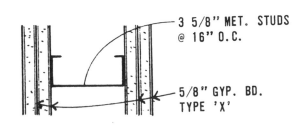

3 5/8" MET. STUDS
@ 16" O.C.

5/8" GYP. BD.
TYPE 'X'

D TWO HOUR

PLAN INDICATION

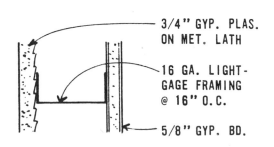

3/4" GYP. PLAS.
ON MET. LATH

16 GA. LIGHT-
GAGE FRAMING
@ 16" O.C.

5/8" GYP. BD.

H [H] CORRIDOR

 PLAN INDICATION

193

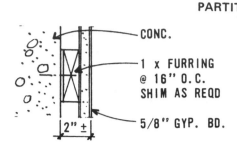

CONC.

1 x FURRING
@ 16" O.C.
SHIM AS REQD

5/8" GYP. BD.

2" ±

A TYPICAL FURRING

PLAN INDICATION

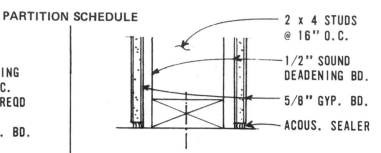

2 x 4 STUDS
@ 16" O.C.

1/2" SOUND
DEADENING BD.

5/8" GYP. BD.

ACOUS. SEALER

E SOUND RETARDANT

PLAN INDICATION

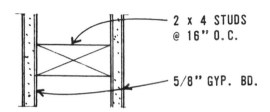

2 x 4 STUDS
@ 16" O.C.

5/8" GYP. BD.

B TYPICAL

PLAN INDICATION

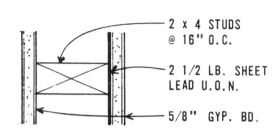

2 x 4 STUDS
@ 16" O.C.

2 1/2 LB. SHEET
LEAD U.O.N.

5/8" GYP. BD.

F LEAD LINED

PLAN INDICATION

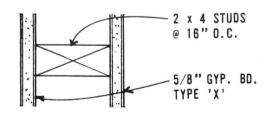

2 x 4 STUDS
@ 16" O.C.

5/8" GYP. BD.
TYPE 'X'

C ONE HOUR

PLAN INDICATION

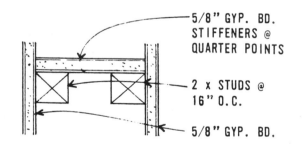

5/8" GYP. BD.
STIFFENERS @
QUARTER POINTS

2 x STUDS @
16" O.C.

5/8" GYP. BD.

G ☐ G ☐ CHASE

PLAN INDICATION

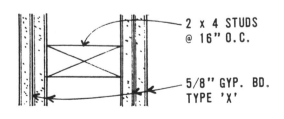

2 x 4 STUDS
@ 16" O.C.

5/8" GYP. BD.
TYPE 'X'

D TWO HOUR

PLAN INDICATION

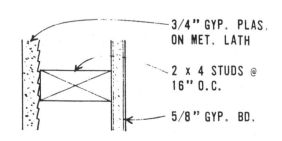

3/4" GYP. PLAS.
ON MET. LATH

2 x 4 STUDS @
16" O.C.

5/8" GYP. BD.

H ☐ H ☐ CORRIDOR

PLAN INDICATION

CHAPTER FOURTEEN

SIMPLIFIED SCHEDULES:

FINISH SCHEDULES,
WINDOW SCHEDULES,
DOOR SCHEDULES, ETC.

THE EASIEST TO USE FINISH SCHEDULE

DOOR SCHEDULES

WINDOW SCHEDULES

FINISH HARDWARE

EQUIPMENT SCHEDULE

FIXTURE AND FIXTURE HEIGHT SCHEDULES
INSTEAD OF INTERIOR ELEVATIONS

THE EASIEST
TO USE
FINISH FORMATS

The floor plan segment to the right shows a streamlined approach to Finish Schedules, one that cuts the work to a fraction of traditional formats.

What makes this method unique is that instead of listing room names down the left hand side of the schedule, only the finish key reference letters are listed . . . as below: A, B, C, etc.

The A, B, C codes designate the entire range of finishes in a particular room. This reduces what might have been a large Finish Schedule covering much of a drawing sheet down to a small block that can be included next to the floor plan.

The pages that follow show short and easy approaches to Door, Window, Finish Hardware and Equipment Schedules. (Courtesy of the Northern California AIA Production Office Procedures Manual.)

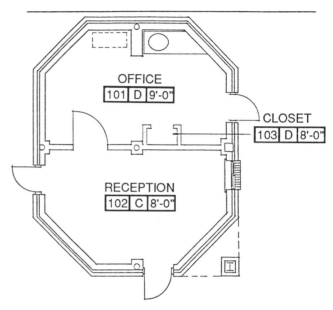

FINISH SCHEDULE					
FINISH KEY	FLOOR	BASE	WALLS	CEILING	REMARKS
A	CONCRETE	NONE	CMU	NONE	ROOM 102 NOT TO BE PAINTED
B	CONCRETE W/HARDNER	VINYL	CMU	24" X 24" ACOUSTICALBOARD	OMIT HARDNER IN ROOM107
C	CARPET	VINYL	GYPSUM BOARD	ACOUSTICAL TILE	
D	CARPET	VINYL	GYPSUM BOARD	GYPSUM BOARD	
E	VINYL	VINYL	GYPSUM BOARD - CMU	GYPSUM BOARD	
F	TERRAZZO	TERRAZZO	GYPSUM BOARD	GYPSUM BOARD	
G	ACCESS FLOORING	VINYL	WOOD PANELING	GYPSUM PLASTER	SEALER ON CONC. BELOW ACCESS FL.
H	BRICK PAVERS	WOOD	WOOD PANELING	ACOUSTICAL PLASTER	
J	CERAMIC TILE	CERAMIC TILE	VINYL WALL COVERING	24" X 48" ACOUSTICAL BOARD	
K	QUARRY TILE	QUARRY TILE	CERAMIC TILE	ACOUSTICAL TILE	

DOOR SCHEDULE

Depending on the complexity of the project, a door schedule should be a list of all doors or a list of groups of identical doors. Some projects having typical doors and few variations do not require listing all doors. However, some projects having extensive variations require a listing of all doors. The recommended schedule (type 1) can be used for both types of projects. It is desirable to have the schedule appear on the sheets with floor plans as is the case with the room material schedule.

An alternate schedule (type 2) is a schedule that has an identical master list for every job and covers most door type conditions.

RECOMMENDATIONS

The following should be used in preparing the door schedules for type 1.

1. Items tabulated on the schedule can be added or deleted to fit a firm's requirements. As an example a hardware group column can be added, if desired, which will eliminate paragraph 3 below.

2. Determine the typical items, such as door construction, facing and finish, etc. and verify with schedule notes. The more items that are typical, the simpler the schedule.

3. Schedule 1 does not list hardware. The hardware group is shown in the door symbol on the plans. The hardware groups are covered in the specifications.

DOOR SYMBOL, TYPE 1

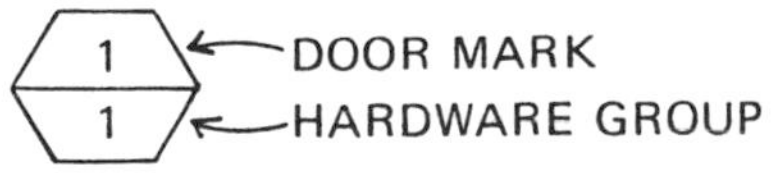

TYPE 1

DOOR SCHEDULE								
DOOR MARK	OPENING SIZE	TYPE (NOTE 2)	THICKNESS (3)	CONSTRUCTION (4)	FACING & FINISH (5)	GLASS (6)	RATING (7)	FRAME (8)
1	3'-6" x 7'-0"	A	✓	✓	✓	CW	1½	26

1. "✓" SHOWN ON SCHEDULE INDICATES TYPICAL
2. DOOR TYPES:

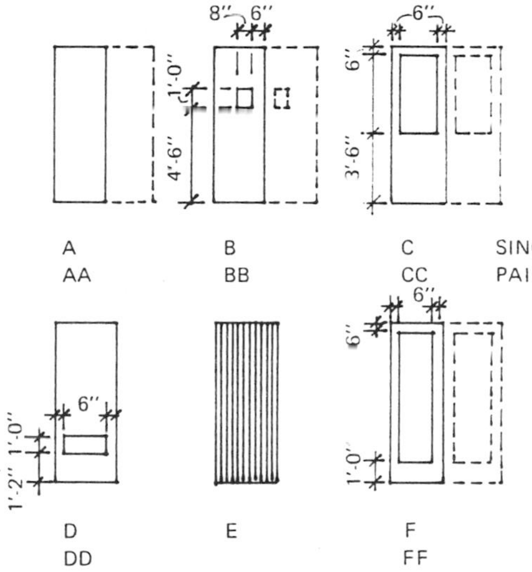

3. ALL DOORS 1¾" THICK UNLESS OTHERWISE NOTED
4. DOOR CONSTRUCTION:
   TYPICAL = SOLID CORE
   HC = HOLLOW CORE
   HM = HOLLOW METAL
   AL = ALUMINUM & GLASS
5. FACING & FINISH:
   TYPICAL = RED BIRCH, TRANSPARENT
   PT = PLASTIC LAMINATE, TEXTURED
   MP = METAL, PAINTED
6. GLASS:
   TYPICAL = CLEAR PLATE
   SG = SHEET GLASS
   CW = CLEAR WIRE
   TP = TEMPERED PLATE
   LG = LEAD GLASS
7. ¾, 1, 1½ ETC. INDICATES HOURS OF FIRE RATING.
8. TYPICAL FRAMES SHOWN "✓". NUMBER INDICATES DETAIL SHOWN ON SHEET______.

WINDOW AND LOUVER SCHEDULE

The window and louver schedule is a summarized depiction of various units throughout the project. It should clearly identify types, dimensional aspects, and operational characteristics. For small and simple projects, where elevations can easily show all window types, no schedule is required.

RECOMMENDATIONS

1. Scale ¼ inch = 1 foot 0 inches.

2. Window wall construction should be indicated and detailed separately.

3. Unit Elevation should always be drawn as viewed from the exterior side.

4. Dimensions shown on schedule should be rough or masonry openings.

5. Directions of vent swings or sliding panels should be indicated on the schedule.

   a. Top hinged always swings out toward exterior, u.o.n.

   b. Bottom hinged always swings in toward interior, u.o.n.

   c. Side hinged always swings out toward exterior, u.o.n.

   d. Pivot hinged always rotates about indicated axis.

6. Indicate symbols on Floor Plans typically. For small projects keying to elevations may be preferable.

7. If a window type is placed into more than one type of wall construction, detail reference marks should be placed on the building elevations where each type of wall construction occurs.

8. If a window or louver unit is to be made of more than one material, a type number or letter should be assigned for each material type.

9. Typical glazing should be noted in the specifications. Special glass types and thicknesses other than typical or minimum, as required by referenced glazing standards, should be noted on the schedule.

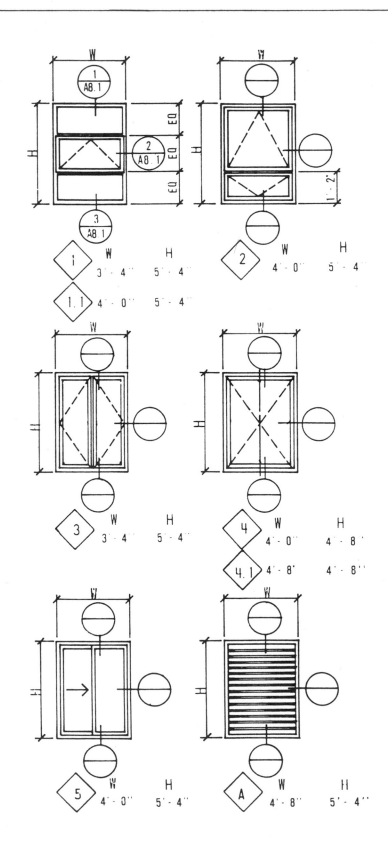

198

FINISH HARDWARE SCHEDULE

A simple to use standard for scheduling Finish Hardware is mandatory for a concise contract between owner and contractor. The system should lend itself for an office standard for all projects large or small and should also serve as a "shopping list" for selection of each hardware item wherein the group number or code identification for the item remains always the same. Soon the draftsman, specifier or estimator will easily identify many commonly-used numbers with specific hardware functions.

RECOMMENDATIONS

Itemizing of finish hardware is best accomplished by establishing groups in the specifications and indicating the group number in the door symbol on the plans or in the door schedule. The type 2 finish hardware schedule system must be used with type 1 door schedule.

TYPE 1

Large projects, with many similar doors.

Finish hardware group number is placed in lower half of door symbol and shown on the floor plans for every door.

The specifications for finish hardware shall:

Type 1a

have necessary preamble for determining quality and procedures and list all hardware for each group.

Type 1b

have necessary preamble for determining quality, procedures and code for each hardware item. The hardware group shall then be listed on the schedule. This codified system is recommended for projects that have many finish hardware groups.

DOOR SYMBOL

Door Mark

Hardware Group

Type 1a

(EXAMPLE)

Group 3 Each Door to Have

1½ pr. butts	T2314 McKinney
1 Lockset	D80PD Schlage
1 Door Stop	FB13 Glynn Johnson
1 Door Closer	4010 LCN

Type 1b

(EXAMPLE)

CODES

1. Lockset Types:

    L1   D60PD Schlage
    L2   D51PD Schlage
    L3   D80PD Schlage

2. Closer Types:

    C1   4010 Regular Arm Series LCN
    C2   4110 Parallel Arm Series LCN
    C3   930 with magnetic hold open Rixon

HARDWARE SCHEDULE   (with type 1b codes)

Group No.	Hinge	Lockset	Closer	Exit Device	Push Pull	Stop Holder	Bolt	Kick Plate	Thresh	Weather Strip	Misc
1	H1	L2	C3			S1	B1				M3
2	H2	L1	C2	E3		S2		K3		W1	
3	H1	L6	C4		P2	S2			T3		

TYPE 2

For small or Specialized Projects the finish hardware group is placed in an extra column of the door schedule. The specifications shall be as recommended in Type 1a or Type 1b above.

DOOR SYMBOL

Door Mark

EQUIPMENT SCHEDULE

In the past, many architects never felt the need for scheduling equipment. The type and size of equipment was indicated on plans with additional information given in the specifications. For small, non-repetitive projects this may be still the best method to convey to the Contractor the requirements of the Owner.

On larger or repetitive projects a simple coding system will facilitate not only production of documents but future changes or quantity take-offs and purchasing.

RECOMMENDATIONS

The equipment schedule is best included in the specifications wherein each item has a MARK CODE which is the key to each item on the plans.

1. Equipment is grouped by TYPES:

   Example: XR = X-Ray
   RF = Refrigerator
   FS = Food Service
   FI = Film Illuminator

2. Each type is then listed with a specific MARK code indicating a particular model or kind. Same models will have the same mark code.

3. Each piece of equipment shown on drawings must have an equipment symbol.

4. Specifications should list all equipment groups and mark code numbers with a detail description of name, model number and contract responsibility (who shall furnish and install).

SYMBOL

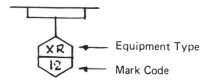

Equipment Type

Mark Code

FIXTURE AND FIXTURE HEIGHT SCHEDULES INSTEAD OF INTERIOR ELEVATIONS

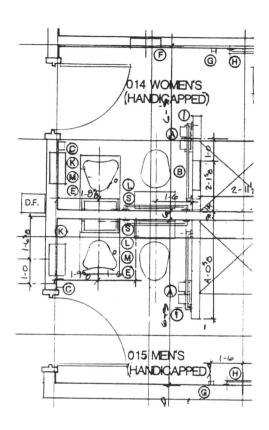

Fixture Heights Schedules do the work of most Interior Elevations. They show the basic fixtures and equipment and their heights in a simple, consolidated format. Thus a grab bar won't be shown over and over in Interior Elevation drawings, it's shown once on the schedule. The locations of the grab bar are shown on plan.

Many sheets of interior elevations can be eliminated through the use of Fixture Height schedules such as that shown below and on the next page.

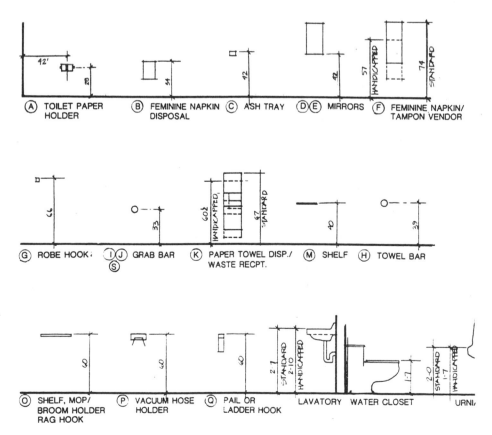

201

Most offices prefer a diagrammatic fixture heights schedule such as the one shown on the previous page because it's clear and explicit about appearance of fixtures and mounting heights. Some firms swear by the list format shown below in two examples because of its simplicity and ease of editing.

These standardized schedules aren't just limited to bath fixtures. They can be used for all kinds of interior design data for all building types such as office, school, and medical equipment.

MISC EQUIP SCHED

1. TACKBOARD (NIC) - PROVIDE BACKING PER UNL STANDARD.
2. CHALKBOARD (NIC) - PROVIDE BACKING PER UNL STANDARD.
3. PROVIDE PLAM WITH METAL TRIM 3'-0" H x 3'-0" W (2) WALLS. PROVIDE BACKING FOR FUTURE MOP STRIP AT 4'-6" AFF.
4. FEMININE NAPKIN VENDOR BOBRICK NO. B-282.
5. FEMININE NAPKIN DISPOSAL BOBRICK NO. B-270
6. TOILET PAPER DISPENSER FORT HOWARD NO. 571-98
7. TOILET STALL GRAB BAR (HC) BOBRICK NO. B-62061X48 MOUNT 33" AFF.
8. 16"x30" MIRROR (HC) BOBRICK NO. B-294-1630 MOUNT 40" AFF. TO BOT.
9. 18"x24" MIRROR BOBRICK NO. B-290-1824 MOUNT 44" AFF. TO BOT.
10. PAPER TOWEL DISPENSER FORT HOWARD NO. 574-10 MOUNT 48" AFF. TO BOT.
11. SOAP DISPENSER AMERICAN DISPENSER CO. NO. 82 "LATHURSHELF" MOUNT 40" AFF.
12. 18" TOWEL BAR BOBRICK NO. B-6737X18
13. SHOWER GRAB BAR (HC) BOBRICK NO. B-6265 MOUNT 36" AFF.
14. SHOWER SEAT (HC) BOBRICK NO. B-506 OR B-507 MOUNT 19" AFF.
15. SHOWER CURTAIN ROD BOBRICK NO. B-204 CURTAIN: 8 GA VINYL WITH RUSTPROOF GROMMETS AT 6" O/C AND HEMMED EDGES. MOUNT ROD 7'-0" AFF.
16. SOAP HOLDER CERAMIC TILE THIN SET HOLDER MOUNT 3'-6" AFF.
17. 30" x 24" MIRROR BOBRICK NO. B-290-3024 MOUNT 44" AFF. TO BOT.
18. TOILET PARTITIONS
19. MOUNT EXG. BRONZE PLAQUE WITH (4) - EXP. BOLTS. USE BRONZE BOLTS TO MATCH EXG. OBTAIN PLAQUE FROM U.N.L. INVENTORY DEPT.
20. 24" x 24" MIRROR BOBRICK NO. B-290-2424 MOUNT 38" AFF. (SEE DETAIL 1V/A15 FOR TYPICAL INSTALLATION).

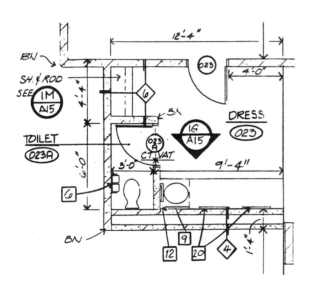

FIXTURE MOUNTING HEIGHT SCHEDULE

ALL DIMENSIONS ARE FROM FINISHED FLOOR LINE.

FIXTURE	STANDARD	HANDICAP	NOTES
WATER CLOSET	15" TO RIM	19" TO RIM	
URINAL	24" TO RIM	19" TO RIM	
LAVATORY AND WASH FOUNTAIN	31" TO RIM	34" TO RIM	
MIRROR	48" TO BOTTOM	40" TO BOTTOM	SUPPLY ONE PER LAVATORY
SOAP DISPENSER	*1	40" TO BOTTOM	FURNISHED AND INSTALLED BY OWNER
PAPER TOWEL DISPENSER	48" TO TOWEL OUTLET	40" TO TOWEL OUTLET	FURNISHED AND INSTALLED BY OWNER
WASTE RECEPTACLE (WALL MTD.)	42" TO RIM	40" TO RIM	FURNISHED AND INSTALLED BY OWNER
TOILET PAPER DISPENSER	24" *2	24" *2	FURNISHED AND INSTALLED BY OWNER
SANITARY NAPKIN DISPOSAL	27" TO TOP	27" TO TOP	
DRINKING FOUNTAIN	40" TO SPOUT	34" TO SPOUT	
GRAB BAR	N/A	33"	
HAND DRYER	48" TO BUTTON	40" TO BUTTON	

*1   CENTER BETWEEN BOTTOM OF MIRROR AND LAVATORY AT WALL.
*2   MEASURED TO CENTER LINE OF SPINDLE.

CHAPTER FIFTEEN

PHOTODRAFTING: REUSING WHAT EXISTS INSTEAD OF REDRAWING IT

PHOTODRAFTING:
   WHEN SOMETHING IS ALREADY THERE,
   DON'T DRAW IT, PHOTOGRAPH IT

PHOTO SPECS

CAMERAS THAT DATE-STAMP THEIR PHOTOS

PHOTO DETAILS

USE JOBSITE PHOTOS FOR QUALITY CONTROL

PHOTODRAFTING

WHEN SOMETHING IS ALREADY THERE, DON'T DRAW IT, PHOTOGRAPH IT

Examples of basic photodrawing are shown below and on the next page.

The principle is simple, when dealing with existing conditions, take photos (with scale markers if necessary), make copies to add notes and dimensions, and incorpoate them in your working drawings.

CADD allows the inclusion of video and electronic camera images, just a variant on standard photodrafting.

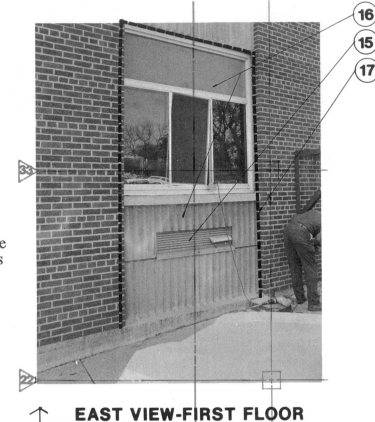

EAST VIEW-FIRST FLOOR

NOTE:
DEMOLITION AT 3RD FLOOR EXISTING
BUILDING OCCURS UNDER ALT.#8.

EAST VIEW-SECOND & THIRD FLOORS

MATCH LINE B
CONTINUES ON
(F)

REMOVE WOOD BLK'G

TYP. 1930 STYLE DOOR FOR FIELD MEASUREMENTS TO REPLICATE STYLE.

SEE DOOR SCHED. FOR TREATMENT THIS DR.

WEST WALL AREA 254

26 **ELEVATION**
SC: 1/2'' = 1'-0''

PHOTO SPECS

Years ago a few architectural firms, such as Jarvis Putty Jarvis in Dallas, introduced photo illustrations to supplement their specifications.

It's still a great idea and it's been used consistently and successfully by a few firms ever since.

The simplest application is to include photos from manufacturers' catalogs or from technical handbooks. Specifying incandescent light fixtures? Why not show the fixture along with the product name and catalog number. Sometimes all the data is ready to photocopy in one shot from a catalog.

The same goes for plumbing fixtures, manufactured casework, wall-mounted equipment . . . anything that's going to come from a catalog can be photocopied and plopped into drawings or specifications. That way when the contractor pulls something out of a box, it'll either match, or not match, the photo. The photos are a double check against errors and typos in model and catalog numbers that might be input to specifications.

Very successful application: Nothing beats a simple photographic image for showing samples of fine stonework masonry, or textured wood, or textured concrete, brushed concrete pavement, fabric patterns, etc. in drawings or in words.

Photos or drawings are photocopied direct from catalogs or handbooks or taken from the world around us. Those who use photo details and photo specifications say you can find almost everything you'll ever need in the way of complex items to illustrate in just one city block: all manner of finishes of concrete, tile, brick, block, wood, window sun screens, handrails, curbs, signs, furnishings, and on and on.

Interior designers have long used photos and catalog copy in their presentations. Only a few have thought to carry the process over to actual working drawings and to add photodrafting, photodetails, and photo specifications to their repertoire.

Landscape architects get similar advantages by copying small images of plant drawings from the manuals onto their site plans along with the formal plant names. That helps avoid the all-to-common problem of intentional or accidental plant substitution by the nursery companies.

Structural engineers have relatively simplistic detailing to worry about and it's very easy to photograph generic footing, column, bearing wall, post, and beam connections during construction for easy photodetailing or specifying.

Mechanical and electrical engineers also use much the same fixtures and equipment again and again. Much of it is in the handbooks and catalogs just waiting to be photocopied to become part of drawings and specifications.

Sometimes it pays to show photos of what you want and don't want in terms of workmanship. For example, what does "clean and neat" mean in a spec? Photos can show well-kept vs. not well-kept jobsites; well-cleaned and not well-cleaned interiors; "clean and true" brick masonry courses, vs. unacceptable; examples of required finish workmanship in slabs, stucco, tile work, etc.

Other illustrations that can help clarify specifications include key plans, and mini-elevations, cross sections, and wall sections. They break up the monotony of the printed spec pages and are handy for showing bidders and subcontractors the extent and location of work.

CAMERAS THAT DATE-STAMP THEIR PHOTOS

Construction progress is photographed to the hilt by Upshaw Architects -- five or six rolls of film on every visit. *"One of the best things I've ever done is to add a date-stamp device on the back of the camera,"* says Upshaw. You can use a simple automatic focus camera with zoom and a date-stamp device. There is no better way to assure accuracy in documenting jobsite work, accuracy which is often lost in the delay of processing and manually dating jobsite pictures.

Self winding, self focusing cameras come with zoom and wide angle options -- they're great for fast, almost automatic, complete picture taking as you roam through a jobsite.

Date-stamp devices can be added to the backs of standard 35 mm cameras. Check with your dealer.

PHOTO DETAILS

A St. Louis architect once told us: *"Virtually all the construction details that you will ever draw are already all around you."*

His conclusion: Why draw them? Why not take photos instead?

This architect started with sitework photos. He started by taking construction site photos as a check on contractor performance. Now he goes several steps further:

1) He selects suitable photos, then copies and enlarges them on his enlargement/reduction copier.

2) He "crops" the photos by laying plain paper over sections where he wants to add new drawing linework and notes. The paper is adhered with a couple of small pieces of magic mending tape at the top edges.

3) Then he draws the cutaway sections by hand. For example, if the photo is of exterior stairs, he cuts the photo at the close edge of treads and risers, draws in the thickness of concrete and crushed rock, adds reinforcing bars, plus typical notes, and the detail is complete.

4) He's done this with planters, retaining walls, handrails, standard parking appurtenances, handicap signs and ramps, etc. Many such items don't need any additional drawing, just the photo and a few notes.

Some simpler photo details, such as stepped slabs, aren't any faster than drawing them from scratch, but they're much clearer because they show the expected final result in perspective as well as the detail cross section. There's no mistaking the architect's intentions with the "real world" photo views.

USE JOBSITE PHOTOS FOR QUALITY CONTROL TOO

The Rule of the Day at many firms for anyone who visits a construction site:

**TAKE A CAMERA
AND
TAKE PICTURES.**

When the contractor makes an error, photograph it. We're told that just having the camera with you makes the contractors more alert even if you don't take pictures.

When your office has made an error, do the same, make prints, and share them with your staff. It's the only way they can learn real quality control, by seeing the tangible results when there are errors or oversights in design and working drawings.

Videos are useful too for construction records, especially for more complex jobs. It can be easier to see potential problems on video than when subject to the confusion and distractions of the construction site.

CHAPTER SIXTEEN

CADD AND CADD MANAGEMENT

WHY CADD STILL HASN'T PAID OFF
FOR MOST OFFICES

CADD -- THE MOST PROFITABLE USES
AREN'T WHAT A/E'S EXPECT

HERE'S WHAT A/E'S SAY ABOUT THEIR
COMPUTERS AND CADD

CADD SHOPPING CHECKLIST

THE COMPUTER AS A PERSONAL
PRODUCTION MACHINE

CHECKLIST: HOW TO SET UP AN A/E OFFICE
COMPUTER FILING SYSTEM

CADD LAYERING AND FILING

WHY CADD STILL HASN'T PAID OFF FOR MOST OFFICES

Throughout the early 1980's, hundreds of A/E offices learned how they could cut working drawing production time and cost by 20 to 40% below the norm.

It happened in small offices, big offices, engineering offices, interiors firms . . . it didn't matter what kind of work they did, as long as they adopted a few key timesaving steps. Those primary timesavers became known as the "Five Stages of Systems Drafting" and were widely publicized by your author.

By the mid-80's many A/E firms stopped or postponed those timesaving reforms. The reason was ironic. They realized they would be doing drawings in the future with CADD systems and CADD would make Systems Drafting obsolete. So why not skip the bother of Systems Drafting, just wait awhile, and let the computer revolution carry them along?

Enter 1993; 98% of A/E firms have computers and well over half have some CADD capacity, **but nearly half the CADD users are not saving much time or money.** Most are not even close to reaching that 20 to 40% time and cost savings that Systems Drafting firms have enjoyed for the past ten years.

The reason? They haven't adopted all the major production timesavers, and haven't integrated the ones they do use in a systematic fashion.

So, one more time, here are the primary stages for putting it all together whether you're on AutoCAD, Macintosh, Integraph, or whatever.

If you introduce these stages, one by one, CADD will make a BIG dent in your work load at very little cost. If you don't, it won't. It's that simple.

The Five Stages:

1) **Housecleaning.** Eliminate the unnecessary waste and redundancies in drafting **no matter how trivial they are;** establish standards for clear, bold, high-contrast drafting (large lettering, etc.); introduce checklists and a quality control system; and establish a simple drafting standards manual to inform new drafters, and remind all staff, of office policy. If you don't cut out the most minor time wasters, like using double lines to draw doors on plans, all the others which are far more time consuming will creep right back into your drawings.

2) **Reuse simple data** once the best drafting policies for clarity, simplicity, and readability have been established. Identify and checklist the redundancies that have to occur from drawing to drawing or project to project, especially notation and construction details. Dozens of CADD managers have told us their computer systems only started to pay off when they introduced their standard detail library or a standard notation and keynoting system . . . or both.

3) **Methodically introduce, or reintroduce, the timesaver tools:** Cameras for photodrafting, video camera for site surveys and computer input, enlargement/reduction copiers, fax machines, talking tape measures, lettering machines, vertical drafters, light tables, car phones, laser printers, desktop scanners, drafting cubicles; and especially, microcomputers for the widest use of project database management and word processing by all staff.

4) **Use pasteup and overlay drafting as early as possible in design development.** Never let drafters redraw repetitive floor plan units instead of pasting up copies of the units.

--**Drafters often argue that it takes as much time to do pasteup as redrawing the simple items.** They may be right in some immediate cases. But they're wrong overall. If you don't insist that they become habitual reprographic systems users, you'll lose thousands of dollars over the long haul. And, waste begets waste. A wasteful practice often makes it harder to implement reforms in related practice. A minor compromise at one stage makes it easier to justify another compromise elsewhere.

--If your plotting time is as slow as most offices (30 to 50 minutes a sheet), take advantage of the photocopier and do pasteup for detail sheets.

--Also, get a 30 to 40% plotting timesavings by printing out floor plan base sheets and overlay sheets separately. One background architectural base sheet may be combined with dozens of auxiliary architectural drawings and consultants' drawings. Your choice: Plot the same floor plan information again and again on each drawing, or plot it once, and merge bases with overlays reprographically. Besides cutting at least 30% of plan plotting time, you get clearer output with high-contrast screened background images combined with solid line drawings. (This advantage will disappear with the introduction of large-format laser or ink-jet plotters. But as long as you use a pen plotter, do all you can to minimize the bottleneck of plotting time.)

--If you're not using CADD, get a running start on production by persuading your designers to use overlay throughout the design development and presentation drawing process.

5) The whole shebang. By spending a little more time on mockups, checklists, and preplanning, you'll get the most uses out of CADD, photodrafting, keynoting, standard details, paste-up and overlay--each in its place, each where it will do the most good and best complement the other systems.

CADD -- THE MOST PROFITABLE USES AREN'T WHAT A/E'S EXPECT

It surprises some A/E's to learn that having a CADD system doesn't mean that they have a SYSTEM that will do any work. Having some hardware and software means having the tools, but a system for **using** the tools for efficient design and production? That's up to the user to figure out and many still haven't.

Most users of CADD mainly imitate the formats and standard procedures of traditional drafting. That's the only guide they have. And that in turn explains why so few have found CADD to be a time saver . . . they just follow old time-consuming procedures in a more complex way.

There are a few specific ways in which CADD is fundamentally different from drafting and if these differences are exploited to the maximum, users can get their money's worth:

1) The complexities of layering, filing, updating, and coordinating data creates an opportunity -- each job MUST be planned and monitored with exceptional care. That's a plus since there's very little such planning in traditional drafting -- an endless source of errors, omissions, redundancies, and contradictions in traditional drawings.

2) Everything drawn on CADD can be stored forever and combined with any other data. The computer makes the use of standard interchangeable parts in drawings almost mandatory. Thus standard details and standard notes, which most firms know they should use anyway, can be stored, filed, and retrieved relatively easily.

3) The filing and retrieval systems inherent in good CADD management expedite the integration of file numbers for notation, de-

tails, and specifications. In other words, it's as easy to have a file number system to link related notes, details, specs, and other data than it is not to. Thus every primary element in a drawing can automatically refer to and be coordinated with every other related item.

4) The reuse and coordination factors make it relatively easy to create a valuable building facilities management database to sell to the client. As most A/E firms know now, this is the biggest selling point of CADD services to many clients. Most know this, but only a minority actually offer facilities management related services.

Successful and not-so-successful CADD users who speak at the A/E/C Systems conferences confirm these observations. Those who do best with CADD are those who are most aware of the above four areas of departure from traditional practice and use them to good advantage.

The most productive users of CADD at the A/E/C Systems conferences exploit all the differences just listed. For example:

--CADD working drawings are roughed out towards the end of design development and outlined in miniature on computer. Mockups of working drawing sheets are laser printed on 8-1/2 x 11 sheets and bound as project books. Besides helping gain an overview of the job for scheduling and supervision, the mockup identifies file and layer numbers and helps maintain control of the project.

--Working drawing content is printed in separate layers to simplify content.

For example:

--Plans with walls and architectural data such as room names, doors, partition types, finishes, architectural details and notes.

--Floor plans with dimensions and no other data.

--Floor plans with equipment and special fixtures with walls as background data.

--Reflected ceiling plans in the usual manner.

This unclutters the sheets and makes them more readable and useful to the contractors. It allows plotting at smaller scale to avoid match lines, and makes reduced size laser-printed check and reference prints more practical to use.

--Standard detail and note libraries are not optional among the successful CADD users, **they're mandatory.**

--Standardized layering and building component identification systems are strictly followed. Layering usually follows a simplified version of the AIA standard. CSI numbers are used to file details and CSI numbers are tagged onto standard notes.

--Decision checklists are still not used in a majority of these firms, but the MOST systematic of them consistently follow some operational checklist. Typically a list of materials and components for sitework, construction plans, fixture plans, etc. is checked off as decisions are made. This list is cross-coordinated with all drawings, and changes are referenced back to a decision sequence log that's built into the checklist. This operation hasn't been widely computerized and integrated with CADD software yet, but is expected to be widely used in the years ahead. (See the chapter in this book on MANAGEMENT BY CHECKLIST.)

--A graphic database to supplement the record drawings is provided by the more sophisticated offices. Most clients for larger commercial and institutional buildings expect to receive a facilities management graphic database, and government agencies will soon all require such "CADD deliverables." There is no general standard on the content and format of such documents, but those offices that haven't systematized their documents as we're describing here are up the creek in this department.

Another unique quality of CADD is the opportunity for direct interactivity with building clients and users during planning and design. Only a handful of the most advanced users take advantage of this option.

It means sitting down with a client at the computer, pulling up an assortment of building components, and creating plans or building sections in response to immediate, direct client feedback. The client meeting opens with an explanation of basic program assumptions and schematic site and floor plans as "starters." Then the designer and client follow the essential steps of planning, side by side.

Users say this process can save an enormous amount of time by cutting out the usual multi-step cycle of having a meeting, then laying out several options for a plan, having another meeting to review those options, revising the plan, having another meeting, etc. It compresses the multi-step feedback process. AND, since the clients are in on the logic of the decision making process, it's somewhat less likely that they'll come up with arbitrary suggestions for changes later on.

HERE'S WHAT A/E'S SAY ABOUT THEIR COMPUTERS AND CADD

Early in 1993, we did a survey of computer use among architects and their consulting engineers. The results showed major improvements in CADD use over the previous two years.

No surprise to learn that IBMs and clones are the dominant CADD hardware -- 57% of survey respondents use DOS machines, 28% Macintosh, and 14% "other."

No surprise that AutoCad is the dominant CADD software -- 52% AutoCad, 21% varied Mac programs, and the remainder a mix of DataCad, Microstation, AutoArchitect, Isi-Cad, VersaCad, etc.

Most newsworthy: A substantial majority say their CADD systems are saving them lots of time. For years most users told us they felt lucky to break even. Now 68% say it's faster than manual drafting. Only 31% say it's the same as manual drafting or slower -- a nearly two-to-one positive rating. Most of those who commented on improved productivity use DOS and AutoCad, not Macintosh.

Aside from saving time, the primary benefits of CADD in descending order are:

1) Coordination, especially with consultants. (72% agreed.)

2) Brings in the work. (55% agreed.)

3) Speeds up consultants' work time -- faster turnaround. (37.5%)

4) Facilities management database documents. (30%) Most of these benefits were reported by the IBM-clone and mini-computer users.

The reported benefits matched the dominant reasons for choosing IBM-clone/ AutoCad systems in the first place, namely client acceptance or demand, and a general industry standard among consultants and joint venture partners.

A/E's are quickly upgrading their machines to Intel 486 chips. About 48% of the DOS machines run on Intel 386; 50% are 486. Most users are eager to upgrade and probably do so annually or at least every other year throughout the '90's.

Primary applications of CADD in descending order:

1) Floor plans, including layered output for consultants.

2) Construction details.

3) Presentations.

4) Design. Individual firms vary widely in specializations. Some use their systems for 3-D presentation output 90% of the time, whereas most firms barely use 3-D at all. A few firms use CADD almost 100% exclusively for details.

Fifty-one percent of users run their CADD systems 90% to 100% of the time, 21% run them over 50% of the time, and the remainder from about 5% to 20%. Again, Macintosh users scored low in CADD use, using their computers mainly for other functions, such as desktop publishing.

Thirty-two percent of CADD users charge for CADD time in addition to operator time. The rates range from a high of $60 per hour to a low of $8. The average rate is $27 per hour, down from an average of $35 last year. Since so many firms charge an hourly CADD rate, all should at least consider doing so. A nominal $10 to $15 hourly rate isn't likely to encounter much resistance.

CADD
SHOPPING
CHECKLIST

As a shopping guide, and to avoid previous errors, here's an update of the best CADD evaluation checklist we know of:

1) Begin with a wish list of desired work functions. What functions and features do you want to be able to readily utilize?

Don't be surprised if a well-known CADD system you're considering buying has little or no direct A/E applications built in. Those are often provided as add-ons by others so ask your dealer about what they do and how much they cost.

Use this checklist to note what you might want, and how different programs you consider actually compare. Rank their relative importance.

__ Multiple use of CADD workstation for word processing, spreadsheets, book keeping, etc.

__ Multiple use of CADD workstation for personal management functions such as phone index, phone dialing, personal fax modem, mailing list manager, etc.

__ Multiple use of CADD workstation for desktop publishing and laser printer output for brochures, proposals, office manuals, etc.

__ Design criteria and programming checklisting

__ Spatial allocation and bubble diagraming

__ Sketching and paint for renderings

__ Photo and still video image input and output for photodrafting

__ Construction detailing, existence of standard detail libraries

__ Finish/door/window/fixture/equipment and other schedules

__ Keynoting

__ Standard notation files
__ Convenient word processing for notation on drawings

__ Specifications (and spec links, by CSI number, to keynotes and detail numbers)

__ Structural engineering design, calculation, and diagramming

__ HVAC and energy use design, calculation, diagrams, and symbol library

__ Electrical design, calculation, diagrams, and symbol library

__ Plumbing design, calculation, diagrams, and symbol library

__ Sitework, contours, property maps and surveying calculations

__ Civil engineering, and cut and fill diagrams

__ Facilities database management

__ Multimedia, computer animation, and video editing

__ Quick access to macro files and command files

__ User-defined sub-routines

__ Ease of scanning input of existing drawings

__ Multitasking

__ Quick backup of data

__ Printer and plotter spooling

__ Networking with other workstations

___ File security through code names

___ Job tracking of operator times

___ Fast refresh of screen image

___ Working drawing floor plan functions including:

    ___ Automatic double-wall and wall materials indication

    ___ Automatic wall intersection corrections

    ___ Automatic door and window inserts and reversals

    ___ Database link to doors, windows, other fixtures and materials

    ___ Automatic extrusion for elevations

    ___ Automated extrusion from plan to isometric or simple perspective drawing

    ___ Automatic area computation and dimensioning

    ___ Automatic coordination and updating of layers

    ___ Coordination with work by other disciplines who may work on other systems

    ___ User defined symbols

More considerations:

___ Cross linkage and data transfer to spreadsheet and database programs

___ System for recovery of lost data

___ Professional looking symbol libraries

___ (Others)

2)   Estimate how many new workstations you need to start with.

3)   Once you identify desired CADD software options, the software vendors will tell you how much computer memory (RAM) and hard disk storage (ROM) and speed you'll need. Ask vendors for names of user groups and call members for their evaluations. (Just make sure those you're referred to are not working part-time as vendors; some A/E's are vendors on the side and their advice may not be objective.)

4)   Review hardware enhancements -- cost and potential value: Graphic cards, larger screen monitors, extra-large and/or portable hard disk drives; archival backup. You may be able to upgrade much of your original hardware satisfactorily rather than trade in for new goods.

5)   Review input/output devices the same way: Scanners, laser printers, plotters, etc.

6)   Make a worksheet to compare competing software and hardware:

___ Estimate annual and monthly costs of hardware, software, and maintenance contracts

___ Include interest cost of financing.

___ List non-CADD software you'll also want to install and add cost

___ Estimate the cost of training each new operator, and getting up to working speed (CADD teachers at your local community college can give you a good working estimate)

___ Include loss of production time (it's not only that you lose the pay you provide an employee during training, there's also the loss of the billable work time)

7)   Estimate your cost benefit (assume 20% time savings in CADD related tasks over current productivity within one year for MS-DOS or IBM and clone machines, and 25% improvement in three to six months for Windows or Macintosh)

216

THE COMPUTER AS A PERSONAL PRODUCTION MACHINE

How do you get the most out of your computer? *"Personalize it,"* says a pioneering A/E AutoCAD user. Computers will do hundreds of jobs, but only a few uses will suit you or save you time or money. To find the highest and best uses, find a few primary special uses and go from there.

--Most CADD input/output consists of floor plans. Floor plans and elevations go on the computer, and the rest of the drawings are a combination of hand drafting, paste-up, overlay, and photocopier stickyback. *"It gets us started and that's all I want. Otherwise we get bogged down in computer management,"* says an architect whose staff is just getting used to CADD operations.

--A few offices have input their mini-working-drawing formats so they can diagram (or "Cartoon") the contents of a working drawing set in a jiffy. They also input short-form checklists to print the essential contents of each working drawing sheet on the mock-ups as guidance for the drafters and CADD operators. (Laser-printed images look like miniatures of the real thing and are a great plus when discussing total scope of work with clients.)

--Some offices make detailed plots of exterior elevations to paste-up on models. Some do their plots right on foam-board to draw cut lines, doors, windows, materials patterns, the works. A few firms **only** use their computers for such purposes.

As some offices have discovered, the Macintosh or other Graphic User Interface (GUI) machine is easiest to personalize and to fit in with any existing combination of computers. There are a lot of reasons why.

For example:

--Mac Powerbooks are portable and popular, so office staff willingly get on a waiting list to check one out to take home for self teaching. Learning is far faster than on MS-DOS computers and most of the software is self-explanatory. When a user learns to use a program, such as word processing, he or she learns most of what's needed to use any other such program.

--Data created in one program can be transferred to another. This has special meaning to A/E's who might want to have a working drawing sheet which has video-digitized photo-drafting, a finish schedule from a data base or spreadsheet program, keynotes from word processing, and standard details from a drawing or CADD package.

--Portability means it can go to job shacks and clients' offices. Ease of use means it can be used on the spot by and with clients and contractors. The machine is used by some firms only for personal marketing, for client relations and on-screen participation by others. Those who have built-in modems Fax their drawings and text from their computer screen to any other fax machines.

Those who zero in on special applications get some remarkable productivity too:

Ten hours per drawing for large apartment complexes, as achieved over the past few years by the Connecticut office of Dennis Davey who uses a highly individualized checklist and keynote system.

. . . 45 minutes for complete client-required revisions of residential working drawings, by Canadian architect Barry Pendergast.

. . . Four hours apiece on large drawings, especially structural details, by a mid-sized Northern California office.

. . . Ten to fifteen minutes for a quick residential "builder's" floor plan with dimensions, elevations and a simple perspective, by Southern California contractor Craig Savage.

That shows some of the productive possibilities. In every case the users have personalized the computer, trained it to do exactly what they want, the way they want it. Then it becomes their servant and not vice versa.

A funny thing happened to James Adams who runs a six-Macintosh office in Missouri. Someone needed tracing paper and, try as they might, they couldn't find any anywhere in the office. Nobody could remember when they had last bought some.

It's not a "paperless office" by a long shot. We don't know any office that's truly "paperless," but at least the Adams firm is "tracing paperless." ALL their project drawing is done on the Macintosh.

There's no tracing paper and no plotter. Their design and working drawings come off standard Macintosh printers.* That includes complete documents for bank buildings, office interiors, small hospitals, and recent projects such as a 40-unit housing project and a 22,000 sq. ft. industrial building.

James uses Predesign and Planning checklists with what he calls "on-off" toggles. When items on the checklists are decided to be included in a project, the "on" toggle of the electronic checklist activates and links up related entries for keynotes, cost estimating, specs, and working drawings.

Detail files include a detail history database. The database experience is leading them into facilities repair and maintenance management for local utilities companies.

Interrelated checklists are among the most powerful tools there are for coordinated production and project management.

Checklists are also the basis for creating project design databases for facilities management. The office database checklist first guides the sequence of decision making during a project and helps assure everything is done that needs to be done. As the decisions are recorded, that creates a new list that documents everything that goes into the project. Finally, the lessons learned in the creation of new projects become part of the original office database.

The Adam's firm drawings are printed in segments on a laser printer, pasted onto carrying sheets and reproduced on on a large-format photocopier to make full-size presentation or working drawing sheets.

CHECKLIST
HOW TO SET UP
AN A/E OFFICE
COMPUTER FILING
SYSTEM

Here's a recommended sequence of drawing identification:

__ **Client:**

Client name abbreviation such as IBM, ITECH, JB, etc.

__ **Job Number or Name:**

This may be a date or the sequential number of a job for a particular client. JB 01, for example, would be the first job for the JB Corporation. Sometimes the client name and job name are the same.

__ **Job Phase:**
(following the AIA Scope of Designated Services):

01 __ Predesign Services
02 __ Site Analysis
03 __ Schematic Design
04 __ Design Development
05 __ Working Drawings
06 __ Bidding
07 __ Construction Contract Administration
08 __ Postconstruction Services
09 __ Supplemental Services

__ **Discipline:**

			__ Architectural
			__ Site Analysis
01	or	AR	__ Structural
03	or	CI	__ Mechanical HVAC
05	or	ST	__ Mechanical Plumbing
07	or	HV	__ Electrical
09	or	PL	__ Interiors
11	or	EL	__ Landscaping
13	or	IN	
15	or	LS	

__ **Drawing Group** -- usually a division of construction, building contract, or discipline such as:

01	or	SW	Sitework
03	or	FN	Foundations
05	or	FR	Framing
07	or	RO	Roofing
09	or	EX	Exterior Enclosure
11	or	IN	__ Interior Partitions
13	or	CE	__ Ceilings
15	or	FI	__ Finishes
17	or	FX	__ Fixtures and Equipment
19	or	HV	__ HVAC
21	or	EL	__ Electrical
23	or	CM	__ Communications
25	or	LS	__ Landscaping

__ **Drawing Subsection:** An individual portion of a drawing sheet such as a schedule or construction detail file numbers.

__ **Drawing Layer:** The computer separation of data for control of plotter line widths and for separation of various discipline's work and building components. Layering differentiations vary enormously from office to office but here's one simple version for building plans:

01	or	EX	__ Exterior perimeter walls
03	or	IN	__ Interior partitions
05	or	RM	__ Room names and numbers
17	or	DR	__ Doors and door identification
09	or	FN	__ Finish schedule keys
11	or	DE	__ Detail and section keys
13	or	NT	__ Identification notes
15	or	CN	__ Construction assembly notes
17	or	DI	__ Dimensions
19	or	PL	__ Plumbing fixtures
21	or	HV	__ Mechanical equipment HVAC
23	or	EL	__ Electrical power
25	or	LI	__ Lighting
27	or	CE	__ Reflected ceiling plan
29	or	SP	__ Sprinklers
31	or	AL	__ Alarm system
33	or	FF	__ Floor framing
35	or	CF	__ Ceiling and roof framing
37	or	FX	__ Fixtures and equipment
39	or	FU	__ Furnishings
41	or	CM	__ Communications

__ **Drawing Sheet:** The final drawing number which may summarize codes for much of the previous categories of information.

__ **Here's an example,** the job is for JB Corp., it's the first one of the year for JB (01), the Job Phase is Working Drawings (05), it's Architectural Drawing (AR or 01), combined with an Electrical Plan (EL or 11), showing the Reflected Ceiling Plan layer (CE or 27) and the Lighting Layer (LI or 25).

JB Corp.	Abbreviation: JB		
Job #1	01		
Job Phase --Working Drawings	05		
Discipline --Architectural	AR	or	01
Drawing Group --Electrical	EL	or	21
Layers (by discipline group)	CE	or	27 &
	LI	or	25

As you see, the final computer drawing file coding number is rather long.

JB   01   05   AR   EL   CE   LI

or

JB   01   05   01   21   27   25

__ **A complex code is OK,** it's just a way to break down graphic data into its components for easy retrieval and reuse on the CADD system. The actual sheet number on the final drawing to be plotted would be much simpler, perhaps something like: E - 7.

__ **Other breakdowns** that have to be added to some drawing identification codes include:

   __ Building Floor or Level
   __ Building Wing
   __ Drawing Revision Phase

__ **Each plotted drawing sheet has to include a composited identification code** somewhere in a "coding block." This is a section on a sheet, perhaps in the title block, where numbers or letters appear depending what document and layer is being composited.

__ **Some offices prefer to use a consistent number for each discipline** regardless of whether referring to Discipline, Drawing Group, or Drawing Layer. Thus they might always refer to Mechanical by a CSI correlated number.

__ **You have total latitude in creating your drawing file codes for use with CADD.** The main thing is that it be easy to remember, enforced consistently through the office, and that it be only as elaborate as needed for your particular practice or discipline.

CADD LAYERING
AND FILING

The AIA has published recommendations for layering and filing systems. This was the product of the Task Force on CAD Layer Guidelines and is an extremely helpful product as you'll see from the two pages reproduced here.

CAD Layers and Drawing Strategies

Any analysis of CAD layers must consider the methods available for sharing graphic information. Most CAD systems support two basic approaches.

Both methods require coordination among members of the design team to ensure integrity of information. For example, if both architects and electrical engineers were to place light fixtures at the same time, the resulting drawings would likely be in conflict.

1. Single File–Multiple Drawing Approach

With this approach, multiple drawings are produced from one CAD file by turning selected layers on or off. The method is straightforward and ensures good coordination of information. The drawbacks to the technique are that CAD files can become very large and that only one person can work on a file at a time.

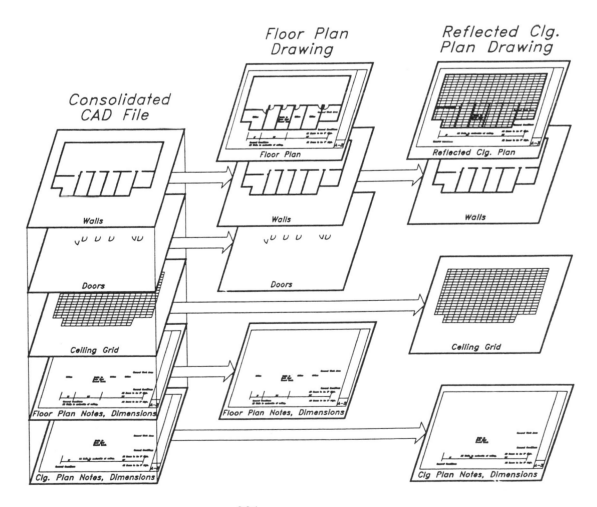

221

2. Multiple File Approach

This approach combines selected layers from several CAD files to produce a drawing. The technique is the converse of the first method. It is accomplished either by merging drawings or by using reference files. CAD systems that support reference files allow several files to be "attached" to an active file for viewing and plotting. It accommodates a team approach better than method 1 but requires a greater degree of coordination.

CAD File 5
Floor Plan

Sheet Border, Notes
Arch Walls
Structural Plan

CAD File 6
Reflected Clg. Plan

Sheet Border, Notes
Arch Walls
Arch. Ceiling Grid
Elec. Light Fixtures
HVAC Diffusers

Reference File 1

Architectural

Reference File 2

Structural

Reference File 3

Lighting

Reference File 4

HVAC System

Example of a Multiple File Strategy

CHAPTER SEVENTEEN

SUPERFAST DRAFTING WITH PHOTOCOPIERS

PHOTOCOPIER DRAFTING AND
11" X 17" WORKING DRAWING SHEETS

LARGE FORMAT COPIERS FOR
SYSTEMS DRAFTING

COPIER DRAFTING: HOW IT WORKS STEP BY STEP

PHOTOCOPIER DRAFTING AND 11" X 17" WORKING DRAWING SHEETS

You can get extraordinary time and cost savings from a combination of simple drafting tricks. In fact, the combination is so simple, it may be hard to see the full range of advantages until you try it.

The first step is to stop using large drawing sheets for small and moderate-sized buildings.

Think small. Firms of all types and sizes have successfully used 11" x 17" sheets for years. They say it's a minor first step that leads to big savings.

The second step is to start making maximum use of photocopier drafting.

The office copier is one of the great drafting time savers, but you have to go beyond the use of copier stickybacks to get the full benefits. To get all the advantages, the copier should have enlargement/reduction capability and up to 11" x 17" sheet size capacity.

Since most phtotcopiers still don't make direct 50% reductions, you may have to use the double-step reduction trick -- that is, punch in a .707. The product will be within a hair of half size. The two-step number to get a double-size reproduction is 1.414. The two-generation process is a nuisance, but quality holds up well on most current copier models.

What to expect when combining copier drafting and small-sheet formats:

--10% to 15% improvement in time and cost from using smaller sheet sizes. We'll explain why later.

--30% or more improvement by using opaque paste-up drafting with the copier.

--Significant quality control improvements by applying "poor person" versions of Systems Drafting techniques with the copier.

For example:

--Make paste-ups of all repetitive data -- symbols, standard notes, titles, details, and repeat elements in plans, elevations, and sections. Paste-up drafting achieves the most spectacular time savings of any Systems technique. Paste up sharp, crisply-drawn original components, and you can "burn out" edge lines and tape marks in the photocopier by slightly increasing the photocopier's light and exposure time. Use Scotch repositionable tape for best results.

--Photocopy onto tracing paper. That allows you to transform opaque original material from diverse sources such as prints, technical literature, catalogs, photographs, etc., into translucent second originals for regular diazo ammonia-developed printing.

--Use typewritten or computer printout strings of notes or keynotes. You can type directly on the small-sheet sizes with large carriage typewriters or printers. Or better still, use paste-up keynote strips.

--Combine specifications with drawings. Photo-reduce typed or computer printout specification pages to about 2/3rds-size, splice and tape them on 11" x 17" sheets, and photocopy them on the backsides of the working drawing prints. That allows you to have sitework specs right beside the drawings, excavation and concrete specifications adjacent to the foundation plan, etc.

--Proprietary specifications are even clearer when you include photocopied pictures of the products you're specifying: light and plumbing fixtures, cabinets, etc.

--Include engineering printout or preprinted tables for building department reference -- with photocopy paste-up, of course.

--Use paste-up and photocopy project photographs for rehab and remodeling work. When photocopied on tracing vellum, you can erase away unwanted portions and draw in the revisions.

--If you have a small engineering scale plot survey, photocopy it at modified size onto tracing paper, to avoid redrawing it at a revised scale. Photo-reduce a copy of the ground floor plan, and paste it up as a building floorprint on the site.

--Want the best looking trees, foliage, and other entourage on the site plan? Just photocopy pages from Ernest Burden's McGraw-Hill book, Entourage, or other drawings, cut out the images, and tape them into place.

--Freehand details always sharpen up when photo-reduced. Freehand and design sketches are often perfectly usable and don't need to be finish drafted. The copier reductions combined with paste-up often give perfectly acceptable results, even from rather rough originals.

There it is -- many or most of the benefits of Systems Drafting and reprographics, at extremely low cost and with standard office equipment.

Some further notes on small-sheet sizes:

--They make for easy drafting. Small sheets are far less time consuming to manipulate and draft on than are regular drawing sizes.

--Smaller sheets force the use of smaller working drawing scales. 1/8" scale is fine for virtually all floor plans and elevations. 1/4" scale has been common for small buildings in the past, only because the traditional sheets were so oversized. The smaller-scale drawings are perfectly readable and noticeably faster to draw and revise.

--Small-scale 11" x 17" drawings are cheaper to print and much more convenient for bidders, contractors, and workshop people on the jobsite. Subcontractors and workers will keep copies in their hip pockets and consult them directly, instead of leaving the prints in the truck or job shack.

LARGE FORMAT COPIERS FOR SYSTEMS DRAFTING

Plain and simple: Paste-up drafting is still the primary time and money saver for the most productive design firms.

Nothing else will let two drafters produce 54 drawings for a housing project in three and a half weeks, or create permit drawings for five towers with 1,800 units in ten days with seven people. That's routine for the high-output Systems Drafting offices.

Of all Systems Drafting tools and methods -- paste-up is still the hands-down winner for doing high-speed design development and working drawings. Not only are the original drawings fast to produce, they're even faster to revise. And revisions, as any drafter will tell you, are the way most drafters spend most of their work time.

Now large-format copiers make paste-up easier and more efficient than before. A typical large-format copier will accept original drawings, prints, or paste-ups up to 36" wide and will make a same-copy on plain paper, tracing paper, or polyester ("Mylar").

The main obstacle to widespread paste-up work in the past was that it required a vacuum frame printer, translucent paste-ups, and a diazo print developer. That meant special equipment, special diazo print films, and special work habits that many firms were unable to manage.

Now it's direct. Mount opaque paste-up elements from virtually any source (as long as they're reasonably sharp and clear) and you can make clear, crisp copies on plain paper or draftable transparencies. Anyone can catch on to the technique quickly, and it won't be long before drafters and designers start coming up with numerous special applications.

Here is an example of the possibilities in three steps:

1) One firm "builds" their buildings from their detail library. Once the overall construction system is decided, they select construction details for footings, floors, walls, roofs, etc. and photo-reduce the details to 1/2" scale. They compose and paste up the details as components of foundation-to-roof exterior wall section drawings.

2) They recopy the assembled wall sections at a reduction to 1/4" or 1/8" scale, and paste up left-right wall sections into outline building cross sections. After making a reproducible, they connect the roof and floor lines of the wall sections. Presto, instant overall building sections.

3) They copy the building cross sections and reprint them as "screened" background images as the base for exterior elevations drawings. They do this by photocopying the sections with a "white dot" screen between the cross section paste-up and the copier copy board. The white dots don't appear except to break up line work in something like a "halftone" image. (Art supply stores sell these as "copy-dot" or similar name screens. You can make your own by buying white Zip-a-Tone dry transfer dot sheets and transferring the white dot pattern to clear plastic sheets.)

Net result: Exterior elevations, longitudinal and lateral building cross sections, and wall sections -- all built from details with very little original drafting. They're composed by combinations of photo-reduction and large-format drafting photocopying.

Here are other examples of handy large-format copier uses:

--BLOCKOUTS. This is the easy way to make large area erasures of existing work. Suppose you have to test out several possible variations of large areas of a building floor plan. Then lay out sheets of opaque blank white paper to size and adhere them over the old work with repositionable Scotch Tape. Apply copies of the variation drawings in position and make new copies. It goes **very** fast.

--PLOTTER ENHANCEMENT. Roller-ball pen plots on plain paper are faster and more trouble-free than regular ink pen plots. Copy the plain paper plots onto vellum or polyester to make reproducibles and you'll darken and enhance the linework in the process of making faster and lower cost final computer plots.

All drawing types are done most quickly and economically when you create a file of reusable elements. For **architects** that means everything that's drawn including construction details, blocks of standard notes, drawing titles, and door/window/finish schedule forms. For **interior designers** it means furnishings and partitions. For **engineers**, it means all the myriad symbols, details and notes used day in and day out. For everyone it can get down to small symbols and components such as room names and North arrows, as long as they're filed conveniently.

Two rules:

1) **Keep all reusable stock drawing components clearly sorted and filed for easy retrieval and convenient reuse.**

2) **Have someone watch over the graphic work area where paste-ups are done to be sure file stocks are maintained, materials are available, and equipment is in working order.**

227

COPIER DRAFTING: HOW IT WORKS STEP BY STEP

Here are some of the jobs A/E's are doing with their enlargement-reduction copiers:

--They modify oddly scaled drawings to more usable common architectural scales.

-They use pasteup to assemble repetitive parts of floor plans and exterior elevations during design development.

--They use economical in-house photo-drawings of existing conditions for rehabs and restoration jobs.

--They copy repeat elements of every sort for pasteup and reuse -- from door swings to room names to drawing titles (some firms print crisp titles and notes of all sizes with their computer laser printer for office copier paste-up drafting).

--They do paste-up strips of repetitive standard notes and keynote strips.

--They run paste-up finish, door, and window schedule computer spread sheet printouts.

--They photo-reduce specifications at reduced scale and print on back sides of working drawing sheets.

--They make background shadow sheets by using white dot screens.

--They draw details or other elements freehand in large scale, then photo-reduce to give a tool-drafted appearance.

The first step in creating a pasteup of a construction detail sheet. (18" x 24" enlargement-reduction copier courtesy of OCE Corp.)

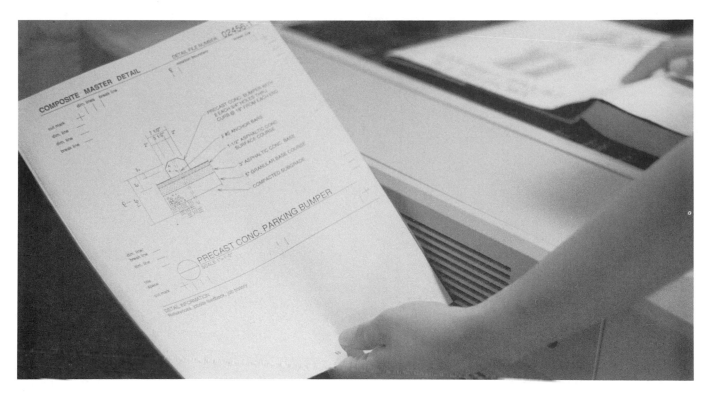

The construction detail can be enlarged or reduced, or
edited by blocking out unwanted portions.

Details are cut to size to fit the working drawing sheet
detail module.

Once details are sized, they can be assembled for
paste-up on a regular size working drawing sheet.

This process is faster and less expensive than having a
CADD system plot out a full sheet of detail drawings.

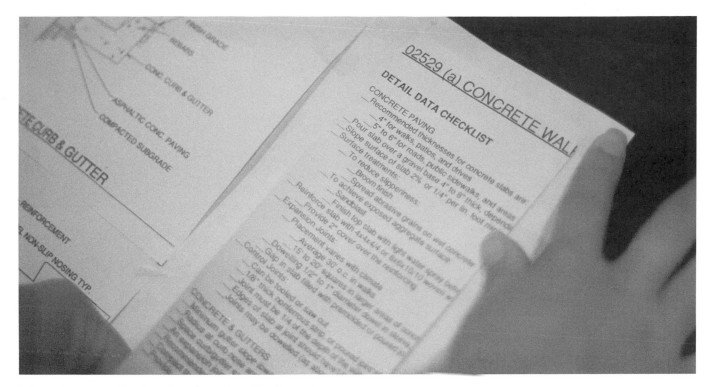

Keynotes, general notes, data from handbooks and
building codes can all be photocopied at any desired
size.

This kind of opaque paste-up can be done on standard
drawing sizes such as 24" x 36" but it especially lends
itself to 11" x 17" format.

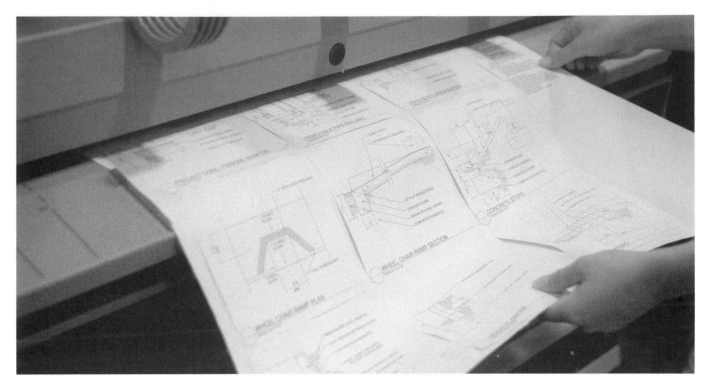

A large sheet of opaque paste-up is easily copied with a large-format engineering copier such as this OCE unit.

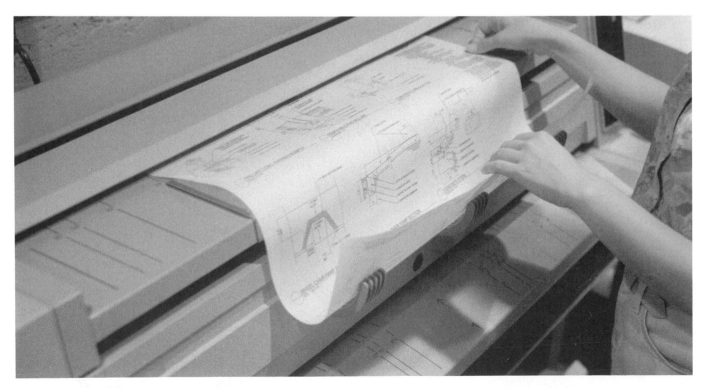

The original can be on paper or pasteboard and the final copy can be on opaque paper, tracing paper, or polyester drafting film.

In this example, the final transparency made from the large format OCE engineering copier is used as an original transparency for running standard diazo prints.

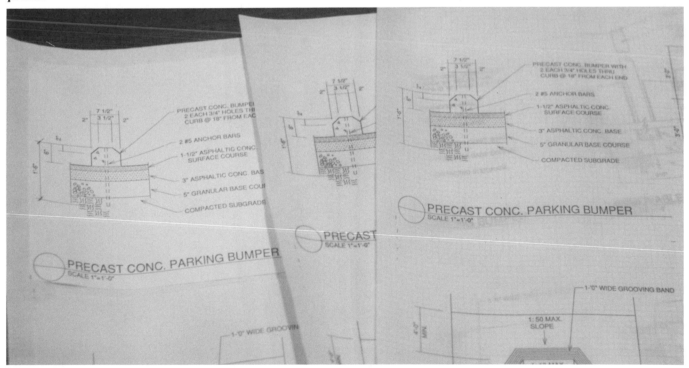

A reminder of the sequence: Paste-up of photocopier prints, engineering copier transparency, final dizao blue-line print.

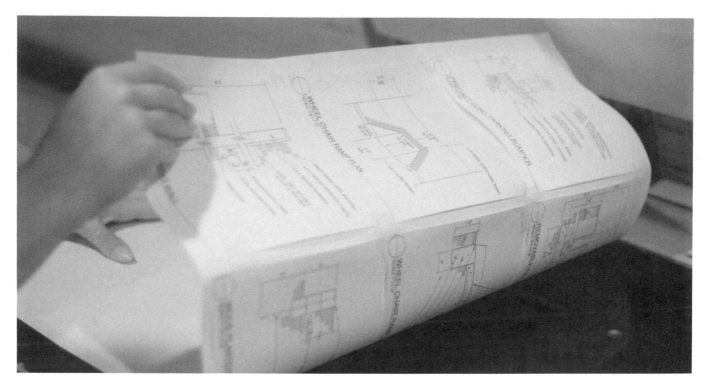

A large format original of any type -- CADD output, paste-up or drawing -- can be photoreduced on the enlargement-reduction copier.

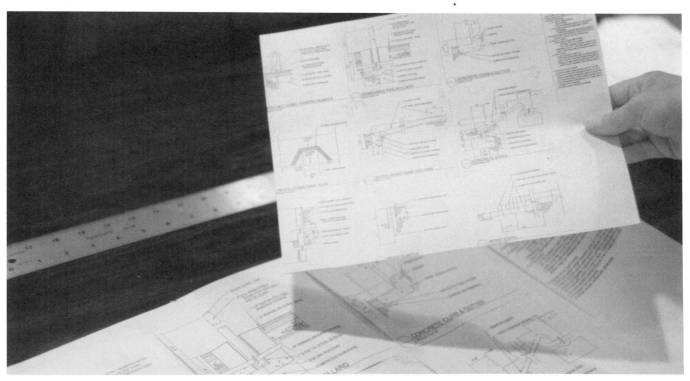

The reduced size version is very handy for review and reference sets. Vital rule: The original lettering and dimension numbers must be in large type.

Components of details can be assembled from generic
standard details and resassembled into whole new details.

One of the benefits of using the photocopier for doing
job prints -- you can copy on both sides of each sheet.

Thus you can work at 11" x 17" format, but then double the effective sheet size by printing on both sides.

The work horses:

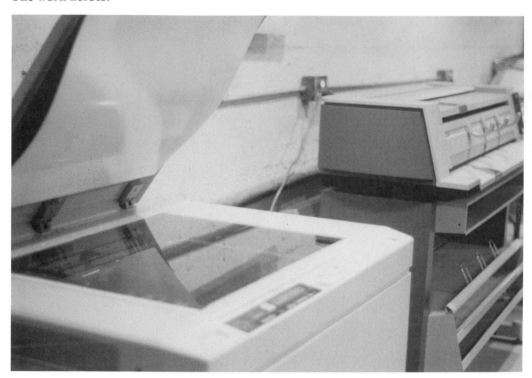

Enlargement-reduction copier (18" x 24" is best although the standard 11" x 17 ' size is very useful for
any design office).

And the large-format engineering copier --
hundreds of uses in any drafting room, especially as an
enhancement to CADD.
(Copiers and printers courtesy of OCE Corp.)

CHAPTER EIGHTEEN

NEW, BETTER WAYS TO COORDINATE THE CONSULTING ENGINEERS

A RADICAL IDEA:
ARCHITECTS SHOW ENGINEERS WHERE
THEY'VE MADE CHANGES IN THE DRAWINGS

WORKING DRAWING SCOPE OF WORK
CHECKLIST FOR CONSULTANTS

ARCHITECT AND CONSULTANT
PROPORTIONS OF WORK AND FEES . . .
SOME RULES-OF-THUMB

BRINGING CONSULTANTS TOGETHER IN
A FREEHAND SKETCH CHARETTE

CONSULTING ENGINEERING DRAWINGS
CROSS COORDINATION

THE LAST WORD ON ENGINEERING COORDINATION

A RADICAL IDEA: ARCHITECTS SHOW ENGINEERS WHERE THEY'VE MADE CHANGES IN THE DRAWINGS

Some customs go on so long that nobody questions them, no matter how odd they might seem. For example, a young architectural drafter might wonder why, when progress prints of work are sent to the engineers, there's no indication of the latest architectural changes.

The engineering counterpart might wonder the same since it's an obvious nuisance, never knowing for sure if he or she has the latest architectural changes.

After seeing the same process time and again and hearing no complaints, the wonder might cease and the drafters will probably adjust to it and proceed to mind their own business.

A few firms have decided it IS a problem. After meeting with engineers to improve coordination, they discovered ample reasons to red - mark the areas of latest architectural changes.

Why not ask your consultants if it would help them? But don't let the answer come just from the principals; they may not know. Get the word from the people on the boards who have to deal with your work. MOST whom we've talked to over the years agree that YES, it would help them a lot for the prime design firms to show where changes have been made.

Establish how much and how precise the change information should be. For smaller jobs, it may be adequate just to draw red circles around areas of change on the check prints. The guiding principle is to make less work for everyone, not more.

The best method we know requires that transparencies be made of all check prints. The checkprints which show the work in the previously checked phase are then compared by layering with the newest work. The differences, the changes between the old and the new, show up clearly and guide the consultant's drafters right to the changes without back-and-forth comparisons of drawings.

ENGINEERING DRAFTING CAN BE ELIMINATED . . . LET THE ARCHITECT DO IT

Architects often make deals with engineers to do the engineering drafting. That's happened where architects have an especially well-run Systems Drafting or CADD operation going. They save up to 50% of the engineer's fees by doing the final civil, structural, and mechanical hard line drafting.

In one notable case, the architect asked for freehand structural details which he was to redraft in finish hardline. As in the case cited above, he realized the flimsy-paper freehand drawings were good enough to use as is, so he mounted them on paste-up sheets and reproduced them with a large-format copier.

"The engineer was mad at first," said the architect, *"but later he decided it was a good idea, and he's doing all his final details freehand now, too.."*

238

WORKING DRAWING SCOPE OF WORK CHECKLIST FOR CONSULTANTS

Not all architects and engineers have a shared understanding of how much work is included in a standard engineering consulting contract.

That leads to disputes, claims and counter-claims late in the working drawing phase, at a time when cooperation is most crucial.

As a guide for you to edit and adapt to your consulting engineering contracts, here are some model checklists of CONSULTANTS' SCOPE OF SERVICES.

CIVIL ENGINEERING WORKING DRAWINGS

__ **Comprehensive**
__ **Partial**
__ **Schematic or outline drawings**

__ Soil tests and boring plans
__ Irrigation
__ Utilities systems, on site and off site
__ Grading
__ Drainage
__ Fire protection systems
__ Site-related plumbing work
__ Soil erosion control
__ Site-related electrical work
__ Structural engineering
    such as retaining walls
__ Cut and fill
__ Excavations
__ Construction cost estimates
__ Final pre-bid check set review
__ Preconstruction meeting
__ Construction consultation
__ Direct coordination with other
    consulting engineers

__ Specifications:
    __ Outline
    __ Comprehensive
    __ Other:

STRUCTURAL ENGINEERING WORKING DRAWINGS

__ **Comprehensive**
__ **Partial**
__ **Schematic or outline drawings**

__ Structural grid or system key plan
__ Structural cross sections
__ Retaining walls
__ Wall sections
__ Foundation plan
__ Construction details
__ Framing plans
__ Code-required calculations
__ Framing sections
    Other calculations
__ Slab reinforcing plans and schedules
__ Beams, lintels, and other materials
    schedules
__ Construction cost estimates
__ Final pre-bid check set review
__ Preconstruction meeting
__ Construction consultation
__ Direct coordination with mechanical
    engineer
__ Specifications:
    __ Outline
    __ Comprehensive
    __ Other:

LANDSCAPE WORKING DRAWINGS

__ **Comprehensive**
__ **Partial**
__ **Schematic or outline drawings**

__ Site planting and landscaping plan
__ Planter details
__ Site-related plumbing work
__ Landscape furniture planning and details
__ Site-related electrical work
__ Fencing, yard walls, pavement grate
    details, etc.
__ Planting schedules
__ Plant support and related construction
    details

__ Landscape cost estimates
__ Final pre-bid check set review
__ Preconstruction meeting
__ Construction consultation
__ Direct coordination with other consulting
    engineers
__ Written instructions for landscape care and
    maintenance
__ Specifications:
    __ Outline
    __ Comprehensive
    __ Other:

MECHANICAL ENGINEERING WORKING DRAWINGS

__ **Comprehensive**
__ **Partial**
__ **Schematic or outline drawings**

__ Sitework and landscape irrigation and
    plumbing plans and details
__ HVAC system type and standard
__ Plumbing supply and drain types and
    standard
__ Fire protection systems
__ Mechanical equipment spatial
    requirements in plan
__ Mechanical equipment spatial
    requirements in section
__ Chase sizes and locations
__ Duct sizes and locations
__ HVAC support systems construction
    details
__ HVAC and plumbing noise and vibration
    control details
__ HVAC heat load and cooling calculations
__ Plumbing supply and drain calculations
__ Plumbing supply and drain sizes and
    locations
__ Plumbing supply and drain diagrams and
    vertical sections
__ Plumbing fitting support and other
    construction details
__ Roof drain calculations and plan
__ Equipment and materials schedules
__ Construction cost estimates
__ Final pre-bid check set review
__ Preconstruction meeting
__ Construction consultation]
__ Direct coordination with other consulting
    engineers

__ Mechanical systems operation and
    maintenance instructions
__ Specifications:
    __ Outline
    __ Comprehensive
    __ Other:

ELECTRICAL ENGINEERING WORKING DRAWINGS

__ **Comprehensive**
__ **Partial**
__ **Schematic or outline drawings**

__ Sitework and landscape lighting and
    power
__ Reflected ceiling lighting plans
__ Power and switching plans and sections
__ Chase sizes, locations, and clearances
__ Communications equipment, chases,
    and outlets
__ Fire and smoke detection and alarms
__ Security systems
__ Electrical power duct and chase sizes and
    locations
__ Electrical equipment sizes and locations
__ Electrical vaults, transformer room sizes
    and locations
__ Equipment and materials schedules
__ Electrical fixture supports and other
    construction details
__ Electrical equipment noise and vibration
    control details
__ Specifications:
    __ Outline
    __ Comprehensive
__ Code-required calculations
__ Construction cost estimates
__ Final pre-bid check set review
__ Preconstruction meeting
__ Construction consultation

ARCHITECT AND CONSULTANT PROPORTIONS OF WORK AND FEES . . . SOME RULES-OF-THUMB

Middle-range architectural fees for simple projects currently range from 4% to 10% of construction cost.

More complex projects and higher square foot cost projects, such as custom luxury homes, are charged at from 10% to 19% of construction. (Construction cost in these averages does not include sitework -- just the building.)

These averages include average-range engineering fees -- structural, mechanical, and electrical -- which are administered and passed along by the architects. The averages do not include civil engineering or landscape architecture, which can vary considerably.

About 25% of the gross architectural fee goes to engineering fees in a typical complex residential job, or an average commercial or institutional project.

Here is the average national range of percentage of gross architectural fee that is passed along for consulting services:

--**Structural** = 5% to 9% of gross fee. (This can double in seismic sensitive California.)

--**Mechanical** = 4% to 12% (These fees tend to be on the high side in California and other states that require special energy calculations.)

--**Electrical** = 2% to 5%

--**Civil** = less than 1% to 3%

--**Other** (interior design, landscape architecture, acoustical, kitchen consulting) = .05% to 3%

The percentages will vary widely for specific projects and especially for complex or technical building types.

241

BRINGING CONSULTANTS TOGETHER IN A FREEHAND SKETCH CHARETTE

Freehand drafted working drawings have been discovered and rediscovered numerous times during high-pressure rush projects. The designers do properly scaled and well-drawn freehand or semi-freehand design sketches and someone realizes that they're good enough to use in the working drawings. It helps when such drawings are done with grid background guidelines and drawn at double size. When they're reproduced at half-size on a reduction copier they come out looking like straight-line tooled drafting.

Same-size freehand work can be perfectly acceptable too and has included designer elevations, cross sections, interior elevations, and even complete floor plans. (Same-size freehand drafting is usually done with the aid of grid sheets and long scales. The scales are used as straight edges for rapid drafting of longer lines.)

One of the most intriguing uses of freehand drafting: complete consultant participation in the planning and mockup of working drawings.

Here's how it works:

1)   After preliminary floor plans are essentially decided and tool drafted, they're pinned on a conference room wall.

2)   Representatives of the major consultants, especially structural and mechanical, are brought in to meet together and work directly on overlay sketch sheets. They and the architect have to work out all construction conflicts and interferences on the spot. This proves far faster than doing it long-distance, piecemeal, with back-and-forth checkprints during working drawing production.

3)   The architect and engineers continue to semi-freehand draft their work throughout the project and meet to repeat the coordination process as the building is firmed up.

4)   When production is at 80-90% completion and all major decisions are finalized, the freehand drawings are assembled and all the consultants' plans and sections are finish drafted. The finish drafting is no more than copying and, since it's not held up by questions and decision making, it goes extremely fast. Smaller drawings, especially details, that have been freehand drawn are reused directly without redrafting when possible. If they're not good enough for reuse, they're copied by hand or on CADD, but such straightforward, uninterrupted copy-drafting goes at high speed.

There are two important secrets behind the success of this method:

1) MOST DRAFTING CONSISTS OF MAKING CHANGES.

2) DESIGN AND DECISION MAKING IS TOTALLY DIFFERENT, AS A PROCESS, FROM DRAFTING. Drafting is documentation of design decision making. When documentation is mixed up with design, it takes forever.

When drafting or inputting is ONLY that, it can go like lightning. When design drawing is used for decision making only and not slowed by the rigors of drafting, it can go considerably faster than usual. So the object is to separate the two different processes and delay the finalization, the finish drafting, until the last phase of work.

We first learned of an office making full-size mockups of their working drawings in 1969. It was a small firm that had received a large hotel commission as a rush project. Three architectural and three engineering personnel completed the multi-million dollar project in six weeks. Three months would have been normal (with added personnel) using the draft-erase-draft and redraft system.

242

CONSULTING ENGINEERING DRAWINGS CROSS COORDINATION

These CONSULTANT/ENGINEERING DRAWINGS CROSS-COORDINATION CHECKLISTS provide cross-coordination data and recommendations regarding which disciplines are responsible for which aspects of drawings and specifications.

MECHANICAL ENGINEERING CROSS COORDINATION -- PLUMBING

CONSTRUCTION COMPONENT:	DRAWINGS OR DWG. SCHEDULE BY:	SPECIFICATIONS BY:
____ Catch basins, meter traps, plumbing manholes, and related gratings, covers, and ladders.	____ MECHANICAL ____ ARCHITECTURAL	____ MECHANICAL
____ Underground tanks, concrete tunnels, and supports. Coordinated with:	____ MECHANICAL ____ STRUCTURAL	____ MECHANICAL
____ Sleeves for plumbing pipe, other mechanical work.	____ MECHANICAL ____ ARCHITECTURAL	____ MECHANICAL
____ Ceiling and wall access doors/panels to plumbing not otherwise accessible.	____ MECHANICAL ____ ARCHITECTURAL	____ MECHANICAL
____ Hangers, anchors, or other supports for plumbing related equipment. Coordinated with:	____ MECHANICAL ____ STRUCTURAL	____ MECHANICAL
____ Plumbing vents and flashing.	____ MECHANICAL ____ ARCHITECTURAL	____ MECHANICAL
____ Final painting of exposed plumbing.	N/A	____ ARCHITECTURAL
____ Wall- or floor-mounted bathroom/rest room accessories.	____ ARCHITECTURAL	____ ARCHITECTURAL

CONSTRUCTION COMPONENT:	DRAWINGS OR DWG. SCHEDULE BY:	SPECIFICATIONS BY:
___ Floor plumbing trench. Coordinated with:	___ ARCHITECTURAL ___ STRUCTURAL	___ ARCHITECTURAL
___ Floor plumbing trench covers and gratings.	___ ARCHITECTURAL	___ ARCHITECTURAL
___ Floor drain. Coordinated with:	___ MECHANICAL ___ ARCHITECTURAL ___ STRUCTURAL	___ MECHANICAL

MECHANICAL ENGINEERING CROSS COORDINATION -- HEATING, VENTILATING, AND AIR CONDITIONING

CONSTRUCTION COMPONENT:	DRAWINGS OR DWG. SCHEDULE BY:	SPECIFICATIONS BY:
___ Floor mounts, bases, and pads for mechanical equipment. Coordinated with:	___ ARCHITECTURAL ___ MECHANICAL ___ STRUCTURAL	___ ARCHITECTURAL
___ Floor trench. Coordinated with:	___ ARCHITECTURAL ___ MECHANICAL ___ STRUCTURAL	___ ARCHITECTURAL
___ Floor trench covers and grating.	___ ARCHITECTURAL	___ ARCHITECTURAL
___ Wall and ceiling grilles, registers, and vents. Coordinated with:	___ MECHANICAL ___ ARCHITECTURAL	___ MECHANICAL ___ ARCHITECTURAL
___ Exterior wall louvers. Coordinated with:	___ ARCHITECTURAL ___ STRUCTURAL ___ MECHANICAL	___ ARCHITECTURAL ___ MECHANICAL
___ Door grilles. Coordinated with:	___ ARCHITECTURAL ___ MECHANICAL	___ ARCHITECTURAL ___ MECHANICAL
___ Sleeves for HVAC work.	___ MECHANICAL	___ MECHANICAL
___ Ceiling and wall access doors/panels to HVAC work not otherwise accessible.	___ MECHANICAL ___ ARCHITECTURAL	___ MECHANICAL
___ Hangers, anchors, or other supports for HVAC-related equipment. Coordinated with:	___ MECHANICAL ___ STRUCTURAL	___ MECHANICAL ___ MECHANICAL

CONSTRUCTION COMPONENT:	DRAWINGS OR DWG. SCHEDULE BY:	SPECIFICATIONS BY:
___ Gravity roof ventilators and power ventilators. Coordinated with:	___ STRUCTURAL ___ MECHANICAL ___ ARCHITECTURAL	___ MECHANICAL
___ Cooling tower supports.	___ STRUCTURAL	___ STRUCTURAL
___ Roof-mounted mechanical equipment supports.	___ STRUCTURAL	___ STRUCTURAL
___ Pitch pan or pocket flashing for roof-mounted mechanical equipment supports. Coordinated with:	___ ARCHITECTURAL ___ STRUCTURAL	___ ARCHITECTURAL
___ Sheet metal and flashing for concrete bases at roof-supported mechanical equipment.	___ ARCHITECTURAL	___ ARCHITECTURAL
___ Curbs, nailers and flashing for roof-mounted fans, vents, and ductwork.	___ ARCHITECTURAL	___ ARCHITECTURAL
___ Roof penetration counter-flashing for hot pipe, stacks, and ductwork.	___ MECHANICAL ___ ARCHITECTURAL	___ MECHANICAL
___ Final painting of mechanical equipment.	___ ARCHITECTURAL	

ELECTRICAL ENGINEERING CROSS COORDINATION

CONSTRUCTION COMPONENT:	DRAWINGS OR DWG. SCHEDULE BY:	SPECIFICATIONS BY:
___ Covers for underground electrical ducts and conduit.	___ ELECTRICAL	___ ELECTRICAL
___ Transformer vault.	___ STRUCTURAL ___ ARCHITECTURAL	___ STRUCTURAL
___ Electrical manholes, pull boxes, and ladders. Coordinated with:	___ ELECTRICAL ___ ARCHITECTURAL	___ ELECTRICAL
___ Electrical floor trench. Coordinated with:	___ ARCHITECTURAL ___ STRUCTURAL	___ ARCHITECTURAL

___ Electrical floor trench covers and gratings.	___ ARCHITECTURAL	___ ARCHITECTURAL
___ Electrical floor ducts. Coordinated with:	___ ELECTRICAL ___ STRUCTURAL	___ ELECTRICAL
___ Concrete base for switchboards, other electrical equipment. Coordinated with:	___ ARCHITECTURAL ___ STRUCTURAL	___ ARCHITECTURAL
___ Electrical conduit sleeves. Coordinated with:	___ ELECTRICAL ___ ARCHITECTURAL ___ STRUCTURAL	___ ELECTRICAL
___ Ceiling and wall access doors/panels to electrical equipment not otherwise accessible.	___ ELECTRICAL ___ ARCHITECTURAL	___ ELECTRICAL
___ Hangers, anchors, or other supports for electrical-related equipment. Coordinated with:	___ ELECTRICAL ___ STRUCTURAL	___ ELECTRICAL

STRUCTURAL ENGINEERING DRAWING AND ARCHITECTURAL COORDINATION

This coordination list is adapted from the Northern California Chapter, AIA Production Office Procedures Manual.

ARCHITECTURAL DRAWING

___ Structural grid with dimensions and coordinates.

___ Floor elevations at tops of slabs.

___ Foundation outline with overall building foundation dimensions.

___ Slab depressions, raised floors, and curbs, with locations and size dimensions. Details of mat frames, edge strips, etc.

___ Openings in slabs and walls: location and size dimensions, details of frames, sills, jambs, edge strips, etc.

STRUCTURAL DRAWING

___ Same as architectural.

___ Same as architectural.

___ Detailed foundation dimensions. Elevations of slabs, slab variations, and bottoms of footings.

___ Details of slabs and curbs showing location, and connection or integral design with other structural elements.

___ Details of structural elements related to openings such as integral beams, lintels, headers, etc.

____ Waterproofing membranes at slabs on or below grade, and below-grade walls.

____ Outlines of structural elements such as columns, piers, pilasters, etc.

____ Exterior construction: walks, ramps, slabs, steps, trenching, etc.--locations, dimensions, and architectural details.

____ Steel frame fireproofing identification, notation, and detail key references.

____ Miscellaneous metal details such as metal stairs, railings, ladders, etc.

____ Details of anchors, hangers, brackets, etc.

____ Elevator shaft and machine beam support. (Location of beams supplied by elevator manufacturer.)

____ Show relationship of membranes to structural members.

____ Identification and detail key or schedule key reference of structural elements. Locations and size dimensions of all structural elements.

____ Outline of architectural features and details of structural components. Details of connection of exterior construction to structural members.

____ Identification of primary, secondary, and non-support members to establish required fireproofing.

____ Structural element connections to miscellaneous metal.

____ Structural element connections, sleeves, and inserts for anchors, hangers, brackets etc.

____ Locations and sizes of all elevator shaft and machine room supports.

THE LAST WORD ON ENGINEERING COORDINATION

Just because it's so simple, so effective, so important, and so commonly not done in coordinating the consultants' drawings, we quote from the chapter on **CHECKING THE DRAWINGS:**

"Side-by-side checking of drawings that have to be coordinated is still the dominant method of checking and it's the worst way to do it."

"Multi-story plans, elevations and cross-sections, architectural and engineering drawings SHOULD ALL BE PRINTED AS TRANSPARENCY CHECK PRINTS AND OVERLAYED ON A LIGHT TABLE FOR ONE-ON-ONE COORDINATION COMPARISONS."

CHAPTER NINETEEN

FASTER, BETTER SPECIFICATIONS

"ARCHITECTS PRETEND TO WRITE SPECIFICATIONS
AND WE PRETEND TO READ THEM."

CUT OUT POINTLESS, WASTEFUL REDUNDANCY
IN THE SPECS

COMBINING CHECK PRINT OVERLAYS WITH
SPEC COORDINATION

HOW TO WRITE YOUR SPECIFICATIONS AS CHECKLISTS

YOUR BEST SPEC DATA SOURCES

SAVE TIME AND MONEY WITH A SPEC SYSTEM MAKEOVER

"ARCHITECTS PRETEND TO WRITE SPECIFICATIONS AND WE PRETEND TO READ THEM."

A/E's are issuing specifications that are riddled with omissions, the wrong data, and uncoordinated data.

The result: Higher bids, more extras, more change orders, more claims. And when claims go to court, those specs provide contractors and owners with glaring evidence of faulty documents.

Contractors who specialize in low-ball bids look to specs to find the extras they'll need to make up for their low bids. So do the marginal contractors who use lawsuits as a wedge to worm out of their contract obligations. A/E's make it too easy for them.

This is the word from a recent survey of contractors about the specifications they're supposed to use; a survey conducted by the University of Florida School of Building Construction and **Engineering News Record** magazine.

--OVER 84% OF CONTRACTORS POLLED SAY SPECS HAVE MAJOR OMISSIONS.

--52% OF ALL SPECIFICATIONS ARE RATED FAIR TO POOR.

"Remove boiler plate - it only raises prices and serves no useful purpose other than to employ lawyers," says a contractor quoted in the **Engineering News Record** survey.

Others report that specifications commonly have sections that have no relationship to the project; they're just cut out of other project specs with no attention to their relevance.

From the ENR survey report:

"Robert Dean, vice president of Heery International Inc., Atlanta, says he has seen firms *'not only using office masters without editing or updating them, but also using commercially available masters without editing or modifying them from job to job.'"*

The specs are biased, too, according to 87% of respondents who say specs favor the interests of the designers, owners, or suppliers. That makes the specs something to resist instead of use.

Perhaps the worst news in the survey: 55% of respondents say specifications *"generally or often conflict with other contract documents."*

That shows some of the depth and extent of the problem. Read on for solutions.

CUT OUT POINTLESS, WASTEFUL REDUNDANCY IN THE SPECS

quirements with materials and materials specs. Not only do they lead to a false sense of security -- since the performance requirement may not be enforceable -- they have also generated successful contractor claims based on time and expense incurred in attempting to meet them after materials and methods requirements had been satisfied."

If you specify materials and workmanship and also specify performance, you're wasting time and negating yourself according to construction attorney Robert J. Smith in the *Building* newsletter for the University of Wisconsin College of Engineering.

He says:

"When materials and workmanship are specified in detail, the specs are subject to what the courts call an 'implied warranty.' This implied warranty says if the plans and specs are followed, a satisfactory result will be obtained. Thus, even in the face of a spec which requires a no-leak roof or foundation, if the contractor can conclusively establish he/she complied with the materials and methods requirements of the plans and specs, that will be deemed 'full performance.'*

"Another way that the courts have looked at this problem is to call it a "defective specification.' That is, since the roof still leaks even after the materials and methods requirements have been scrupulously followed, the problem must be with the spec.*

"To those who would argue the contractor promised to provide a new, no-leak roof or basement, the courts respond by noting that the contractor also promised to follow materials and workmanship requirements, and in the event of a conflict between the descriptive and performance specs, descriptive requirements will likely govern since the contractor had to comply with those first.*

"Thus design professionals should think twice about combining final performance re-

251

COMBINING CHECK PRINT OVERLAYS WITH SPEC COORDINATION

It pays to have spec writers do the overall quality control checks of working drawings. They have to do their own checkset reviews of items to specify anyway and are usually the most technically knowledgeable people in the office. So quality control checking is usually a good role for them. (This isn't always practical because spec writers are often overloaded as it is, but it's worth consideration.)

But whether or not the spec writers do quality control checks, they can also benefit by using the overlay checksheet technique.

Here's how:

1) Make check prints in the usual manner and attach a blank sheet of tracing paper or thin polyester on each print.

2) Draw a red dash line or check box on the overlay beside items on the check prints that may be specifiable. Don't mark every individual item, of course. **One mark for each type of specifiable item is sufficient.**

3) After or below each red dash line or box, write the appropriate CSI Broadscope Division number: 02 for sitework, for example, 03 for concrete, 04 for masonry, etc.

In this process, you're creating a graphic shorthand checklist of all your spec items. It's integrated with the drawings. It doesn't require a lengthy, separate list-writing process, and it greatly reduces the chances of overlooking something in the drawings that should be specified.

Use overlays for doing the marks. That way you can update your overlay checksheet spec checklist at each subsequent checking phase by re-attaching the overlays in registration with the up-to-date prints. (See the chapter on CHECKING THE DRAWINGS for more details on the timesaving "overlay checking" system.)

In the final phase of checking, add check marks beside all the items you've written specs for. This final check will show if you missed anything in specifications or have any extraneous sections. It's all far easier than the old style back-and-forth, sheet-by-sheet comparison of specs and drawings.

HOW TO WRITE SPECIFICATIONS AS CHECKLISTS

You can make your specs far faster to write and much easier to read with one simple move: Turn them into checklists.

Despite the best efforts of reformers at CSI, AIA, Sweets, etc. and despite years of effort at Outline Specifications, Streamlined Specs, and Short Form Specs, most such documents are still written to read like law books. Most specifications are still written in **paragraph** form in a style that combines the worst aspects of legal contracts and technical manuals.

But specifications aren't legal contracts and technical manuals; they're more like To-Do lists and shopping lists. They're lists of things for the contractor to do, things to buy, things to install, and the standards by which you'll judge the contractor's work.

Here's a simple sample sentence from a published master specification:

"Provide temporary piping and water supply and, upon completion of the work, remove such temporary facilities."

It's not that bad a sentence but compare it's <u>usefulness</u> with a rewritten checklist format version:

Provide temporary:

__ Piping

__ Water supply

__ Remove them when work is completed

It's the same information in each version but the checklist format has major advantages over traditional sentence/paragraph writing:

--It can be checked off point by point to simplify the original writing process. (Anybody can write an intelligible list.)

--It's easily adapted and revised when doing future specs.

--It's readily checked off and costed by subs and suppliers doing takeoffs for bids.

--It can be checked by the contractor as the work is done.

--It's checked off by the designer when visiting the site to assure compliance.

As a multi-use checklist, specs become a working tool for everyone involved instead of just being a paperweight in the construction shed.

Here's another example of standard traditional spec writing:

"Provide and maintain for the duration of construction all scaffolds, tarpaulins, canopies, warning signs, steps, platforms, bridges, and other temporary construction necessary for the completion of the Work in compliance with pertinent safety and other regulations."

To change this paragraph to a checklist, first mark the "checklistable" items with asterisks like this:

*"Provide and maintain for the duration of construction all *scaffolds, *tarpaulins, *canopies, *warning signs, *steps, *platforms, *bridges, and *other temporary construction necessary for the completion of the Work in compliance with pertinent safety and *other regulations."*

Then rewrite it in checklist format as follows:

__ Provide and maintain during construction:

    __ Scaffolds

    __ Tarpaulins

    __ Canopies

    __ Warning signs

    __ Steps

    __ Platforms

    __ Bridges

    __ Any other temporary construction needed to complete the Work

__ Temporary construction shall comply with:

    __ All safety regulations

    __ All other applicable codes

The checklist version has the same data as the first version, but is much easier to read, understand, and interpret. And as noted before, the specification becomes doubly useful as a clear construction site punchlist.

Here is an example of checklist format for a portion of specifications for sitework:

02220   EARTHWORK--BACKFILL AND COMPACTION

__ Perform backfill and compaction in a systematic pattern, to assure complete and consistent work.

__ If any overexcavation accidentally occurs, correct it with well-compacted backfill.

__ Fill and thoroughly compact holes from root and stump removal pits.

__ Place termite and other soil poisons with backfilling.

__ Perform testing and inspection of backfill and compaction.

__ Layer backfill in 6 inch to 12 inch increments.

__ Compact all fill.

__ Use stabilized fill material of an approved type and from an approved source.

__ Test and approve fill material delivered from other sites.

__ Do not allow any debris to be mixed with fill.

__ Cure concrete foundation or retaining walls to reach required strength before backfilling.

__ Brace foundation or retaining walls as necessary to assure structural stability and prevent damage from backfilling.

__ Thoroughly waterproof basement foundation walls before backfilling.

__ Notify the design firm of below-grade waterproofing and arrange for observation of the work.

__ Do not allow any damage to waterproofing from backfilling.

__ Alternately place backfill at two sides of a wall to avoid unbalanced loading.

__ Correctly replace boundary markers, monuments, and stakes if they are moved or damaged during excavation, backfilling, and grading.

YOUR BEST
SPEC DATA
SOURCES

The best source of all, of course, is the Construction Specifications Institute. Send for a complete list of all their publications, including their Masterformat, SpecText, ManuSpec, Spec Data, Manual of Practice, etc.

___ **The Construction Specifications Institute (CSI),** 601 Madison St., Alexandria, VA 22314. 703-684-0300.

Sources will change, of course, as time goes by but as of this publication date your best current sources are:

___ **AIA MasterSpec.** AIA Service Corporation, 1735 New York Ave., NW, Washington, DC 20006. 800-424-5080.

___ **Sweet's Building Selection Data** section of Sweet's catalogs. McGraw-Hill Information Systems, 1221 Avenue of the Americas, New York, NY 10020. 212-512-4450.

___ **Library of Specifications Sections.** Over 250 master specification sections covering all 16 CSI divisions. This is an outstanding source of text in four volumes by specifications expert Hans W. Meier. Prentice-Hall Business & Professional Books, Englewood Cliffs, NJ 07632. 201-767-5937.

___ **American Concrete Institute Publications Catalog / Architect's Guide to ACI Publications.** American Concrete Institute, P. O. Box 19150, Detroit, MI 48219-0150. 313-532-2600.

___ **ASTM Publications Catalog.** ASTM, 1916 Race St., Philadelphia, PA 19103. 215-299-5585.

___ **Federal Construction Guide Specifications. NAVFAC Guide Specifications. Military Specifications.** Write for price list from: Federal Construction Council, Naval Publications & Forms Center, Code 3015, 5801 Tabor Ave., Philadelphia, PA 19120. 215- 697-2000.

___ **Federal Specifications.** GSA Business Services Center, 300 N. Los Angeles St., Los Angeles, CA 90012.

___ **National Forest Products Association,** 1250 Connecticut Ave., N.W., Suite 200, Washington, DC 20036. 202-463-2700.

___ **American Plywood Association,** P. O. Box 11700, Tacoma, WA 98411. 206-565-6600.

___ **Expansion Joint Manufacturers Association,** 25 North Broadway, Tarrytown, NY 10591. 914-332-0040.

___ **American Institute of Steel Construction,** 101 Park Ave., New York, NY 10017.

___ **NRCA Roofing & Waterproofing Manual,** National Roofing Contractors Association, 8600 Bryn Mawr Ave., Chicago, IL 60631.

SAVE TIME AND MONEY WITH A SPEC SYSTEM MAKEOVER

The main barrier to carefully written and edited specs is simple time and cost. The production budget is often nibbled down by overruns during the design phase. By the time someone gets around to specs, there may be nothing left to pay for them. If it happens often enough, the last-gasp, last-minute, cut-and-paste system becomes "normal" office custom.

At some point, somebody has to say *"ENOUGH!"* and get the office to start over.

The office has to create its own master spec and spec writing system. The system has to be clean, rational, and up-to-date. When it is, the specs that follow can be put together much faster and cheaper than the error-prone rush jobs done in the old way.

A small firm's investment will amount to a couple of weeks of part-time effort. A mid-sized firm may spend a couple thousand dollars worth of time -- a fraction of what the next spec blunder might cost. Offices that do it will make their investment back in time and cost savings in the first few projects that follow.

Here are the steps:

1) If yours is a small (1 to 6 people), varied practice, get the latest spec section and technical publications list from the CSI. They publish most of the sections you would want to work with and you can rework their text and input it into a standard specifications database on your computer.

If yours is a mid-sized firm, consider using the AIA or CSI Master Spec on computer for a year. As you go along, adapt and rewrite the most suitable sections of those master specs to create your own permanent in-house Master.

2) Send for all the best and latest technical reference data and toss out the old.

3) Write your specifications text as checklists. For all the reasons noted previously, written as checklists are far easier to write, read, edit, and use than the traditional sentence/paragraph format.

4) Use the CSI Masterformat specifications division and section numbers as the basis for keynoting and reference notes.

CHECKLISTS -- DAY-TO-DAY

As noted previously, checklist specifications are comparatively easy to write mainly because anybody can write a list. And checklists are easy to read because they segment and divide information in easy-to-grasp doses.

A/E's aren't limiting checklists to office operations manuals and specifications. For example:

--Correspondence. If you have a list of items, then make it a list and not a long paragraph peppered with commas. Most people will not finish reading a list of items in a paragraph, but they will read and check off the same items if they're listed in a vertical column.

--Memos and instructions. Want to give instructions or make some points in the most effective way possible? List them boldly and if it's the type of memo that needs confirmation of receipt and feedback, include checkmark spaces beside the main points for the user to mark.

CHAPTER TWENTY

EXPEDITING BIDDING AND CONSTRUCTION CONTRACT ADMINISTRATION

BIDDING -- TOO OFTEN A CAULDRON OF FRAUD AND EXTORTION

HOW TO GET THE BEST CONTRACTORS

HOW CONTRACTORS TURN A/E ERRORS & OMISSIONS INTO A PROFIT CENTER

A "JUSTIFICATION OF CHANGES" CHECKLIST

HOW TO AVOID THE "BAD NEWS" LOW BIDDER

APPROVING PROPOSED SUBSTITUTIONS DURING BIDDING

BIDDING --
TOO OFTEN
A CAULDRON OF
FRAUD AND
EXTORTION

First let's define the terms:

--Bid cutting: A prime contractor wins a contract based on low bids from suppliers and subcontractors, then forces them to cut their prices even more as a condition for getting the work.

--Bid peddling: A subcontractor or supplier gets information from the prime contractor about other bidders' quotes in order to shave his or her bid enough to get the job.

--Bid shopping: A prime contractor gathers estimates and bids and gives the information to others to persuade them to bid lower.

--Low-balling: A prime contractor determines what sort of bid will beat the competition, submits it, and then uses bid shopping to get costs in line. (Not to mention aggressively exploiting every opportunity for change orders and claims for extras during construction.)

--Owner bid-shopping: The owner puts a job out to bid, rejects the bids, and makes superficial changes as a pretext for second-round negotiations to knock down the original bid figures.

The results of these practices:

--Cheap substitutions, as suppliers and subs struggle to make up for the losses they'll sustain for underbidding.

--A battleground of adversary relationships on the jobsite, as everyone in the chain watches for every chance to pass on a cost to someone else.

--Espionage and cutthroat tactics that crowd out contractors who are primarily concerned with just doing a competent job at a reasonable price.

Here is as vivid a description of the situation as you could ask:

"On a major job you may spend 80-100 man-hours estimating that job and putting off others. If you give the numbers to one person who's not honest, the competition has it in five seconds and you have wasted all your time." (David Miller, electrical contractor in Newark, Ohio and former president of the American Subcontractors Association; quoted in Engineering News Record, *"Shop Till They Drop,"* pg. 26, March 9, 1992.)

"Some developers and architects think it's the way to get the best deal, but it's just the opposite," says New York general contractor, Danny Torrello.

Torrello continues: *"As a sub, when you know you're dealing with a bid shopper, you just give him the numbers he wants first and give him the shaft later. There are a thousand ways to make up losses in this business if you really have to."*

On some projects the shaft starts at the top where the owner and prime design firm are counting on unrealistically low construction costs to carry the project.

The implication is that somebody at the bottom is going to take the losses, go under, and carry the debt. Sometimes dozens of subs and suppliers get chased off the job on various pretexts after their work is in, and they never get paid. The owner doesn't intend to fully pay off the prime contractor, the prime doesn't intend to fully pay the subs, the subs don't intend to fully pay suppliers.

That, of course, is why the lien laws exist -- these practices go back for centuries, and generations-old legal protections do have teeth. So in most cases, everyone in a chain of bid-shopping and low-balling may plan to pay just enough to forestall legal action; not enough to kill the participants, just enough to exploit them to the maximum they'll tolerate.

What can be done? On the legal front, not much. Bid shopping and low-balling are themselves not illegal and the Justice Department has let trade associations know they won't tolerate any cooperative effort to stop such practices. What is illegal are the cheap substitutions that are made by the suppliers and subs to cut their costs.

On the A/E side, three recommendations:

1) Use a Contractor Performance Evaluations system to build a file of the most reliable contractors, subs, and suppliers. (See the recommendations in this chapter.)

2) Use a system for approving proposed substitutions that keeps all other bidders informed. (Also in this chapter.)

3) Establish clear cut rules and procedures for handling shop drawings and shop drawing changes and substitutions (elsewhere in this chapter).

HOW TO GET THE BEST CONTRACTORS

Some excellent advice on choosing contractors from Production Consultant, Banks Upshaw of Scottsdale, Arizona. A formal screening process is normal in commercial work but, he says, it's surprising how little contractor screening is done even on multimillion dollar residential jobs.

"It saves you a lot of time," he says. *"Good contractors take only a third as much of your time during construction as someone who's run-of-the-mill. There are fewer questions, things get done right the first time, they don't get hung up with indecision because of inexperience."*

When new contractors call to see if there's any work (which is common these days), he asks for the following:

--Financial statement

--Client references

--Two to three major subcontractors in each category of construction

--Financial references

We would add:

Check that their license is current. Run a TRW or comparable personal credit rating service check on personal finances; and a Dun and Bradstreet report on business finances and stability. Ask the candidate to pay for the reports.

Upshaw goes to look at their work and he calls or visits the people the contractor has worked with before. If the contractor passes muster, he or she gets on the office list. When a job goes to bid, Upshaw can give the owner five or six truly qualified contractors to do the job, and not have to worry about it. Bids are read with care and if any of the numbers seem out of whack -- particularly if a sub is bidding way too low for the work -- Upshaw asks that the bid be checked and resubmitted.

All in all this is some of the best problem-prevention insurance a small office can provide . . . and an excellent value to the client.

HOW CONTRACTORS TURN A/E ERRORS & OMISSIONS INTO A PROFIT CENTER

First, a look at specifications. Most A/E's have never considered what contractors know or think about the spec writing process. These quotes, from a book that's been a best seller* among contractors, are eye opening:

*"The technical specifications are an architect's **secondary priority**. Throughout the process of design development, specific thought had been given to the precise materials desired and their configuration. The design process is what the architect is best at. It is the process of bringing the initiated concept through the detailed design that gives the architect the greatest satisfaction. It is only when the design is complete that the technical specifications and the contracts themselves are finally assembled. It is not the priority, but an unfortunate, tedious necessity, as far as the architect is concerned. It's a second effort. It is for this reason that cut-and-paste is looked to not as the most effective method of comprehensive preparation of an important legal document, but as a fast, cheap way to get it over with.*

*"The technical specifications and the front-end documents are **assembled at the eleventh hour**. The design process itself usually continues until the last possible moment. It is only when the deadline of the bid solicitation looms that serious attention is diverted to completing the specification. The specifications then become likely to be 'written' during evening and weekend sessions and to be fit in between the*

*Contractor's Guide To Change Orders -- The Art of Finding, Pricing, and Getting Paid for Contract Changes and the Damages They Cost, by Andrew M. Civitello, Jr., Prentice Hall, ISBN #0-13-171588-7.

'real' design work. What is worse, junior architects and even clerks may be enlisted to prepare certain portions of the documents.

"It's bad enough that the documents you've been working with had originally been written for somebody else and made to 'fit' your job. Add to this the hectic second effort just described, and it is easy to see how the probability of errors and confusion increases."

You could hardly find a more accurate, or more damning description of how specs are actually prepared in most design offices.

It's not just specs, as you'll see below; the author has plenty to say about working drawings. We have never seen a more useful guide to working drawing errors and oversights. If anyone in an office thinks specs and working drawings are not top priority items, this book will snap their eyes wide open.

Contractors who know what to look for can spot and plan for change orders and extras from the earliest stages of construction. Just the ordinary level of care given to an average set of working drawings and specs offer a bonanza to the well-prepared contractor.

On finding lapses in coordination by comparing the spec index with the finish schedule:

"A comparison of the Finish Schedule with the Specification Index can expose potential duplications and/or omissions. It can indicate whether or not the design process seems to have been completed in a coordinated manner, or if it was performed by different individuals who didn't speak with each other. Any discrepancies discovered in this review will be fairly obvious."

On inadequate level of detail:

"Providing an inadequate level of detail in a design necessary to complete a particular construction is an attempt by a designer to:

1) Avoid spending the proper amount of time necessary to complete the design.

2) Shift the burden of any remaining design

onto the contractor under the guise of 'coordination.'

"Missing design information can take many forms, and can range from the obvious to the subtle. Examples include:

--Mounting or fastening details (do you want stainless steel brackets or rubber bands holding up the limestone?).

--Not enough dimensions to allow even an elaborate calculation to locate the work properly.

--Complete descriptions (is the blocking to be continuous, 24" on center, or eliminated altogether?).

--Vague descriptions of special shapes, angles, and so on (what is the exact angle of the spandrel glass?).

--No precise layout given (what is the radius of the curved stone in front of the building? Do you have to lay it out in the parking lot to find out?)"

On consistency in plan orientations and match lines:

"The plan orientations of all drawings should be identical. . . . If you . . . observe a casual attitude on the part of the designers toward this most basic design procedure, you can expect problems at every level. Not only will it be likely that there will exist errors directly caused by careless match line and plan orientation practices, but that design attitude will be sure to permeate all other design considerations as well. In other words, if the designers have not properly taken care of the most basic design/layout considerations, you can expect that the more complicated responsibilities were likewise improperly completed, if at all."

On conflicts in ceiling spaces:

"There probably hasn't been a project designed and built that hasn't had some kind of problem with everything that was supposed to fit so neatly into tightly restricted spaces above the ceilings. . . . The smaller the area, the more complex the building systems and the greater the probability of conflicts and change orders.

" . . . Invariably the architect located ceiling recesses in spaces that looked clear -- like corridors. Those same spaces had looked so clear in the design's beginning that the plumbing, fire protection, and the H.V.A.C. designers all had the same idea. So everything had been located on the individual drawings all in the same space."

The author recommends using a light table and overlaying architectural and engineering plan sheets to better find the conflicts that may require change orders. To this day most A/E's still don't follow this eminently logical practice in coordinating their drawings. Money in the bank for the astute contractor.

He provides a checklist that should be posted in every A/E drafting room:

"Overlay the lighting plans. Check each location of ceiling light fixtures, emergency lights, soffit lights, exit lights, undercabinet lights.

"Overlay the architectural plans. Check for discrepancies in the locations of walls, soffits, and cabinets that will affect light locations, ceiling patterns, and exit light layouts.

"Overlay the H.V.A.C. plans. Check that each register, grille, and diffuser location is consistent. Confirm that the equipments' actual sizes are accommodated in the layout and that everything misses the lights.

"Overlay the sprinkler layout. Do the heads miss the lights? Do the heads fall in the center or quarter-center of the ceiling tile in each instance (if specified as such)? Is there an architectural pattern in the ceiling tile that will change location preference?

"Overlay the electrical power plans. Do the smoke detectors miss the lights and everything else?"

Similar checklists are provided for reviewing structural plans, sitework, etc. and, of course, the total process of prospecting for change orders.

A summary of the author's intent should make every design firm get deadly serious about its quality control:

"The real talent for prospecting for changes lies in the ability to expose extra cost items in good time to allow their incorporation in stride. If we can learn to take the initiative to actively search and discover potential extra cost items, there will be time enough to work the process through in a less hurried manner. We'll have the luxury of being able to contemplate the best strategies and to consider those tactics that will get our change orders approved with a minimum of friction. It's the old business of making it happen instead of letting it happen."

In other words, your errors, omissions, and "clarifications" which used to an inconvenience to contractors will henceforth be a profit center for them.

A "JUSTIFICATION OF CHANGES" CHECKLIST

When the bids and stress are high, the pressure is on to accept any proposed changes that might expedite the work and save some money. These are the changes -- last-minute changes and substitutions -- that are involved in about half of all design, construction, and materials failures.

In response to the problem, some firms have created "Justification Checklists" to document the process of analyzing and approving changes. They're written as standard approval forms to be filled out and signed before any important change orders can be written. They help assure that some attention is given to the implications of a change before approval

Checklist information includes:

__ Reason(s) for the change:

    __ Problem with the existing requirement.
    __ Improvement to result from the new requirement.
    __ Proposed change complies with codes and regulations.

__ Proposed substituted material, product, equipment, fixture, finish, construction standard, or system is genuinely equal or superior to the original.

__ Comparable quality is validated.

__ Design disciplines affected, additional design work required.

__ Additional A/E documentation required and estimated additional design:
    __ Time. __ Cost.

__ Estimated additional construction:
    __ Time. __ Cost.

HOW TO AVOID THE "BAD NEWS" LOW BIDDER

The low bidder is often the worst choice for a construction contractor.

We all know the problem of low-balling, and during hard times we know that even good contractors get desperate and submit unrealistically low bids. A low bid contractor who gets into deep water will try to recover losses one way or the other. That means disputes, extras, faulty work, costly repairs, extra jobsite observation, and endless other problems after construction. The low bid ends up costing far more than the high bid.

There are systems for dealing with this hazard -- systematic ways of screening contractors as much on the basis of reliability as dollar bid. Systems have been evolved by professional clients who contract out millions of dollars of construction each year; clients such as General Motors, Union Carbide, major utility companies, etc.

The system is based on formal Performance Evaluation.

As a first step, use an evaluation form, as shown on the next page, to record how well your previous contractors did their job. For contractors you don't know, call their previous clients and design firms. Write down and then average out their evaluations.

You can write your own list of weighted evaluation criteria, to fit your needs. A sample list follows on the next page. **The list should be weighted so that you gauge the relative importance of each factor and come up with a final score you can use to multiply against the contractor's bid.**

Thus you'll multiply a bid from a contractor with a low Performance Evaluation by a factor number such as 1.2 to increase the actual poten-tial dollar amount of the bid. The bid of a contractor with an excellent Performance Evaluation record will be multiplied by perhaps .8 which will lower the actual potential dollar amount of the bid. (See next page for the source of factor numbers.)

For example . . . suppose an "Excellent" contractor bids on the high side at $1,000,000. You multiply that bid by .8 for a weighted bid of $800,000. This is called the "Management Discounted Bid" value. An "Average" contractor bidding $900,000 would get a weighted factor of 1.0, and the "Management Discounted Bid" would be $900,000.

By comparison, a low-rated contractor's bid of $700,000 would be multiplied by a weight factor of 1.2 with a "Management Discounted Bid" value of $840,000.

The "higher bid" by the best contractor computes as the best deal.

On the next page is a suggested contractor Performance Evaluation Worksheet.

In the Worksheet, if an accumulation of ratings under #1 Contractor Capabilities averages out as "Good," then that's a rating of 0.9. The weight of importance of that rating is 3. So the total Category Performance Value would be 0.9 x 3 = 2.7.

The weightings are from contractor performance worksheets used by 17 major corporations, as compiled by professor George S. Birrell, of the School of Building Construction at the University of Florida, and published in the June 1988 issue of Commercial/ Renovation magazine. If you disagree with the % weightings shown here, feel free to use your own.

The obvious value of this system is that it helps objectify the selection process and enhance the criteria well beyond the simplistic "low bid."

It has other benefits, not the least of which is that when contractors and subcontractors learn that their current work is being evaluated on this basis, they tend to improve.

CONTRACTOR PERFORMANCE EVALUATION WORKSHEET

Performance Levels
Category

Contractor Performance	Excel	Good	Satis	Poor	Terrible	Category	Performance
Evaluation Categories	0.8	0.9	1.0	1.1	1.2	% weight	Value
1. Contractor Capabilities	_____	_____	_____	_____	_____	X 3.0 =	_____
2. Overall Performance	_____	_____	_____	_____	_____	X 7.5 =	_____
3. Site Staff Management	_____	_____	_____	_____	_____	X 9.5 =	_____
4. Safety & Housekeeping	_____	_____	_____	_____	_____	X 9.0 =	_____
5. Labor Management	_____	_____	_____	_____	_____	X 10.0 =	_____
6. Cost Management	_____	_____	_____	_____	_____	X 4.5 =	_____
7. Time Management	_____	_____	_____	_____	_____	X 12.0 =	_____
8. Claims by/against Contr.	_____	_____	_____	_____	_____	X 3.0 =	_____
9. Cooperation w/Owner	_____	_____	_____	_____	_____	X 6.5 =	_____
10. Cooperation w/ Other Contractors	_____	_____	_____	_____	_____	X 6.0 =	_____

TOTAL VALUE OF
PROJECT PERFORMANCE _____

To quote Professor Birrell:

"The system is a method of moving the contractor bid evaluation from subjectivity to objectivity. Also, it can be used both by owners with a long history of development, and those with a limited background. The system allows an owner to use both past performance and bid value as criteria for awarding contractors.

"Furthermore, an owner can expect improved performance by current contractors when the system is implemented, as was evidenced by the reaction when a large U.S. utility began to implement its own approximate system of objective, written contractor performance evaluation. The contractors realized that their current performance on projects for that owner would be a factor in future bid evaluation, i.e., their competitiveness for future work. As a result, the quality of their performance took a clearly observable upward jump."

APPROVING PROPOSED SUBSTITUTIONS DURING BIDDING

Contractors can often improve on your drawings and specifications. They may have access to non-specified products that are cheaper and as good as, or better, than items you're familiar with. So it's reasonable to use the "or equal" phrase in specs to allow for that likelihood.

On the other hand, "or equal" opens you up to chiseling. Low-ball bidders specialize in coming up with low-quality substitutions. It's during the latter phases of bidding, and during construction that ill-informed, flawed, last-minute decisions are likeliest to be made by A/E staff.

The "or equal" risk factor goes way up if you, your client, or any of your staff accept bidder substitutions that the other bidders don't know about. At that point, you're properly subject to lawsuit by contractors who didn't have a chance to offer similar substitutions and bid on the same basis.

Some top design firms enforce a procedure that cuts through the dilemma nicely. The key is a special section on "Approvals" in the GENERAL CONDITIONS of the contract.

It says the following:

1) Your specifications name the materials, products, and manufacturers required for the job.

2) Bidders may submit other materials, products, etc. for consideration as substitutions.

3) Submittals of proposed substitutions have to be:
--In writing.

--Received ten (or other suitable number of) days before the bid opening.

--Made in good faith, that is, as verifiably equal or superior to the specified items.

4) Submittals have to include all the data that would be in construction drawings and specifications:

--Complete names and descriptions.

--Dimensions.

--Performance figures.

--Latest catalogue numbers.

5) If new materials are named, data is provided on required laboratory tests, standards, etc.

6) If a new fabricator is named, data is required on capabilities and experience.

7) If the design firm approves a bidder substitution submittal, copies of the submittal will be delivered to all other bidders so that all bids reflect the same options.

NOTE:

For more on how to handle changes and proposals for changes, see the section on SHOP DRAWING CHECKING AND COORDINATION in the chapter on **CHECKING THE DRAWINGS.**

CHAPTER TWENTY ONE

HOW TO REUSE EVERYTHING YOU DO: THE DATABASE AS A LIFE-LONG ASSET

TURNING YOUR LIFE'S WORK INTO A LIFELONG ASSET

STANDARDIZED ROOM DESIGN -- A BOON TO PRODUCTION

WHAT ARCHITECT'S SAY ABOUT ROOM DESIGN STANDARDS

A SAMPLE SET OF ROOM DESIGN STANDARDS

TURNING YOUR LIFE'S WORK INTO A LIFELONG ASSET

A/E's face the same two choices as they start every new project:

1) **Keep throwing their life's work away.**

2) **Or start turning their work into permanent, life-long assets.**

Most architects and engineers continue on track number one.

They work through a job for months or years, put the finished drawings in storage, and start all over with the next project. Time and again, up to 50% of the design decisions, managerial actions, technical research, drawing components, and even half the errors, are repeats from previous projects. The work is repeated, but the results aren't reused.

"Turning your work into life-long assets" sounds great, but how do you actually do it?

We're learning from those who do it at least in part -- in marketing, for example, or design, production, project management, and office administration. No one does it all, but here are some hints of what's possible:

--A Bay Area firm made a radical cut in jobsite problems by introducing a simple list of procedures for reviewing contractor-proposed changes and substitutions. The list forces the users through step-by-step analysis of a proposed substitution, and helps prevent hasty, ill-considered decisions.

--The same office put an end to budget-busting designs years earlier with a checklist of cost constraints their designers had to follow. Several categories of building types were listed, along with acceptable materials and cost ranges for building systems, fixtures, and equipment.

That list represented a condensation of years of experience -- the same experience available to any office, but rarely put in writing.

Both examples illustrate the same point: You can compile isolated expertise within an office and make it accessible to everyone. That gives you two big advantages: it multiplies the value of past project time by allowing you to reuse instead of redo previous work, and it saves time while reducing errors in new projects.

Some further examples from database-oriented offices:

--When one person discovers something good or bad about a building product, everyone gets the word via an office's private version of Sweet's Catalog. Most firms have no formal method for sharing such information.

--When a project manager discovers a short cut in the permits process at City Hall, that becomes part of the office's codes and permits checklist. In many firms, such helpful information is not only not shared, the discoverers tend to treat it as private, privileged information.

--When A/E's find construction or design errors at the jobsite, they take photos and write Jobsite Feedback Memos that everyone can learn from. In most offices, only those responsible for a problem learn about such things. That leaves everyone else sufficiently ignorant to repeat the same mistakes.

The examples reveal a near-universal problem -- along with an opportunity. The problem is the insular nature of most design offices. Individuals and teams have great latitude in the way they do their work. That's as it should be. But it also allows people to refuse to learn from others while keeping their own experiences and expertise to themselves.

One way to consolidate different types of information created by staff members is the OPERATIONS MANUAL.

Almost all well-established A/E firms have created Operations Manuals to guide project staff members on what has to be done and when, in all aspects of work. The better manuals go further than just providing checklists of action; they have space for people to write in what works well or doesn't work during a project.

Here's how that's done:

The project manager makes a new photocopy of the Operations Manual for each job and pulls out various sections as they're needed. It might be time to use a Working Drawing Coordination Checklist, for example. While using those pages to do coordination checks, he or she writes in notes on where the checklist items seem in error, over-done, or incomplete. That information later goes back to whomever is in charge, and the Operations Manual data is reviewed and revised accordingly.

These preceding steps reveal two essential points about creating and managing a successful office database:

1) Someone has to be in charge.

2) The database has to incorporate a feedback system for updating and improving the information.

Checklists and operations manuals wither and die if they aren't improved upon by the experience of the users. Every page of an office procedural checklist should provide space for job identification, dates of use, and especially space for "comments" in which positives, negatives, and suggestions for improvement are recorded.

Database systems become useless if no one has final responsibility and authority to screen and upgrade the design, technical, and managerial data that everyone in the office is to share. That person shouldn't make all the decisions; it's partly an office-wide responsibility too. But one person has to be in charge of making sure the database information is available to all, is being used, and is being improved with feedback notes by the users. The job title can be Production Manager, Quality Control Manager, Technical Director -- the exact words don't matter, just so the job is identified, given a title, and treated seriously.

Here are steps for starting an Operations Manual:

--List office work functions to establish the overall categories you'll be dealing with, such as Programming and Predesign, Design, Production, Contract Administration, Marketing, etc. Decide which to pursue first.

--Have those who perform the work functions list the steps involved -- especially the steps of checking and preventing errors, and of coordinating one's work with others.

--Set mutually agreed upon deadlines for the preliminary checklists. Assemble lists for editing and input; then give the revised lists back to people to use and mark up with additional feedback.

--Encourage people to suggest new procedures. Make it clear you're open to good practices from whatever source.

--Design the feedback system and a schedule for reviewing and utilizing feedback data. It's the only way to keep the Operations Manual alive and ever improving.

--Go for the long haul. This isn't a one-shot project; creating and improving the office database will be a permanent part of office operations.

Think in terms of creating an office database that's so useful, it could be sold to other firms. Think of what kind of information you would be happy to buy from other offices. Then, as per a suggestion from Tucson architect, Frank Mascia, you'll be creating a tangible office asset that can have a hefty price tag on it when it's time for ownership transition. Add those dollars to the savings you accumulated in the process of creating the asset, and you truly come out well ahead of the game.

STANDARDIZED ROOM DESIGN -- A BOON TO PRODUCTION

Every day thousands of drafters and designers repeat essentially the same design research and decision making about rooms in buildings.

They look in the same handbooks, weigh the same considerations, **and come to the same conclusions.**

Since they don't file the data to reuse again, they and others in their firms will repeat the same work on project after project, year after year.

This amounts to 20% to 30% of design and drafting time. Nearly all that time can be saved by using standardized room designs.

"Standardized room design" doesn't mean making rooms that look alike. "Standardized room design" means recognizing the simple fact that design standards for similar rooms in similar buildings are virtually identical. A corporate meeting room in Anchorage is virtually the same as one in Miami; a new Maine school classroom has essentially the same requirements as one in Los Angeles.

Each such room has the same range of commonalities in materials, finishes, substructures, door frames, door types, hardware, cabinets, fixtures, and other accessories.

Engineering requirements are about the same for each room type, too: Lighting, electrical outlets, air changes, plumbing. Building code requirements are very similar for fireproofing, exits, and alarms.

Differences that exist are mainly in cost. A first-class hotel suite will differ from a family motel room, but first-class hotels have essentially the same quality of materials in most any city.

So there are common design standards for every room type with variations based mainly on cost and sometimes on regional custom.

All of this has major time and cost saving implications for any office willing to take advantage of the situation.

Most of our readers have, or certainly know about, standard detail files. Now what would it mean to apply that concept to room types?

Suppose you're doing an upscale condominium and you have a list of room types as part of your program. Rooms might include garage, elevator lobby, laundry, residence rooms, etc. Each room type would be in a file folder, or on a computer, just like a standard detail.

You look up a room type, such as "Kitchen, Residential--Cost Category A," and there would be all the basic design data for that room: Finish schedule, fixtures and cabinet heights, electrical, a list of construction details by file number, plus specification sections that pertain only to that space.

If you specialize in complex building types, such as hospitals, a single room may have fifty or more special attributes on the Room Design Standard sheet.

Whether the buildings are simple or complex, you and your drafters and designers are getting all that reference information from one source in one shot. Your own customized in-house database.

Normally the research required to set stand-ards for a single room might come from different sources: GRAPHIC STANDARDS or TIMESAVER STANDARDS, the building code, SWEETS, specialized handbooks for the building type in question, and one or more previous projects.

And, normally, the finish schedule, detailing, and specifications pertaining to the room would be figured out at different times by different people. **Their research and decisions will often duplicate or contradict what's already been done by others.**

269

The difference is that this time the information that's researched and decided for a project doesn't disappear in file drawers.

It's organized according to a systematic format which expedites the job at hand, and the format becomes a standard -- the ROOM DESIGN STANDARDS reference sheet.

Here's a checklist for compiling room data:

INTERIORS

__ Floors & subfloor
__ Fixtures
__ Counters
__ Cabinets

__ Walls & Substructure
__ Ceiling & Substructure
__ Fire ratings

__ Acoustic ratings
__ NC
__ STC

DOORS & WINDOWS
Recommended
Minimum

__ Doors
__ Frames
__ Hardware
__ Related Fixtures

FIXTURES/EQUIPMENT
Recommended
Minimum

PLUMBING
Recommended
Minimum

HVAC
Recommended
Minimum

ELECTRICAL POWER
Recommended
Minimum

LIGHTING
Recommended
Minimum

__ Lamps
__ Switching

ILLUMINATION
Recommended
Minimum

__ F.C. (& other critera)

Whenever the issue of creating an office design database comes up, the question always arises:

Who has time to do this?

The answer is every office has the time because every office creates and recreates the same information for every project that goes through the office. All that's necessary is to categorize it and file it in an easily retrievable fashion.

WHAT ARCHITECTS SAY ABOUT ROOM DESIGN STANDARDS

Most rooms are highly standardized but most design offices treat each one as if it were the first born on earth.

The duplication of effort is staggering. For every room in every building they do, designers and drafters look in the same handbooks, check the same codes, ask the same questions and get the same answers time after time.

"Since they don't file the data to reuse again, they and others in their firms will repeat the same work on project after project, year after year. . . . This amounts to 20% to 30% of design and drafting time. Most of that time can be saved by using standardized room design data."

Architect Helen Grathaway disagreed with me at one of my workshops: *"I run a one-woman show and I don't need to make a whole file. I just remember it."*

What she means is that she does file the data for reuse but it's in her head and nobody else can use it. For the slightly larger office -- two, three and more people, the data in the boss's head doesn't do that much good. That's when every new staff member starts re-inventing everything they do.

"I kept a room design standards data base for years," retired architect Ray Jones told a University of Wisconsin seminar several years ago. He used standardized finish and fixture schedules for the homes, schools, and clinic buildings he specialized in.

One day in the mid-60's Jones had noticed that the finish and door schedules of two buildings he had done five years apart were identical. He thought there was a mistake -- the wrong prints mixed with the wrong jobs. How could two buildings designed independently have exactly the same finish and door schedules?

"I was so shocked; you could have knocked me over with a feather. I didn't believe it. But the more I looked, the more repetition I found throughout every one of my building types. So I smartened up and started doing what you're always talking about . . . ."

Here's the kicker:

If you establish a room design data base, you can link it to your standard details, standard notation, and standard specifications. That means that when you name a room, you'll automatically identify much or most of the working drawing content that goes with it.

The main difference in room materials, fittings, utilities, etc., is price. Some buildings or rooms are at top budget, and many are bottom of the line. That means that a room standards database has to be divided into a two or three tiered range of quality and cost. Beyond that division, the organization of the data is fairly simple.

The examples that follow are from the Restaurants chapter in the book ARCHITECT'S ROOM DESIGN DATA HANDBOOK*

The book is useful, but you can create your own standards. Especially if you have a smaller practice and specialize in only a few building types

*ARCHITECT'S ROOM DESIGN DATA HANDBOOK, Fred Stitt. Published 1992 by Van Nostrand Reinhold.

DESIGN STANDARDS -- RESTAURANT DINING ROOM

GENERAL	Recommended Standards	Minimum Standards
__ Floor Area	__ Recommended seating areas for dining: __ 4 diners = 130 sq. ft. __ 6 diners = 160 sq. ft. __ 8 diners = 180 sq. ft. __ 10 - 12 diners = 248 sq. ft.	__ Minimum floor space for: __ 4 diners = 90 sq. ft. __ 6 diners = 150 sq. ft. __ 8 diners = 172 sq. ft. __ 10 - 12 diners = 210 sq. ft.
__ Minimum Widths and Clearances	__ Width per place setting = 32" __ Aisle access behind chairs = 36" __ Service aisle = 40" - 48"	__ Minimum width per place setting = 28" __ Minimum aisle access behind chairs = 24" __ Minimum service aisle = 36"
__ Ceiling Height	__ 10' - 14' __ 7'-6" soffits OK	__ 9' __ 7' soffits
__ Special Plan Considerations	__ 4-person tables set at 45° provides maximum seating room and greatest planning flexibility (36" table, 18" seats, 18" aisle per module)	

FINISHES	Cost Categories A	B	C
__ Floor	__ Carpet, glued, non-stain synthetic, static proof (wool not recommended) __ Non-slip ceramic tile, stone tile, or quarry tile __ Hardwood, 3/4" oak parquet or strip, Prime or #1 grade	__ Carpet, glued, non-stain __ Hardwood, standard grade	__ As per Cost Category B __ Vinyl floor OK in family convenience restaurants
__ Base	__ Hardwood __ As per floor	__ Hardwood	__ Hardwood, paint grade __ Vinyl
__ Cabinets, Casework	__ Wood, Premium grade casework	__ Wood, Custom grade casework	__ Wood, paint grade
__ Counter tops	__ Wood, Premium Grade __ Polished and sealed stone __ Ceramic tile __ Plastic laminate, HPDL grade	__ Wood, Custom Grade __ Plastic laminate	__ Plastic laminate
__ Wainscot	__ Hardwood, Premium grade __ Plaster __ Fabric or heavy duty decorative wall cover	__ Hardwood, Custom grade __ Fabric or medium weight decorative wall cover __ Gypsum board, painted	__ Gypsum board, painted
__ Walls	__ Hardwood, Premium grade __ Fabric or heavy duty decorative wall cover __ Plaster __ Gypsum board	__ Hardwood, Custom grade __ Fabric or medium weight decorative wall cover	__ Gypsum board, painted
__ Soffits & Ceiling	__ Wood paneling, exposed beams, Premium Grade __ Plaster __ Gypsum board	__ Gypsum board, painted __ Plywood paneling	__ Gypsum board, painted
__ Acoustics	__ 35-40 STC if sound protection between dining space and adjacent rooms is required	__ 25-35 STC if sound protection between dining space and adjacent rooms is required	__ No minimum STC requirement or recommendation

DESIGN STANDARDS -- RESTAURANT DINING ROOM

DOORS & WINDOWS	Recommended	Minimum
__ **Doors**	__ 3'-0" to 3'-6" x 1-3/4" SC premium to #2 grade, 7 ply __ Traffic door: double acting with push plate and view panel __ Solid core, or SC with gaskets if sound isolation required	__ 3'-0" x 1-3/8" HC #2 grade
__ **Frames**	__ Wood -- custom or prehung	__ Wood, prehung
__ **Hardware**	__ Knobs/handles -- brass, stainless steel __ Kickplates -- brass, stainless steel	__ Brass plate or chrome __ Plated metal, standard grade
__ **Window light**	__ Equal to 25% to 30% of floor area or more	__ Equal to minimum of 20% of floor area

FIXTURES	Recommended	Minimum
__ **Built-in**	__ Picture hanging rail for artwork display __ Shelving or cabinets for display __ Cashier's station __ Maitre d' station __ Wait stations with storage, water, and ice supply	__ No minimum requirements

FURNISHINGS		
__ **Space allowances for built-ins**	__ Dining booths = 5'-6" long x 3'-6" deep __ Counter seating = 20" deep x 24" per diner with 4'-6" for work aisle behind counter __ Built-in bench seating for booth and waiting area = 24" x 24" per person	__ Dining booths = 5'-6" long x 4' deep __ Counter seating = 18" deep x 20" per diner with 4' for work aisle behind counter __ Built-in bench seating for booth and waiting area = 18" deep x 20"wide per person

PLUMBING	Recommended	Minimum
	__ Supply to ice machine and water station sink __ 1-1/2" drains at water station drains __ Sprinklers, concealed __ Fire hose cabinets as per fire code	__ As per Recommended

DESIGN STANDARDS -- RESTAURANT DINING ROOM

HVAC	Recommended	Minimum
__ Occupancy	__ As per program and restaurant consultant -- density of seating will depend on restaurant type and turnover rate	__ As per program
__ Heating	__ Uniform 70° F heat at all heights	__ 70° F heat at point 3'-0" above floor
__ Ventilation WARNING: Drafts near air supply and exhausts, kitchen door, and entry are a common cause of diner discomfort and complaints	__ 25 - 35 c.f. per minute __ Drafts must be avoided at seating height __ Exhaust fan system in kitchen must prevent odors from reaching dining room	__ 20 c.f. per minute
__ HVAC System & Controls	__ Forced air __ Floor conducted radiant heat __ Thermostat, zone or room controls	__ Forced air __ Radiator or fan coil unit __ Central thermostat and or manual controls at radiator or fan coil units

ELECTRICAL POWER	Recommended	Minimum

__ Power Outlets	__ Grounded outlets:
	  __ As required for janitorial maintenance
	  __ Strip outlets at wainscot height at side board and/or buffet for food warmers, other appliances
	  __ At wait station
	  __ At maitre d' station for podium light
	  __ At cashier station
	  __ For artwork
	__ All outlets with ground fault protection
		__ Grounded duplex base outlets at 10' centers
		__ Outlets at sideboards and wait station locations
		__ Ground fault interruption protected outlets

__ Circuts	__ Separate circuits for appliances rated at 20 amps or more	

LIGHTING	Recommended	Minimum
__ Lamps	__ Downlights or accent lights centered over dining tables __ Recessed incandescent __ Track lights or wall wash lights for artwork	__ Central recessed or surface mounted fixture
__ Switching	__ 3- or 4-way switch at each room access __ Dimmer for chandeliers __ Dimmers for hanging or recessed fixtures, and/or wall wash lights	__ Switches at main entry __ Some dimmer control is very desirable

ILLUMINATION	Recommended	Minimum
	__ 5 to 30 f.c. adjustable lighting	__ 30 f.c. general light

COMMUNICATIONS	Recommended	Minimum
__ Communications	__ Smoke and fire alarms, concealed __ Phone and intercom communication at cashier __ Public phone near waiting area or restrooms __ Intercoms at wait and maitre d' stations for kitchen communication	__ Smoke and fire alarms __ Phone and intercom at cashier station __ Public phone near waiting area or restrooms

CHAPTER TWENTY TWO

MANAGEMENT
BY CHECKLIST

MANAGEMENT BY CHECKLIST

CHECKLISTS, THE KEY TO OFFICE DOCUMENTATION

SAMPLE PAGES FROM PROJECT MANAGEMENT CHECKLISTS:

OUTLINE OF CONTENTS
THE PROJECT MANAGEMENT CHECKLIST

SAMPLE PAGE
THE PROJECT MANAGEMENT CHECKLIST

OUTLINE OF CONTENTS
WORKING DRAWINGS CHECKLIST

SAMPLE PAGE
WORKING DRAWINGS CHECKLIST

OUTLINE OF CONTENTS
CONSTRUCTION ADMINISTRATION CHECKLIST

SAMPLE PAGE
CONSTRUCTION ADMINISTRATION CHECKLIST

MANAGEMENT BY CHECKLIST

"All I need every morning is a list that tells me what I'm supposed to do next. That's all I need for my employees, too."

That was the direct and simple challenge an architect laid down before a panel of leading CADD company representatives in 1985. We're still waiting for their response.

Another architect told an AIA convention panel, *"I haven't gotten my money's worth out of my computer yet, but I'd be happy if it would do one thing: Once we're into working drawings, I'd like to have some way of getting a list of what's to be specified and what's to be detailed without having to go over all the drawings. That shouldn't be so hard, should it?"*

And a young Oklahoma firm threw a challenge at us last year to: *" . . . come up with a way to track and coordinate all the decisions and changes throughout a project, but with just one simple checklist, not the three or four you sell now."*

OK, let's see how to do all of the above, either manually or on computer. First, a restatement of the requirements.

We need a single, central control list of design decisions as to what materials and fixtures go into the job and, from that, a way of getting:

--Lists of instructions to drafters--what goes on each sheet of working drawings

--A list of what should be specified

--A list of what should be detailed

--A revision list to show what's changed along the way

--A list of items that have to be coordinated with other work

And it all has to be readily kept up-to-date and accessible by everyone on the job.

It all comes together if you use a Predesign & Planning checklist in a new way. When you use a design decision checklist, add an abbreviated note as to which drawings are affected by each decision, like "P & E" for "Plans and Elevations," "R" for "Roof Plan," etc. After the preliminary design decision go-arounds, have someone re-compile the lists for each type of working drawing sheet: Floor Plans, Elevations, Roof Plan, etc. As that's done, low and behold, **the list of decisions turns into lists of keynotes for drawings that haven't even been started yet.**

The Design Decision list becomes the Keynote List, Working Drawing Checklist, and an all-around Project Coordination Checklist. One list serves all these functions, and later, a few more . . .

Actually, the basic notation on any sheet of working drawings has always been a kind of checklist of what's to be detailed, specified, and coordinated. And as design decisions change over time, the new notes are the documentation of the changes. The obvious problem: Drawing notes that are scattered all over a drawing are unusable as a list. But when they're listed in a single column on the right hand-side of the drawing, presto: instant checklist.

Going back to the beginning, here's how to turn Design Checklists into Keynotes:

--First, fill out Predesign and Planning checklists for each phase of work (the Guidelines checklist or any list of your devising) to finalize decisions with your client: What paving to use? Wood or brick planters? Metal fence, chain link, or wood? What exterior wall finishes? What roofing? etc. Unless you have a preferable system for coding your decisions, **follow the CSI sequence and use CSI reference numbers with your checklists.** Then make reduced-size copies of the checklist and clip the copies to full-size working drawing mockup sheets, or to the working drawing tracing sheets, or plop them into your CADD drawing files.

--Use the checklists as the place to note the changes that are decided throughout the course of a project. Any items listed that relate to other drawings should be identified early on, so that a change in one will be matched by a change in the other.

--At the finish-up phase of working drawings, print out a final version of the Keynote list. It can be done as a printout stickyback to attach to traditional tracings, a paste-up or overlay component for reprographics, or plotted on CADD.

What goes onto the final Keynote Checklist? Four pieces of information for each item: Keynote number to reference the note to the broadscope drawing, the item name (usually a material or fixture), the item's complete CSI number, and the detail reference number, if any. **Like this:**

Number	&	Detail File Number	Reference
2.3	02515.1	`3-S1	Concrete Paving
2.5	02515.11	4-S1	Construction Joint
2.7	02519	.-- --	Gravel Surfacing
2.9	02447.11	1-S2	Metal Fence
2.11	02448.11	2-S2	Flag Pole

Assembly Notes to elaborate on the identification notes can be added in the final printing like this:

2.3 Concrete Paving 02515.1 3-S1
   Slope to drain

2.5 Construction Joint 02515.11 4-S1
   1/2" joints at 8'-0" o.c., unless
      dimensioned otherwise
   Include joints at all adjacent walls,
      curbs, or pavement

The single checklist used throughout the process accomplishes an amazing amount of coordination work; six tasks in six steps:

1) It's the record of design decisions and changes. (Dates and reasons for decisions can and should be recorded at the side of the original Design Decision checklist before it's reprint-ed in the final version prior to going to bid.). That saves misunderstandings and litigation later when people have memory lapses as to what was decided previously, when, why, and by whom.

2) It's a Working Drawing Checklist to tell drafters what has to be included on a drawing. Later, it's a reminder list for the drawing checker.

3) It's the Working Drawing Keynote list. Most notes on broadscope drawings are no more than simple identification, such as "Metal Fence." If more notation is needed, write it in on checkprints, and later input it for printout on the finished drawings as per the example on the preceding page.

4) As the Keynote list, it's also the reference guide to construction details. Any construction item that should be detailed is easily identified as the original list is checked off. If there's a "Metal Fence," then there should be a "Metal Fence" Detail.

5) No surprise: Since any item checklisted with a CSI number is specifiable, the keynote list is automatically a checklist for the spec writer, complete with CSI Masterformat five-digit reference numbers.

6) When construction nears completion, the keynote list becomes a ready-made construction punch list. (It can also be referenced to a more elaborate checklist, such as the Guidelines Construction Administration Manual or any other published construction checklist as needed for larger projects.)

CHECKLISTS, THE KEY TO OFFICE DOCUMENTATION

Sure, checklists can be bureaucratic and overdone. Sometimes the longer they get, the less information they have. And, in government circles and large corporations they can become mandatory interruptions to work instead of helpful tools. But at their best, there's no management tool that can save you more time and money with less investment than a concise, well thought out checklist.

Checklists are increasingly popular for liability protection. Besides helping prevent errors and oversights, they can be persuasive evidence to insurance companies and courts that you're being cautious and thoroughly professional in your work. If a project has problems and you have a detailed checklist showing that you did all the right things, the attorneys will look elsewhere for people to blame.

Clients like checklists, too. The most common client complaint is that their design professionals don't *listen,* and if they listen, they don't pay attention and follow through on what's said. Checklists which capture client needs and comments are **much** appreciated.

Design management checklists have gained hundreds of thousands of dollars in time and cost savings, liability protection, and higher fees, even for small offices. Larger firms can't imagine functioning without them. Here are some of the benefits and rules:

--Checklists are best when they serve multiple functions. For example:

1) Any checklist is first and foremost **a reminder**. People forget things when distracted by the interruptions of office life and need checklists and to-do lists to keep on track.

2) When a checklist is passed from one person or team to another, it's transformed into **a communications reminder** of what kinds of information people need from one another.

3) After managers check tasks that apply to project at hand, the reminder list becomes **a supervisory and instruction list** for those who implement the work.

4) As a project proceeds, those using the lists can note what's done and not done, what tasks are pending, and what they need to finish the pending tasks. It becomes **a communication system for staff to tell supervisors what they need to finish their work.**

5) At the close of each phase of work, it's **a reminder for supervisors who check the work:** What's done, what isn't done, and what's remaining.

6) At the close of a project, it's **a record of what was done, who did it, and when**--a documentation of the work that can make all the difference when disputes arise.

7) If the checklist is used actively along the way, that is, new points and improvements are added to it, it'll be **a guide for improving the office master operations database.**

8) **Someone must be in charge.** Typically this would be the same person who oversees quality control and/or production management. If nobody is in charge of office master checklists, users may go off on tangents, newcomers won't be informed about how to use the checklists, and the asset will lose value from erratic use or disuse.

--Don't let checklists get loaded down with trivia. That's where constant editing comes in. As staff and management suggest new items to add to a list, many such items will be redundant or too obvious to be of value. Sometimes little commands or admonitions are slipped into checklists which are insulting to staff and don't tell anything specific about the work to be done.

SAMPLES PAGES FROM CHECKLISTS:

PROJECT MANAGEMENT

WORKING DRAWINGS

CONSTRUCTION ADMINISTRATION

The pages that follow show sample contents and pages from three basic types of operating manual checklists.

The first is from a PROJECT MANAGEMENT CHECKLIST, task modules of all that has to be done to complete a construction project.

The second is from a WORKING DRAWING CHECKLIST, basically the steps and content of construction documents from start to finish.

The third is from CONSTRUCTION ADMINISTRATION, all the items that have to be checked in the course of construction.

Samples of other types of checklists such as PREDESIGN & PLANNING, ROOM DESIGN STANDARDS, and SPECIFICATIONS are shown elsewhere in this book.

OUTLINE OF CONTENTS
THE PROJECT MANAGEMENT CHECKLIST

The overall content list follows the AIA's SCOPE OF DESIGNATED SERVICES.

PHASES:

1) PREDESIGN

> MARKETING AND PRESENTATION MANAGEMENT
> FEASIBILITY AND FINANCIAL ANALYSIS
> PRECONTRACTUAL ADMINISTRATION
> PROJECT PLANNING
> PROGRAMMING AND PREDESIGN

2): SITE ANALYSIS

> PREDESIGN AND SCHEMATIC SITE REVIEW
> ENVIRONMENTAL IMPACT REPORT
> PERMITS AND APPROVALS

3) SCHEMATIC DESIGN

> CONSTRUCTION COST ESTIMATING
> SCHEMATIC DESIGN AND DOCUMENTATION
> BUILDING CODE AND FIRE CODE SEARCH

4) DESIGN DEVELOPMENT

5) CONSTRUCTION DOCUMENTS

> WORKING DRAWINGS
> CONSULTANT/ENGINEERING DRAWINGS CROSS-
>   COORDINATION CHECKLIST
> SPECIFICATION WRITING AND COORDINATION

6) PREBIDDING, BIDDING AND NEGOTIATIONS

7) CONSTRUCTION CONTRACT ADMINISTRATION
> SHOP DRAWING CHECKING AND COORDINATION

8) POSTCONSTRUCTION ADMINISTRATION

9) LONG RANG MARKETING PLANNING

SAMPLE PAGE
THE PROJECT MANAGEMENT CHECKLIST

FROM PHASE 1, PREDESIGN

PROGRAMMING -- OCCUPANCY NEEDS AND SPATIAL ALLOCATION

Project Name/No: Notes by:

Dates Checked:

Checkmark each item to be done and cross out the check when completed. Mark with a -- if an item is not to be done.
If an item is in doubt, mark with question mark and add a note of what to do to resolve the question. By:

____ Establish criteria for importance of room functions and relationships, and create a
        User Questionnaire.

    ____ Volume of traffic.
    ____ Frequency of interaction.
    ____ Relative value or cost of the interactions or personnel.

____ Create a Departmental Spatial Interaction Matrix (list of departments that shows their   relationship
        to other departments).

____ Create room by room spatial interaction diagrams showing all room relationships.

____ Identify numerical ratings of the importance of relationships of each room to other rooms.

____ Make link and node diagrams to show departmental and room relationships identified in the
        interaction matrices.

____ Make bubble diagrams showing spaces with relationships indicated and their importance
        rankings. Manipulate bubble diagrams until link crossovers (plan conflicts) are eliminated.

____ Create diagrammatic/schematic building plans.

____ Note relative spatial areas for all departments, rooms, mechanical, vertical transportation,
        service, exit stairs and corridors, and horizontal circulation.

____ Review program and predesign decisions with client.

(Note after each item: When due. When to review. Who is in charge. Information needed.)

OUTLINE OF CONTENTS
WORKING DRAWINGS CHECKLIST

SAMPLE PAGE
WORKING
DRAWINGS
CHECKLIST

SITE DRAINAGE

___   FRENCH DRAIN   (02401)
    ___ materials   ___ dimensions   ___ detail keys   ___ notes/refs
    ___ slopes   ___ depths
    coord check: ___ civil/soil   ___ util   ___ paving   ___ landscape

___   PAVEMENT UNDERDRAIN   (02410)
    ___ materials   ___ dimensions   ___ detail keys   ___ notes/refs
    ___ slopes   ___ depths
    coord check: ___ civil/soil   ___ util   ___ paving   ___ landscape

___   TRENCH DRAIN   (02410)
    ___ materials   ___ dimensions   ___ detail keys   ___ notes/refs
    ___ slopes   ___ depths
    coord check: ___ civil/soil   ___ util   ___ paving   ___ landscape

___   SUBDRAINS   (02410)
    ___ materials   ___ dimensions   ___ detail keys   ___ notes/refs
    ___ slopes   ___ depths
    coord check: ___ civil/soil   ___ util   ___ paving   ___ landscape

___   STORM DRAINS   (02420)
    ___ materials   ___ dimensions   ___ detail keys   ___ notes/refs
    ___ slopes   ___ depths
    coord check: ___ civil/soil   ___ util   ___ paving   ___ landscape

___   AREA DRAINS   (02420)
    ___ materials   ___ dimensions   ___ detail keys   ___ notes/refs
    ___ slopes   ___ depths
    coord check: ___ civil/soil   ___ util   ___ paving   ___ landscape

___   DRAINAGE FLUME/SPILLWAY   (02420)
    ___ materials   ___ dimensions   ___ detail keys   ___ notes/refs
    ___ slopes   ___ depths
    coord check: ___ civil/soil   ___ util   ___ paving   ___ landscape

___   SLOTTED DRAIN PIPE AND SLOT   (02420)
    ___ materials   ___ dimensions   ___ detail keys   ___ notes/refs
    ___ slopes   ___ depths
    coord check: ___ civil/soil   ___ util   ___ paving   ___ landscape

___   ___   DRAIN AT PAVING EDGE   (02420)
    ___ materials   ___ dimensions   ___ detail keys   ___ notes/refs
    ___ slopes   ___ depths

OUTLINE OF CONTENTS
CONSTRUCTION ADMINISTRATION CHECKLIST

SITEWORK

   PREPARATION & SUBSURFACE

   PAVING & SURFACING

   FINISH LANDSCAPING

STRUCTURE

   CONCRETE

   STEEL FRAMING

   WOOD FRAMING

EXTERIOR CONSTRUCTION

   BELOW GRADE WATERPROOFING
   ROOFING

   MEMBRANE ROOFING

   FLASHING

   WOOD SHINGLES AND SHAKES

   ASPHALT, SLATE, TILE SHINGLES

   MASONRY WALL CONSTRUCTION

   WALL JOINT SEALANTS & FLASHING

   WINDOWS & GLAZING

INTERIOR CONSTRUCTION

   SUSPENDED CEILINGS

   INTERIOR FINISHES

FURRING AND LATHING

PLASTERING

GYPSUM WALLBOARD

TILE

FLOORING
   TERRAZZO
   WOOD FLOORING
   ASPHALT, VINYL,
   RESILIENT FLOORING
   CARPETING

DOORS

HARDWARE

FINISH CARPENTRY

FURNISHINGS AND FIXTURES

PAINTING

SAMPLE PAGE
CONSTRUCTION ADMINISTRATION CHECKLIST

SITEWORK CONSTRUCTION - PREPARATION & SUBSURFACE

PRECONSTRUCTION -- SURVEY CHECK   (01330)

Project: Dates: By:

____   The site survey has been checked for errors and corrected.

____   The site surveyor's monument, stake, and flag locations are verified to be correct.

____   Setbacks and easements are verified.

____   Excavation danger points, such as at buried gas and phone lines, have been identified.

Notes:

PRECONSTRUCTION -- SOIL TESTS & SITEWORK PREPARATION   (02010)

____   Test pits, borings, and tests are as required by local regulations.

____   Additional test pits, borings, and tests are provided at questionable locations.

____   Requirements are being met by the contractor to secure necessary sitework and general construction permits and approvals.

____   Owners of adjacent properties have been notified of impending sitework.

____   Public agencies and utility companies have been notified of impending sitework as required.

____   Soil investigation and tests have been conducted as specified.

____   Site photos have been taken, dated, and keyed on a site map.

____   Underground utility lines, pipe, cable, and conduits are identified and clearly marked on site.

____   The contractor is providing construction barricades and signs for public protection as specified.

____   The contractor is providing temporary structures such as scaffolds, platforms, canopies, etc. as specified and with permits required by local authorities.

____   A preconstruction meeting is held with all concerned parties to review sitework preparation, construction, and coordination between subcontractors.

(Note after each item: OK or not. When to recheck. Who is responsible. Information needed.)

CHAPTER TWENTY THREE

TIME MANAGEMENT: HOW TO DOUBLE YOUR PERSONAL PRODUCTIVITY

HOW TO GIVE YOURSELF A HEFTY RAISE THIS YEAR

Chances are that by the end of twelve months you can earn 20% more income than you are right now. That's with some minor changes in what you do. With a little serious additional effort, you can earn 50% to 100% more.

If it's not money you're interested in, the same rule applies to time. Instead of more money, you can increase your discretionary time. Or increase some of both.

Here's how and why:

First of all, you already earn more money per hour than you realize; it just depends on the hour. Suppose your actual net personal income is $20 per hour. That $20 is just an average of your least, and most, productive time. Sometimes you're worth a lot more and other times you're worth a lot less. When you do your best work, chances are your time is worth $80 per hour or more, and when you're doing your least important work it may be worth around $5 per hour or less.

The formula is based on the old, reliable, 80/20 rule.

You get about 80% of your return from 20% of your time . . .
. . . and 20% of value from the remaining 80% of time.

Translated into dollars that would mean that in a 40-hour week where you earn $800 (at $20 per hour), about 80% of those hours are giving you 20% of return -- 32 hours are earning a total of $160. That's $5 per hour. The most productive 20% of your time, 8 hours worth, would be earning 80% of your return or $640. That adds up to $80 per hour.

If you consider this from the standpoint of billable time, and if you're billed at 2.5 times direct hourly wage, then the values increase: a little over $12 for least productive time to $200 for top time.

The "80/20" rule is a simplistic generalization: how realistic is it?

Over the years we've seen hundreds of talented professionals throw away considerably more than 80% of their work time on minimal activities. And we've seen hundreds more with less education or skill become far more productive, not by doing anything exceptionally well, but simply by the way they redirect their hours towards the most productive tasks.

"Redirecting time" on the one hand means dropping the time wasters, like opening one's own mail or habitually blowing the afternoon by overeating at lunch. But there's another, more crucial side to it. That's the side where you find what you do best, find out how many hours you spend on it daily, and double the amount of time you spend on that activity.

If you spend 20% of your daily time getting 80% of your work value, that 20% represents about an hour and a half out of an eight-hour work day. **If you increase your most productive time to three hours daily, you've doubled the net value of your day.**

It's the best deal in town.

Some argue: *"I work for others. I can't control my own time."*

Everybody "works for others" and everybody has to think out the logistical problems that come up when they try to redirect their time.

--Some people realize they can gain discretionary time by coming to work an hour early. (They may gain a bonus of added work time by reducing commute time as well.)

--Some change their lunch hour to 1 p.m. They get an hour of quiet time at noon when everyone else is out of the office.

Equally important to revamping the schedule: Redirecting discretionary time to exploit one's best talents.

The 80/20 rule again:

80% of a person's rewards come from 20% of his or her highest-level abilities. Maybe it's negotiating with people, maybe finance, maybe design, or inventive engineering.

For most design professionals, their highest value time is problem solving and the more of it they apply to higher and higher level problems, the better.

Lining yourself up for your raise:

1)   Apply the 80/20 rule to your income to estimate what your top earning power is when doing your highest value work. Calculate it in dollars per hour.

2)   Estimate how much time you spend per day doing your highest value work. Calculate how much more time you'd have to spend daily on top-level work to raise your true productive value by 20%, 50%, or 100%.

3)   Think out the logistics of modifying your daily schedule to enlarge your discretionary and highest value work time, such as changing commute time or lunch time, working at home one day a week, using a daily "quiet hour," staying late, getting an assistant, etc.

If your best time is worth $25, $60, $150 or more per hour, start treating all your time as the high-value commodity it is. If you treat it with that kind of care, you'll see the rewards that are possible and gain motivation to cut the time wasters and enlarge discretionary high-value time all the sooner.

GETTING TO THE HEART OF PERSONAL PRODUCTIVITY . . .
LIKE: WHAT ARE YOU DOING HERE IN THE FIRST PLACE?

OK, here you are reading this book to help you do your job better.

But what's the point of doing your job better? Personal growth? More money? Better service to clients? Sure, but what else?

You have to be satisfied that your work is truly worth something in order to give full energy and attention to productivity. To be satisfied, you need to know "what else" your work accomplishes.

This is why the single most important step in time management is to identify your personal "Primaries." Primaries are the prime motivators, the things you want to do more than anything else.

These also happen to be the things most people put off the most, which is why a lot of people become dissatisfied with life and work. They take it for granted that they have to put off things that are most interesting and important to them to get other, lesser things out of the way. Then the lesser things dominate daily time and attention for years, and life dreams fade away.

Primaries often change too, without conscious realization. You have to sort them out periodically to make sure you haven't veered way off track, or check that your values are really what you think they are. It may turn out that you're doing more of what you want to be doing than you realize.

Checking up on Primaries is a true low-cost energizer. It only takes about 20 minutes.

Through this process you develop or update your long-range plans. From the long-range plans, you develop short-term plans and tactics.

Once those are established you'll gain control over details of life and work that most people leave to chance. Despite the extraordinary benefits to be gained, the idea of naming or renaming one's Primaries also arouses anxiety, and people tend to avoid it.

There are several reasons why people hide from the exercise of naming their Primaries.

First, Primaries -- the most satisfying and profitable things you do -- may conflict with other tasks that you have to spend most of your time on. Or they may conflict with other activities you've convinced yourself are really more urgent.

Meaningful examination of your Primaries may throw some parts of your present living situation into question. Perhaps you've taken a job for reasons that are obsolete. Now there's a vested position and a pension might be at stake. You may not like the job, but there are "other things" that tie you to it. Or you may sense that to really make the most of your talents you might have to go back to school. And that would take too long and create hardship for those you most care about. Or, to be most satisfied, maybe you should change jobs and locale and that might pit you against your mate's interests. That could be explosive, and the status quo may seem preferable to facing the conflict.

These are all universally common situations, and there's no reason to let them interfere with good time management. You can take another route that doesn't present threats but still tells you the Primaries you need to know about to be most effective in day-to-day planning.

One way to break the impasse is to assume that in lieu of definite, life-long Primaries, you can legitimately name tentative ones. And instead of seeking large-scope, long-range, life plan Primaries, you can seek short term ones that fit within the life and work structure you currently accept.

From that starting point, follow this exercise:

1) Grab a sheet of paper or your computer, and make a list of your main talents. The things you do best. The things you earn the most money for. The things you enjoy most. The things for which you've received the most recognition. Take up to five minutes and try for at least five main talents.

2) Check off the main talents you've listed that combine the most levels of reward: moral satisfaction, adventure, self-expression, recognition, money, family security, and any any other value you hold.

3) Do the next phase of the exercise in three steps. Think ahead 20 years, then 10 years, then 5 years, and visually imagine what you'd like to be doing most in your life. See yourself as if in a movie, living and working on a day-to-day basis. List the activities you see yourself doing and enjoying most in an ideal life situation. Be fanciful and idealistic. You can zero in on more practical aspects later. Spend about five minutes or more. Walk around and think about it if the images don't come clearly.

4) Now take a more serious look. If you knew you were going to die in one year, what would you want to spend your time doing in the meantime? List all the activities that come to mind. Allow five minutes or more. Don't censor yourself. If answers are slow to come, let yourself daydream.

5) Look for overlaps or conflicts in the various lists. Tasks you enjoy most may not match up with the things you're paid to do. You may have to plan on maximizing your earning now, to be able to enjoy some totally non-monetary activities later.

Hopefully you can combine it all -- find Primaries that will give you the best of all worlds and that are, or can be, a part of your daily life now. List the possibilities. Take five or ten minutes, and allow blank spots to be filled in later, if need be.

The net product you can expect from this 20-minute exercise: One, three, five or more specific personal Primaries. Among them will be multi-value activities that are **most** productive, gainful, and personally satisfying.

Now here's the other REAL point of this exercise.

It deals with the absolute fundamental principle of integrated time management. And that is to start to restructure your days so that **you double and redouble the time you spend on your top multi-value Primaries.** Most people spend barely 20% of their time on Primary-related activities. Many people spend far less. So, if you're spending what amounts to an average of one hour a day on your Primaries and double that to two hours, **you've doubled your most productive time.**

If you repeat the rescheduling process periodically, at least annually, you'll double and redouble your net productivity, your income, and your general life satisfaction. <u>Guaranteed</u>.

TIMESAVERS FROM A DESIGN PROFESSIONAL

A/E's often apply their problem solving skills to time management and office productivity with innovative results. For example, consider these tips from Toronto engineer and management consultant Sydney F. Love:

--When you prepare an agenda for distribution to participants prior to a meeting, put people's initials beside the items they'll particularly be responsible for. It's the best way to assure proper preparation and more intense participation.

--Similarly, when sending out a meeting's minutes, they should be action minutes, things people are to do, rather than just a listing of points covered during the meeting. (A meeting that closes without a clear direction of what people are to do next is not a successful meeting.) Have one or more initials listed beside each action decided on in the meeting. Then there's no misunderstanding as to who is responsible for implementation. If responsibility isn't clarified during the meeting, it can be afterwards and then included--action plus initials-- in the final minutes.

--Control the time you spend in appointments by negotiating objectives and time limits with the person you're seeing. If you've initiated the appointment, say: "Here's what I want to accomplish . . . ." Then state a reasonable time objective and negotiate it. "I'd like to wrap this up in 30 minutes. That seem right to you?"

Similarly, if someone else initiates the appointment, ask what his or her objective is, and then work out a time objective to match.

--Mr. Love applies the above technique to phone calls too. When calling, he has a 3x5 card in front of him with a check list of the items he wants to cover. He names his purpose, names a time frame and covers his points.

HOW TO GET THE UPPER HAND IN DEALING WITH TIME WASTERS

"There is no meaningful way to get the upper hand in our lives without controlling our expenditures of time. We may occasionally deal with inconsiderate people who are able to inconvenience us now and then by failing to keep appointments, by engaging us in extended conversations, or by a number of dawdling inefficiencies. But in examining the phenomenon of time-wasting, we are looking for the most part at self inflicted wounds."

In other words, if your time is being wasted, you're responsible for it.

Consultant and author Ralph Charell offers these thoughts for getting the upper hand:

--As a start, Charell suggests you go after your two or three worst time drains. Those are habitual acts which give you little or no return on the time you invest in them, like unused commuting time, random TV watching late in the day, or poking through trivial magazines.

Such habits are hard to break so the trick is not to try to quit some activities and just leave a vacuum but to substitute other, more rewarding, activities. **Instead of changing unproductive habits, crowd them out.**

Today's commuters, for example, make the most of it with phones for doing business, or educational tapes for listening, or pocket recorders for verbally sorting out ideas and plans for each day. Random readers change habit by stacking their most important possible reading on top of their magazine piles.

--Eliminate time wasters seemingly imposed by others by taking as much control as is open to you. For example, confirm ALL appointments, especially airlines, hotels, and car rentals. Reconfirm all appointments, meetings, and delivery times. The people you've communicated with the shortest time ago are the least likely to be late.

--When you deal with a stranger on the phone for reservations, services, deliveries, complaints, etc. always get their name and write it down (plus an extension number, if any). That will encourage them to expedite action. And it will save you time in tracking down results later on.

--To get what you want in troublesome situations, stay cool. We've all seen people having fits, banging on counters and yelling demands. People in the service industries have code words for the yellers and, while getting on the phone and pretending to "do everything possible," they don't. Just like you and I, they respond to courteous, clear statements of need . . .

When analyzing office operations over the years, this problem has always boiled down to issues of prioritization and delegation. The over-extended boss is using junior people to mock-up even minor design and technical problems and then trying to solve them all on his or her own. Instead of having staff create mockups that show the problems or test the bosses ideas for solutions, the boss needs to have staff create mockups of optional solutions. That extra dimension to work assignments can add the breather space the boss thinks he or she needs, will reduce the decision making time, and can lead to more and better use of more streamlined production methods.

--The independent staff and "old timers" problem. Most offices face this to some degree and most deal with it more or less successfully by providing the best in office tools for those who want to use them. Those who are most comfortable pushing pencils will switch to faster methods when pressed to do so by tough deadlines. After a few small-scale successes, they often change habits on their own. (Just be prepared to humor them when they discover the benefits of some system or another and then claim credit for inventing it.)

291

ANOTHER FORM OF INTERRUPTION PREVENTION

The American Management Association made a remarkable discovery about work emergencies and interruptions several years ago.

They did a survey of executive time and found that those who didn't systematically plan and monitor their daily time spent 14% of their work hours on emergencies and interruptions.

Those who did work with daily work plans spent less than 5% of their work time on brush fire problems.

Fourteen percent of a fifty-hour week is seven hours and 5% is 2.5 hours -- quite a spread. Daily work plans, TO-DO lists, are the key to it.

Here are the best ideas we've found in the management world on how to get the most out of the daily TO-DO work plan:

--Plan objectives first; specific tasks second. What one objective do you want most to accomplish by the end of the day? Let that answer guide your specific secondary tasks -- the tasks that support your primary objective.

--Daily "TO-DO" lists are less effective if they don't relate to a weekly plan. Weekly plans have little meaning without monthly, quarterly, yearly, etc. plans. If you don't have up-to-date long-term plans, now's the time to sit down and work them out. Your daily plan will be vastly more effective when it's set up within a long-range context.

--Set your watch or a kitchen timer to go off at least every hour, perhaps every half hour. That will raise consciousness about what you're doing at those moments relative to what you're supposed to be doing.

--Tempting as it is, don't crowd the TO-DO list. Most execs enter on their daily plans far more than they can accomplish in the day. Anyone is lucky to do two or three major discretionary tasks in a work day. Why plan so many that doing any one task seems like an interruption or a barrier to the others on the list?

--Do planning at the end of work days, as well as at the beginning, when what you've done is most fresh in memory. Then modify the plan first thing the following day.

--The earliest part of your work day or work period is likely to be the most under your control. Use that period for the most important business. Then you won't be so bothered when the rest of a day goes to pieces.

--Watch TO-DO lists for tasks that are frequently left over and carried to another day. If they can't be delegated or scratched, give them some special brainstorming. The chronically undone tasks indicate a barrier, a lack of motivation, or lack of data or resources. Schedule a special thinking time slot, time not to do the procrastinated tasks, but a time to figure out why they're not getting done.

GREATER PRODUCTIVITY WITH LESS WORK

At Guidelines we have one of the largest collections of information on time management and productivity in the world. That information includes the most common questions and problems, and the best answers and solutions as provided by top production managers and consultants over the years. Here are some examples:

Question. "Plain and simple, how can you get the most physical work done the fastest?"

Answer. Work--all work--consists of moving or changing things or parts of things. So, to get more done faster means **moving** things faster.

Many workers out-perform others for no other reason than that they simply move themselves around faster. Besides moving yourself faster, you want to move things, parts and tools faster. And move more of them at any one time. If you want to improve your output, or improve your work station, **examine everything from the standpoint of movement.** Then you'll see that some items should be moved closer together or closer to you. You'll see you need more powerful tools to augment motion. That's it, plain and simple.

Question. "We hear the word 'efficiency' all the time. Just exactly what does the word mean? Isn't there a danger in trying to be overly efficient?"

Answer. "Efficiency" has the same root as the word "effect" and originally pertained to bringing something about, to do, or make. In modern usage it also denotes doing or making without loss or waste. The key to doing things without loss or waste is to constantly think in terms of **simplicity, consolidation, reusability, or multiple use.** Moving the things you work with closer to yourself reduces or simplifies physical motion. You eliminate a waste of motion that you didn't perceive as wasted before.

Here's an example of consolidation:

Typically, there are three separate operations to document a meeting: the agenda for a meeting, the notes taken during a meeting, and the minutes of the meeting.

But if the agenda is designed as a checklist with space to note the results of each item as it's handled, then the agenda and notes can be photocopied at the end of a meeting as the summation of the event.

Most of your actions in work have the potential of consolidation or multiple purpose or value. You can often link them up with other actions.

Suppose you're making a trip and ask: "What other things can I accomplish at the same time?" If you ask that and start a list, you'll think of people you can see or call along the way, or special work or reading you can do while waiting in transit. If you pursue the questioning, dozens of items will come to mind. But you have to raise the question to get the ideas flowing.

Being "overly efficient" would mean being overly productive. We have yet to see such a thing. The true question centers around being overly demanding, or fussy about details, and the "right" way to do things. Some folks are very efficient at doing things that shouldn't be done at all.

Question. "What's the very best way to simplify a task?"

Answer. The best simplification is elimination. As just mentioned, some things shouldn't be done at all. Most tasks people set up can and should be eliminated. Most meetings don't need to be held. Most memos, letters and reports don't need to be written. And if they **do** need to be written, most words in them are irrelevant. When you get to higher levels of professional and management work, it may not be **most**, but certainly **many** tasks can be cut out with no problem.

A good exercise in this "elimination" context: Ask what major things that you do are most dispensable. Then dispense with them. Turn that time over to your most important activities.

Question. "Is there any particular tool or guide that's most important in time management?"

Answer. One in particular: Your plan-- the long-range plan that identifies what's most important to your life. Don't think a plan requires putting the most important things off as a distant goal. That's what planners often do, and it's deadly. A good plan gets you a maximum amount of what you want most, **right away.** Preferably **now.** Plan to be doing what's most valuable, profitable, enjoyable, etc. as a life style **this week or this year**--not 5 or 20 years from now.

Plans come in two styles: Strategic or long-range, and tactical. The most immediate form of the tactical plan is your daily schedule. That's the secondary tool that implements the primary tool of strategic planning. Without it, the long-range plan will get lost in a maze of trivial actions and sidetracks. Without it, you don't have a daily constant reminder of what you should be doing that's most important. Without it, you don't have a map for prioritizing, consolidating, simplifying, delegating, or eliminating the jumble of possible tasks you face each day. Any tool that would do this much for you should properly cost thousands of dollars a year. But the actual daily cost is a few minutes of your time and the price of a sheet of paper.

PUT YOUR GRAPHICS SKILLS TO WORK IN DATA GATHERING

When you take notes at lectures, meetings, and/or while reading, here are some tricks that can make them more effective.

--Label each portion of the notes as you go along. Use note paper with wide left-hand margins, and use the margin space to identify, in a few words, the substance of each note / paragraph written on the right-hand side. This will help you keep your mind on the gist of the data while you're getting it down. And it makes the notes much easier and faster to review later. Remember, the left-hand notes are like a topical index: just one or two words to identify each major piece of data that's noted.

--Enhance your notes even more by using drawings. Use an open notebook so that left-hand pages are exposed and open for sketching. Sketches are an addition to -- not a replacement for -- written notes.

--Diagrams, charts, and sketches are all ways to compile data in shorthand form. That's why they're used so often to convey information. When they're not used, you can make your own graphics and record information more efficiently than it's being presented.

--The sketching idea has another purpose. It's a way of taking advantage of our combination of "left brain" and "right brain" capacities. The left side of the brain learns and functions verbally, the right side prefers to work with general overall ideas, and images. It's best to get both hemispheres involved to assure that you gain the most comprehensive understanding of the data you're dealing with.

HOW TO KEEP MEETINGS FROM DESTROYING YOUR DAYS

As the saying goes: "Everyone knows half the time spent in meetings is wasted. The problem is in predicting which half."

Meetings start late, run too long, accomplish too little, are dominated by the wrong people, and are filled with trivia and side tracks.

Here's a list of the greatest tips of all time for cutting wasted time in meetings and getting results:

--Corporations that have studied time waste in meetings now commonly enforce two rules:

1) **Never allow more than six participants.** A group of six or more almost always splits off into smaller groups who get off in small conversations and mini-meetings within the larger meeting.

2) **Always start on time.** When you cater to stragglers, everyone will allow for that and start arriving later and later.

--**"Give structure to recurring situations,"** said William W. Doge of Avery Products Corp. Doge faced a time problem of dealing with weekly formal meetings plus frequent, unscheduled, one-on-one confabs with managers of ten different divisions. Meetings were often less than satisfactory.

An answer to the problem: A desk-top portfolio with ten compartments, one for each manager under his supervision. All memos, notes minutes, correspondence, etc. that came across his desk are placed into the appropriate compartments. **This pidgeonholing automatically compiles background data and an agenda for each future meeting.** When a meeting is on, Doge grabs the proper folder and is ready to go. Added benefit: Others who see that method in use will copy it, and themselves eliminate last-minute preparation time, yet arrive at meetings better prepared than ever before.

--**One of the great double-purpose time-savers: The Agenda to Minutes system.** In four parts:

1) **If calling a meeting, write a checklist outline of the contents with date, time, place, and SUBJECTS of the meeting.** Include a list of participants, and the estimated time limit.

2) **Name the problem/issues to be dealt with and the purpose of having the meeting.** (This little exercise will often bring to mind alternative solutions that will let you drop the meeting idea altogether.)

3) **List the topics to be covered as a checklist**, with lined spaces to note the conclusions. The point of most meeting topics is to identify who is going to do what and when. Write those decisions down or note the lack and postponement and next step when decisions aren't made.

4) **Have your agenda and the notes you added transcribed as a memo of the minutes of the meeting.**

--**Always rephrase or restate any question** someone asks you before you answer it. This serves three purposes:

1) **Everyone will hear the question** (usually a substantial number of participants will either not hear a question or will have been thinking about something else).

2) **When you restate a question, it allows the questioner to clarify** what he or she wants to know before you answer. That avoids giving an answer, then going through the: "Oh, that's not quite what I meant . . . ." interchange.

3) **The time you spend restating the question lets you think and sort out the answer a bit more carefully.** It makes for more thoughtful answers and speedier exchanges of data.

--The timing of meetings can save or cost time. Mid-morning Monday meetings will probably run all the way to noon.

Lunch time is popular among managers who need to exchange data. The advantages:

--Few or no interruptions since everyone else is out to lunch.

--Regular work hours aren't used up.

--There's a built-in deadline that helps keep such meetings on track and on time.

--Reciprocal benefit for staff and the office: Staff gets a free lunch, the office gets the benefit of extra work time.

--When people raise problems and objections:

1) Ask them to suggest a couple of solutions.

2) If they say it doesn't seem solvable. . . rephrase the problem . . usually it's the way a problem is stated that makes it seem beyond solution.

--Be aware that whenever somebody writes or calls you, chances are it's because they want something from you. Since people don't want to seem like "takers," they tend to beat around the bush before saying what it is they want from you. You can phrase your introductory hellos and remarks to make it easier and faster for them to get to the point. After saying hello, say: "Good to hear from you, what's on your mind?" or, get right to it and say: "What can I do for you." It's cordial and it gets the conversation right on track.

--The value of getting to the point quickly goes both ways. It saves everybody's time when you start right in with "I" statements: "I need to know . . . ." "I'd like to . . . ." "I want . . . ."

TIME SAVERS, TIME SAVERS EVERYWHERE

Design professionals have invented thousands of ways to do their work better. Most ideas arise as solutions to minor problems. But over a period of time, these small solutions prove to be very important time savers. Here are a few from some of the better run design firms:

--Folders of correspondence to be checked and signed often get misplaced or buried on desks. Best solution we've seen: Outgoing letters awaiting signatures are carried in odd-sized bright-color folders. There's no missing them, and no losing them.

--Don't let staff members pile work on the secretary without setting priorities. One architect's secretary set up two "IN" trays on her desk. One for "Priority Rush" and one for "Medium Rush." Some offices require all work going to secretaries to be market with #1, #2, or #3. #1 means "most urgent, need ASAP." #2 means "need by end of today." #3 means "need early next day or by date and time noted."

--You rarely see chalkboards or felt-tip writing boards in A/E offices, but they're among the handiest pieces of business equipment available. They're great for exploring ideas graphically; writing brainstorm notes during private meetings; projecting progress charts and notes; and writing highly visible TO DO lists, appointments, and reminders of all sorts.

--Some A/E's use easels with large scratch paper pads instead of chalkboards. They write ideas, notes, etc. on the sheets, and tack the sheets to the wall opposite their desks. It's no more than a giant note pad but, because of its size, it's very commanding. Some prefer it over chalk or felt-tip board notes, because these notes can be hidden away when necessary, and important items are less likely to be lost by erasing.

AS BUSINESS PICKS UP, MEETINGS START TO EAT UP MORE TIME

Advice on how to hold better meetings that truly get better results in less time, from management consultant Frank Snell:

--The purpose of having a meeting in the first place is to save time. Meetings can be the quickest way to share data and coordinate group action. Problems arise when this purpose is forgotten.

--The best way to assure that a meeting is justified in the first place is to send an agenda to all participants ahead of time. If the agenda can't spell out what the meeting is to accomplish, then don't bother with it.

--The agenda outline may include any or all of the following:

1) Date of memo and source

2) Date, time, and place of the meeting (if it's in an unusual location, include a map)

3) Primary subject of the meeting

4) Estimated length of time required

5) Who will attend

6) Background of the subject, if not self-evident

7) Present situation of the subject (or problem statement)

9) Specific goal(s) and purpose(s) of the meeting

--Control the seating arrangement. Divergent or adversary groups tend to cluster at opposite sides of the table, which creates a natural set-up for doing battle. Seat the most argumentative participants in the same row some distance from one another. They'll have trouble going at each other.

--As leader of the meeting, position yourself in the most prominent position. If you sit to one side, someone else will assume the leadership role.

--Unexpressive leaders with droning voices are a major cause of boredom and inactivity among participants. If you lead meetings, add extra life, energy and variations in tone and pitch to your voice. It's play acting, but it's essential to keep a meeting alive and productive.

--Open your meetings by saying: *"I think we can handle this in 20 (or 30, or 40) minutes."* The promise of a lively, short meeting will spark participants to avoid sidetracks and focus aggressively on the subject(s) on the agenda.

MEETINGS ACCORDING TO AN OLD TESTAMENT

(Published by C. Craig Wright in the Xerox Corporation newsletter for managers. Read slowly and with reverence.)

The Master speaketh:

"And as the conclaves multiplied and lapped one upon another, they were delayed in starting and delayed in ending, and were postponed to be called again when those whose presence was required could be made free. For the message was clear, but the ways were hidden.

"Then, from all sides came voices crying out:

'Though I labour from my coming in to my going out, I cannot attend the meetings for which I am summoned.'

'It concerned me not, yet was I called unto meeting.'

'My need for decisions is great, yet am I denied, for all are in meetings.'

'Can the meeting not start by the mark on the glass?'

'To the meeting for which I made ready, no person came.'

'Is thy servant a fool, that thou summonest him to a meeting to schedule meetings?'

"And then at last the chief, hearing of these things, decreed that the axe be laid to the root of these meetings which brought forth bad fruits. And he sent forth a scroll, saying:

I   Thou shalt not meet if the matter can be resolved by other means.

II   Thou shalt make the purpose of each meeting known to those thou summonest.

III   Thou shalt summon only those whose presence is needful.

IV   Thou shalt start at the time announced.

V   Thou shalt stop when it is right to do so.

VI   Thou shalt not run beyond.

VII   Thou shalt combine into one those which need not be separate.

VIII   Prepare thy thoughts, that the minutes not be wasted.

IX   Schedule not in haste, for the day is short in which to do that which thou hast to do.

X   Fear not to cancel if need disappears.

"And in Time the people learned and obeyed these writings, and followed them, putting to and taking away as suited their needs. For as they foresook their old ways, new hours were given unto them, and they were free to do their things, and they saw that it was good."

(Amen.)

GRASSHOPPING THROUGH THE WORKDAY

(There's an old story about a farmer that gets re-enacted in one form or another daily in every office in the land.

The farmer sets out to do some plowing. On his way to the tractor, he notices the water trough is empty. As he goes to hook up the hose to the trough, he sees a hole in the driveway that needs filling, so he goes off to get a shovel. On his way back with the shovel, he sees a fence that needs mending, so he veers off to get some barbed wire. Then he remembers the water trough . . . .

Some managers and employees follow this path of diversions throughout entire workdays. They interrupt every task they start to start another one. Some management consultants have seen it often enough to name it. It's called "grasshopping."

It's usually easier to respond to a fresh stimulus than it is to stick it out, beginning to end, on a routine chore. It's especially easy if the chore is tedious and there's no overriding goal, no plan for the day's work sequence.

British management consultant Rosemary Stewart recommends a special use for the daily To-Do list to shine a light on the problem. As best you can, check yourself every half hour or so and see if what you're doing at any particular time coincides with your day's plan. Also, make a checkmark every time you interrupt yourself.

Every manager complains that interruptions are one of the worst drains on time and productivity.

But what they don't say is that most interruptions are self-initiated. So, the idea is to note which interruptions are really just part of the job, how many you cause, and how many are caused by others.

Here's a tip: Watch for repetitive types of interruptions. If they're repetitive, they're great potential subjects for delegation or systematization. For example:

--Are people calling and repeatedly asking for the same kind of information?

--Are subordinates asking for the same kind of directions?

--Are you repeating the same corrective actions, responding to the same kinds of emergencies?

--Are you having to interrupt yourself and repeatedly having to seek the same kinds of data or help from others?

--When you have to interrupt yourself to get something, is it located so far away that you're vulnerable to distractions and further interruptions en route?

Each such event indicates a delegation problem or an opportunity for a minor reorganizing or systematization of the working situation.

For one week, make a daily note on your TO-DO list to watch for these items; you'll spot some great potential timesavers for yourself.

OFFICE TIME WASTERS AND TIME SAVERS . . . STARTING WITH MANAGEMENT

We all know from The Peter Principle:

During recession, when the incoming work dries up, the work on the boards seems to take forever. And, for whatever reasons, the interruptions increase and it seems harder to take and keep control and maximize the value of one's work time.

As noted earlier in this chapter, most interruptions in work are self-inflicted. So taking control of work time wasters starts with oneself. Start with any one habit, first or last thing each day, that eats time . . . deal with it. Reading the paper during the first 15 minutes of work, for example, opening mail, anything other than getting deeply into work is a time killer.

Then, for dealing with time wasters from others:

-- When you need information or a decision from someone and they say they'll "get back to you," say *"Good. Tell me when that will be, so I can be sure to be available."*

-- Deadlines for tasks are not just ignored by staff members; more often they're misunderstood or forgotten in the distraction of other work. The GOLDEN RULE of supervisory management is to send a memo to go with every verbal agreement, and then verbally reconfirm the due date and time: *"Oh, by the way, did we set that time at Friday at 11 a.m.?"* "That's first thing Thursday morning, right?" That sort of verbal reinforcement usually locks everything in properly on both sides (supervisors have memory lapses, too).

-- Project managers need to keep in touch with the latest products but scattered, drop-in visits by sales reps are terrible time killers. The same problem applies to interviewing drop-in job applicants.

-- Schedule all sales rep and job applicant visits for the same general time and day each week, such as Friday afternoon after 3 p.m. That consolidates the entire task into one period of time that's not usually productive anyway. The line of people in the waiting room helps assure everyone gets through their interview expeditiously.

-- Provide a no-table, no-chair meeting space for stand-up meetings; just a podium or shelf to review catalogs, samples, portfolios and related paperwork. Stand-up meetings are usually formal, crisp, and brief.

-- Coffee breaks are like old elastic, they gradually get stretched way beyond their original limits. Many offices have eliminated separate breaks. They keep the microwave available all day and let employees have coffee or tea whenever they want while continuing to work. (Some managers are properly worried about spilling coffee on drawings. The rule should be: No cups on drafting boards or where they can spill on any work. Accidents are virtually unheard of where such rules have been clearly stated and enforced.)

ARE BUSINESS PEOPLE REALLY MORE BUSINESSLIKE?

Who are more businesslike -- design professionals or corporate managers?

The obvious and popular answer is wrong, and wrong in a special way. While A/E's, especially architects, berate themselves for unbusinesslike attitudes and practices, surveys show more similarities than differences with the corporate world. For example:

A/E's often accuse themselves of spending too much time doing instead of managing. The accusation is true but it applies throughout business. A 1981 survey by the American Management Association finds an average of only 47% of the corporate managerial time is spent doing management. About 31% of the front office time goes to nuts-and bolts office labor, inactive time, and responding to brush fires. The remaining time is spent instructing subordinates.

A/E's say their business judgment is flawed. Yet business biographies such as **"Up the Organization"** by Robert Townsend report that about two out of three executive decisions are mistakes. While errors are a constant source of self-condemnation among design professionals, studies going back many years find one out of every seven office actions to be in error; and one out of two attempts to remember details of past communications are wrong even among top people.

Some A/E's think that their long work hours are a reflection on their efficiency. But the hours are typical of all self-employed business people, corporate executives, and professionals. The American Management Association survey finds 53% of business executives arrive at work early, usually 30 to 60 minutes before the official office starting hours, and 57% also work after hours an average of 30 to 60 minutes. Nearly 73% spend their later evenings on business-related meetings, travel, social, and civic functions. About 50% do "occasional to regular" work days on holidays and weekends.

A **"Fortune"** magazine survey found executive work weeks followed this pattern: 45-48 hours daytime office work; one night weekly working late; two nights a week working at home; and one night on business-connected entertainment. When you add company trips and special projects, the average executive or professional work week easily averages 70 to 80 hours. Long hours in themselves are not considered a problem either among professionals or business people, but proper use of the hours is a matter of serious concern.

A large number of long office hours aren't spent working. Self-employed professionals throw away 25% of their work time, according to research by well-known management consultants Booze Allen & Hamilton of New York. Consultant O. Mark Marcussen says front-line management people commonly lose 1.2 hours daily for varied personal but unavoidable reasons. They waste another 2.4 hours purely out of managerial inefficiency.

Marcussen says first-level supervisors usually have only a vague idea of the time spans required to finish tasks. That in itself leads to 35% of their time losses.

Unclear assignments, incomplete instructions, and unstated deadlines pass the managerial inefficiencies on down throughout lower-level echelons, accounting for 25% of all general office time waste.

Typical managers start each day with only a generalized idea of what's to be done and they spend the day taking each event as it comes. Consultant Robert Adcock, author of a 70's survey of aerospace administrators, reached this conclusion about executive managers:

"The only principle which seemingly was followed by a majority of those surveyed was the Principle of Muddling Through -- little planning, less organization, and practically no controlling."

301

So A/E's may be no better or worse than other business people. That's rather interesting but what's the payoff?

The payoff comes in looking at those managers and professionals who are NOT typical and who have the most effective work styles. The American Management Association report concluded that *"although the allotment of time cannot be increased, the rate of return can be multiplied."*

That's what managers who have broken away from the ranks have done; they've multiplied the rate of return on their work time. Their management style starts with concerted self management that usually includes three key characteristics:

--Clearly stated long-range goals -- usually quantified and linked to definite dates in time.

--Prioritized To-Do lists that they use in daily pursuit of aspects of the long-range goals.

--Relentless concentration on the most important tasks (those related to the larger goals) starting **first thing** each work period.

Most of the most effective managers found in any survey have done self-audits of how they spend their daily time. That gives a base line of data for resetting all their work time priorities. People commonly learn that they spend an hour or two a day on the most important possible tasks. With that knowledge they can double their time on the most important work and thus double their overall productivity and effectiveness without adding one minute of work time. It's all extremely simple and extremely effective.

APPENDIX

SCOPE OF WORK -- WORKING DRAWINGS

INTRODUCTION
HOW TO USE THIS REPORT

EXTREME DIFFERENCES IN DESIGN SERVICES

CONSTRUCTION DOCUMENTS --
WORKING DRAWINGS

GENERAL INFORMATION SHEETS

SITE PLANS

FLOOR PLANS

ROOF PLANS

EXTERIOR ELEVATIONS

CROSS SECTIONS AND WALL SECTIONS

REFLECTED CEILING PLANS

SCHEDULES: FINISH, DOOR, WINDOW, ETC.

CONTENT CONSTRUCTION DETAILS

PRODUCTION MANAGEMENT
DISCIPLINES COORDINATION AND DOCUMENT
CHECKING

PRODUCTION MANAGEMENT
ARCHITECTURAL DESIGN AND DOCUMENTATION

STRUCTURAL DESIGN AND DOCUMENTATION

MECHANICAL DESIGN AND DOCUMENTATION

ELECTRICAL DESIGN AND DOCUMENTATION

CIVIL DESIGN AND DOCUMENTATION

LANDSCAPE DESIGN AND DOCUMENTATION

INTERIOR DESIGN AND DOCUMENTATION

PROJECT DEVELOPMENT SCHEDULING

ESTIMATING PROBABLE CONSTRUCTION COSTS

PRESENTATIONS

BID ADMINISTRATION

CONSTRUCTION CONTRACT ADMINISTRATION
PROJECT ADMINISTRATION PRECONSTRUCTION

SHOP DRAWING CHECKING AND COORDINATION

PROJECT ADMINISTRATION --
ACTIONS REGARDING CONTRACTOR

CONSULTANTS COORDINATION AND DOCUMENT CHECKING

AGENCY CONSULTING, REVIEW AND APPROVALS

SUBSTANTIAL COMPLETION AND PROJECT CLOSE-
OUT OWNER-SUPPLIED DATA COORDINATION

OFFICE CONSTRUCTION ADMINISTRATION

CONSTRUCTION FIELD OBSERVATION AND
REPORTS

FIELD ORDERS AND CHANGE ORDERS

TESTS

CERTIFICATES FOR PAYMENT

CONSTRUCTION ADMINISTRATION

CONSTRUCTION ADMINISTRATION AT
SUBSTANTIAL COMPLETION

SEMIFINAL AND FINAL INSPECTION

FINAL CERTIFICATE OF PAYMENT

IN-OFFICE CLOSE-OUT

POSTCONSTRUCTION ADMINISTRATION

INTRODUCTION HOW TO USE THIS REPORT

Use this survey report to identify in detail the Scope of Services that can reasonably be provided for any particular fee range for any phase of work.

To see where you stand relative to other design firms, first compare the fees you charge for various projects with those listed on the next few pages. That will tell you if your fees rank as "Low," "Medium," or "High" compared to national and regional averages.

The design services are classed as "Minimum," "Median," and "Comprehensive." For the majority of offices, these are the types and amounts of services that are provided for the "Low," "Medium," or "High" design fees respectively.P

The services listed as being in the "Mini mum," "Median," and "Comprehensive" categories arc those provided by a majority of the offices surveyed. However, there are many individual variations. For example, some firms insist on doing "Comprehensive" amounts of drawing checking and cross-coordination although everything else on the same project might be "Minimum" or "Median" level work.

EXTREME DIFFERENCES IN DESIGN SERVICES

The differences in services provided by different firms can be enormous:

--Five sketchy sheets of working drawings vs. 20 highly detailed drawings for similar shopping center projects.

--Plans and exterior elevations copied outright from a magazine vs. weeks of intensive predesign and design development for a housing project.

--A preprinted sticky-back list of general notes and building code references vs. a well-edited, bound set of specifications for low rise speculative office buildings.

Not that the more complicated way is necessarily the best. We all know that after years of reform, reductions in overdrawing, and streamlined systems graphics And not that some clients, especially contractor-developers, don't prefer and need just the bare minimum. They don't need specs and they're better off with a free reign in making construction detail decisions on the job.

But we're talking about standard bid documents and "full service" by design firms of similar apparent qualifications. The three low-end projects cited above were all low fee, but the clients did not know how bad the service they were getting really was.

As noted previously in this report, we provide detailed lists of design services under headings of "Minimum," "Median," and "Comprehensive." These are the ranges of services that are delivered by over 70% of the design firms in 75% of the projects surveyed. The fees paid for work on these projects reflect the services provided -- that is, in most firms, "High" fees correspond with "Comprehensive" services, "Middle" fees correspond with "Median" services, and so on.

In 68% of the projects surveyed, there is a direct relationship between the fees paid and the quality and quantity of services provided. In nearly 30% of the design projects studied in this survey, however, there was very little relationship between the fees paid and the work done. In 15% of the cases, the clients received considerably more design work than they paid for.

When working drawings for similar projects by different firms could be compared, there were differences of as much as two-to-one in terms of quantity and overall quality of work. But in nearly a third of these, there was no comparable difference in fees. In other words, the firms doing the most careful and comprehensive work were often paid the same, or less, as those who did minimal work. Design time for some similar projects varied as much as five to one with no corresponding differences in fees or in evident design quality.

The firms that provide far more work than they are paid for tend to be small firms with strong design values. They considered it a moral commitment and a matter of pride to give a maximum amount of care to each project. That's excellent for the clients, but unless these firms have outside sources of income, they are not likely to survive. They often have less than half the work flow of comparable offices and well under half the revenue. Even with a low volume of work, the principals tend to be overworked from spending extra time on their projects and are often financially over-extended.

Clients obviously don't necessarily get a bargain when they pay low fees. Clients may think they're getting a deal by paying 2% of construction cost for design documents, but most of the firms that habitually undercharge for services usually put in no more work than one would expect for a 1% fee -- the lowest possible extreme of Minimal services.

Here's one approach to the problem, from an Atlanta architect:

He prepared three sets of lists and sketched diagrams of documents. One set outlines the extent of minimal services, one shows what's entailed in full design service, and the third lists and illustrates additional or specialized services. The three Scope of Services lists and diagrams

not only let clients know what to expect from him, but show what they can or should expect from other firms. He says: "Now when clients check other firms on fees and services they have a more detailed basis for comparison and they almost always come back to me."

Other information aids that design firms have provided for clients . . .

--A list of problems clients commonly experience from minimal service firms: permit delays as documents have to be redrawn; sitework problems from skimpy site examination; low-ball bids on minimal documents that lead to extensive extras during construction; maintenance problems; roof and facade failures; and tenant lawsuits.

--Letters of testimony from contractors on the quality of a firm's specs and drawings.

--A documented track record on efficient permit processing and a record of projects designed and built on time and on budget. This helps any client decide to go for genuinely full service at rational fees.

--A brochure showing photos of not just the A/E firm's work but photos and drawings illustrating the components of full service: the steps of Predesign and Programming, Site Analysis, Design Development, etc. The leading professional societies have produced good pamphlets that describe the full service design process but they're too often used as waiting room table-top literature. Such material is most effective when it's handed to clients with personal emphasis that it is an important clarifier of the design services they'll receive.

CONSTRUCTION DOCUMENTS -- WORKING DRAWINGS

GENERAL INFORMATION SHEETS

MINIMUM CONTENT

Includes one to two-phase drafting and checking: 50% completion and/or final completion.

__ Photocopies of applicable regulations and code charts (these are also often included in Median and Comprehensive jobs as clarifiers as well as timesavers):

    __ Building code standards, such as nailing schedules, minimum joist spans, etc., as required by local building authority
    __ State energy codes
    __ General handicap requirements

__ Project site address

REGARDING CONSTRUCTION NOTATION AND DIMENSIONING

__ Single word or short-phrase material identification notes

__ Center-line partition dimensions and finish-to-finish clearances for wall construction

__ Dimensions rounded as much as possible to nearest foot or six inches

MEDIAN CONTENT

Includes three to five phases of drafting and checking, such as 30% completion, 60% completion, 100% completion with 80% to 90% completion check for larger projects.

Includes previously listed content plus:

__ Symbol and nomenclature keys

__ Lexicon of abbreviations and nomenclature

__ Vicinity map -- drawn, photocopied or scanned from road map

__ Consultant contact names, addresses, phones, fax numbers

__ Space for permit approval dates and stamps

__ Space for client phase approvals and dates

__Lexicon

EXPLANATION OF NOTATION

__ Identification notes, such as "See Specifications " or "See Structural," with assembly notes and general cross references to other relevant documents

EXPLANATION OF CONSTRUCTION DIMENSIONING

__ Finish to finish dimensions

COMPREHENSIVE CONTENT

May include five or more phases of drafting and checking: 15% completion, 30% completion, 60% completion, 90% completion, 100% completion plus final pre-bid or post-bid check. Quality control checking may be provided separate from progress phase checking and may be provided by outside consultants.

Includes previously listed content plus:

__ Building permit authority contact names, addresses, phone numbers

__ Symbol and nomenclature keys

__ General notes explaining special drawing features such as keynotes

__ Detailed vicinity map with "magnifier" enlargement of immediate area of site

__ Sample symbols and their meaning for Architectural, Civil, Landscaping, Structural, Electrical, HVAC, Plumbing, etc.

__ Square footages for interior and exterior areas

__ Explanation of special graphic features:
  __ "Half-size" reference prints
  __ Screened shadow background prints
  __ Fixture height schedules instead of interior elevations
  __ Keynotes
  __ Special notation and dimensioning standards

__ Miniature drawing index on every sheet (expedites drawing cross-checking by all parties)

__ Photocopies of manufactured products, fixtures, etc.

__ Photocopies of pictures of desired surface textures, patterns

CONSTRUCTION NOTATION IN GENERAL

__ Identification notes, such as "See Specifications Section 32105" or "See Structural Dwg. S-2," with assembly notes and specific cross references to other relevant documents

__ Identification notes with CSI Masterformat division numbers to relate notes to specifications and to standard detail file numbers

__ CSI Masterformat number references are often combined with Keynote Legend

CONSTRUCTION DIMENSIONS

__ Substructure and rough opening dimensions at plans and elevations

__ Modular dimensions -- most common: 4' or 4" modular grids

__ Dimension leader line arrow heads differentiated for different types of dimension, such as arrows to finish dimensions and dots to 4" or 4' module

__ Floor elevations on plan keyed to site datum

SITE PLANS

MINIMUM CONTENT

Includes one to two-phase drafting and checking: 50% completion and/or final completion.

__ Property outline only and/or minimum data as specified by building department

__ Easements and rights of way

__ Points of utility connection

__ Indication of property slopes and drainage

__ Storm drains, new and existing

__ Building outline or footprint with exterior wall dimensions to property boundary lines

__ Building overhang dimensions

__ Street access, driveway dimensions

__ Parking space as required

__ Handicap parking, pavement markings, signs, ramps and access as required

__ Utility meters and hook-ups

__ Print of the survey may be bound into the drawings for reference

__ General notes regarding work not shown in detail

MEDIAN CONTENT

Includes three to five phases of drafting and checking such as 30% completion, 60% completion, 100% completion with 80% to 90% completion check for larger projects.

Includes previously listed content plus:

__ Spot check of accuracy of surveyor's drawings and notes of discrepancies

__ Zoning check and confirmed setback limit dimensioned

__ Site reference photos to establish views, neighbor proximity, etc.

__ Existing and new site contours

__ Existing and new finish grades at all corners of building

__ Grade slopes at building line

__ Dimensional allowance for final grading elevations with addition of topsoil

__ Building overhang dimensions

__ Roof plan sometimes shown on site

__ Site drain slopes for all drainage, including roof

__ Driveway and street centerline elevations and side elevations

__ Pavement construction joints and movement joints with detail and specification references

__ Existing and new curb inlets, catch basins, manholes, flumes and spillways

__ Pavement slopes to drain

__ Existing site elements to remain/remove

__ Cut and fill grading profiles

__ Utility meters and hook-ups

__ Underground fuel or other storage

__ Buried cables and main warning signs

__ Soil test reference data

__ Drawings for contractor to assist with construction and temporary facilities planning

__ Retaining walls, fences and gates

__ Signs, kiosks, and other site appurtenances

COMPREHENSIVE CONTENT

May include five or more phases of drafting and checking: 15% completion, 30% completion, 60% completion, 90% completion, 100% completion plus final pre-bid or post-bid check. Quality control checking may be provided separate from progress phase checking and may be provided by outside consultants.

Includes previously listed content plus:

__ Confirmed site dimensions from on-site review and measurements at site

__ Meets and bounds confirmation

__ Aerial photographs and conversion of aerial photos and survey to same scale for overlay comparison

__ Site photo and/or video survey encompassing all major site features and surroundings

__ Separate screened prints of background site work to show:
  __ Grading and building excavation
  __ Drainage excavation
  __ Electrical excavation and overhead work
  __ Site lighting and electrical outlets
  __ Plumbing excavation:
    __ Gas
    __ Sewer
    __ Water
  __ HVAC excavation
  __ Landscaping
  __ Site furniture and appurtenances

__ Construction fence design and plan

__ Temporary roads, parking, storage, other construction facilities

__ Roof drainage and final resolution of drainage

__ Landscaping plan:
  __ Plant selection
  __ Plant purchase and supervised installation
  __ Irrigation and/or sprinkler system planning

__ Existing trees and landscaping protection plan and specifications

__ Comprehensive "as-built" record drawings of final site work or review and OK of contractors' record drawings

__ Additional plans as required:
  __ Percolation test plan
  __ Soils testing boring schedule and profile
  __ Test boring locations
  __ Copy of test profile from geotechnical engineer
  __ Temporary erosion control
  __ Test pit and boring plan
  __ Grading plan
  __ Demolition plan
  __ Excavation plan
  __ Construction temporary facilities
  __ Drainage
__ Photos of major site features -- existing to remain, modify, to remove and store, demolish

__ Photo specs to show desired paving patterns or surface textures

FLOOR PLANS

MINIMUM CONTENT

Includes one to two-phase drafting and checking: 50% completion and/or final completion.

__ Rooms dimensioned with room name -- interior finish to finish size such as "Dining 12' x 14'"

__ Walls and partitions with centerline or surface dimensions

__ Walls and partitions with keyed materials indications

__ Doors and windows with sizes/types noted at each unit, no separate door or window schedules

__ Ceiling or roof framing shown by arrows to indicate direction of frame, with note of member sizes and spacings

__ Stairs:
    __ Arrow down or up with noted number of treads
    __ Finish clear opening dimensions

MEDIAN CONTENT

Includes three to five phases of drafting and checking such as 30% completion, 60% completion, 100% completion with 80% to 90% completion check for larger projects.

Includes previously listed content plus:

__ Dotted-line indications for openings, soffits, etc. at ceiling with notes, dimensions, and detail references

__ New work in solid line in contrast to screened background of existing work to remain

__ Existing work to remain identified with symbols or patterns

__ Door and window symbols keyed to door and window schedules

__ North arrow and reference north

__ Detail keys at junctures of different floor materials

__ Detail or schedule keys at floor saddles and thresholds

__ Door and window symbols keyed to door and window frame or rough opening schedules

__ Door and window frame schedules keyed to details

__ All door and window details included on door/window frame schedule sheets

__ Notes of depressed slabs or framing to accommodate divergent thicknesses of finish floors

__ Sound isolation walls

__ Cabinet work keyed to cabinet schedule

__ Stairs:
    __ Rough and clear opening dimensions
    __ Stair profile section and stair detail reference keys
    __ Noted non-slip treads and other safety elements

313

__ Dimensions of chases, shafts, and furred walls

__ Fixtures such as drinking fountains, fire hose cabinets, etc., keyed to fixture schedule and details

COMPREHENSIVE CONTENT

May include five or more phases of drafting and checking: 15% completion, 30% completion, 60% completion, 90% completion, 100% completion plus final pre-bid or post-bid check. Quality control checking may be provided separate from progress phase checking and may be provided by outside consultants.

Includes previously listed content plus:

__ North arrow, reference north, and symbols at exterior walls keyed to Exterior Elevation drawings

__ Finish flooring patterns or reference to enlarged pattern drawings

__ Datum references to show relative floor heights of balconies, landings, mezzanines, etc.

__ Expansion space notes or detail references for perimeters of expansive flooring

__ Numbered and lettered structural grid coordinates with primary framing dimensions

__ Separate framing plan for larger projects as part of structural drawings

__ Walls and partitions with dimensions to framing and substructure, plus cumulative dimensions across plan

__ Walls and partitions with keyed materials indications linked to materials schedule

__ Simplified finish schedules included on floor plan sheets

__ Keys or symbols to break-out work of different trades or contracts

__ Furniture reference overlay

__ Screened prints showing plan in shadow print, separate trades or disciplines in solid line:
    __ Framing plans in reference to floor plan showing plumbing and HVAC vents
    __ Plumbing supply in reference to floor plan with framing and fixtures
    __ Plumbing drainage plans in reference to floor plan, framing, floor slopes, and fixtures
    __ Electrical plans in reference to floor plan with door swings and electrical equipment

__ Anchor detail keys at walls and floors for mounted equipment and fixtures

__ Interior wall construction keyed to enlarged wall plan and vertical sections showing acoustical and fire-rated treatments

__ Exterior wall construction keyed to enlarged wall plan and vertical sections, showing framing, detail keys, thermal insulation, and waterproofing

__ Floor and ceiling plans keyed to floor and ceiling detail sections showing framing, detail keys, acoustic and fire-rated treatments

__ Noted and detailed vibration and noise control at equipment supports

__ Stairs:
     __ Treads and risers with noted T & R dimensions
     __ Rough and clear opening dimensions
     __ Stair section and stair detail reference keys
     __ Noted non-slip treads and other safety elements

ROOF PLANS

MINIMUM CONTENT

Includes one to two-phase drafting and checking: 50% completion and/or final completion.

__ Roof surface with materials, overhangs and slopes shown in combination with Site Plan

__ Roof overhang edge dimensions to property lines and setback lines

   Drain indications and flashing detail keys

MEDIAN CONTENT

Includes three to five phases of drafting and checking such as 30% completion, 60% completion, 100% completion with 80% to 90% completion check for larger projects.

Includes previously listed content plus:

__ Roof construction plan in same scale as floor plans

__ Roof finish materials, overhangs, and slopes

__ Indication of direction of roof framing

__ Exhaust vents, fresh air intakes (confirmed that fresh air intakes are not near or in wind path of exhaust vents)

__ Scuppers, overflow scuppers and roof drainage slopes in inches per foot

__ Drain types, sizes, and detail keys

__ Dotted-line indications for openings, soffits, etc. at ceiling with notes, dimensions, and detail

__ Detail keys for eaves, parapets, drains, gutter guards, skylights and roof-mounted equipment

__ Notation of roof insulation

COMPREHENSIVE CONTENT

May include five or more phases of drafting and checking: 15% completion, 30% completion, 60% completion, 90% completion, 100% completion plus final pre-bid or post-bid check. Quality control checking may be provided separate from progress phase checking and may be by outside consultants.

Includes previously listed content plus:

__ Roof framing plans as separate sheets from plans showing finish roofing

__ Shadow print of upper floor plan in combination with solid roof plan to show coordination of structural frame, drains, skylights, etc.

__ Shadow prints of upper floor roof plan with combination of solid line roof plan to show coordination of roof with work of engineering disciplines:
__ Plumbing
__ HVAC
__ Electrical

__ Roof plan and details referenced to specifications section numbers

__ Roof railing and parapet heights

__ Utility service entries at roof

__ Types and location of anchors, supports and guys for roof mounted equipment

EXTERIOR ELEVATIONS

MINIMUM CONTENT

Includes one to two-phase drafting and checking: 50% completion and/or final completion.

__ Exterior elevations of primary sides of building labeled as "Front," "Rear," etc. or "North," "South," etc.

__ Exterior material notes or pattern indications

__ Doors and windows with general height dimension for door and window finish openings

__ Overhangs

__ Wall and overhang dimensions to property lines if close to setback limits

__ Finish grades

__ Below-grade construction

__ Retaining walls, paving, planters, curbs, etc. adjacent to structure

__ Floor, ceiling, roof heights

__ Roof slopes

__ Major openings

MEDIAN CONTENT

Includes three to five phases of drafting and checking such as 30% completion, 60% completion, 100% completion with 80% to 90% completion check for larger projects.

Includes previously listed content plus:

__ Notes whether vertical dimensions are to finish surfaces, subfloor, or framing

__ Notes whether opening dimensions are rough or finish openings

__ Air intakes coordinated with HVAC drawings (confirmed as safely away from building or car exhausts)

__ Adjacent fences and yard walls

__ Adjacent site furniture such as decks, pavement, railings

__ Adjacent major landscaping features that might affect construction

__ Site features to remain, features to remove

__ Wall-mounted exterior lights and alarms

__ Flag poles, signs, plaques, other wall-mounted appurtenances

__ Hose bibbs, hydrants, standpipes

__ Downspouts, rain leaders, roof drains and scuppers

__ Roof gutter slopes

__ Catch basins, splash blocks, trench drains for roof drainage

__ Wall-mounted marquees, awnings, canopies, etc.

__ Wall-mounted appurtenances which are N.I.C., shown but drawn dotted line or clearly labeled N.I.C.

__ Overhead utility connection points

__ Demountable or movable roof equipment in dotted line with identification note

__ Roof railings and guards with dimension of required safety height

__ Vertical joints for expansion/contraction and specification reference note

__ Horizontal movement joints for frame shrinkage and specification reference note

__ Flashing and waterproofing indications with specification reference notes

COMPREHENSIVE CONTENT

May include five or more phases of drafting and checking: 15% completion, 30% completion, 60% completion, 90% completion, 100% completion plus final pre-bid or post-bid check. Quality control checking may be provided separate from progress phase checking and may be provided by outside consultants.

Includes previously listed content plus:

__ Rough opening structure and substructure dimensions

__ Sill and lintel substructure heights

__ Vertical working dimensions from an established start point such as finish slab or foundation

__ Background screened image of overall building cross section with exterior wall construction in solid line

__ Coded dimension indications for rough, substructure, and finish dimensions

CROSS SECTIONS AND WALL SECTIONS

MINIMUM CONTENT

Includes one to two-phase drafting and checking: 50% completion and/or final completion.

__ Cross-sections and wall sections are not necessarily provided except to show exceptional construction

__ Standard wall construction sections may be photocopied from handbooks

__ Footings and foundation walls (unless covered in structural drawings)

__ Bearing walls including rough openings

__ Floor construction

__ Balconies and mezzanines

__ Parapets

__ Overhangs

__ Roof framing

MEDIAN CONTENT

Includes three to five phases of drafting and checking such as 30% completion, 60% completion, 100% completion with 80% to 90% completion check for larger projects.

Includes previously listed content plus:

__ Interior partitions

__ Suspended ceilings where appropriate to show relationship to structural framing or mechanical equipment

__ Mechanical equipment where appropriate to show relationship with structural framing

__ Suspended heavy equipment in relationship to framing

__ Wall-mounted equipment in relationship to structural frame

__ Roof appurtenances, such as penthouse and stair bulkhead, in relationship to framing

COMPREHENSIVE CONTENT

May include five or more phases of drafting and checking: 15% completion, 30% completion, 60% completion, 90% completion, 100% completion plus final pre-bid or post-bid check. Quality control checking may be provided separate from progress phase checking and may be provided by outside consultants.

Includes previously listed content plus:

__ Key plan with identification of cross section cut points

__ Special sections to show thru-building shafts and light shafts

__ Special structural systems, bracing, stiffeners, etc.

__ Directly connected blow-up details of special connections

REFLECTED CEILING PLANS

MINIMUM CONTENT

Includes one to two-phase drafting and checking: 50% completion and/or final completion.

__ Columns, posts, walls, ceiling-high partitions

__ Suspended ceiling grids

__ Integrated ceilings

__ Exposed ceiling beams, girders, joists

__ Ceiling construction and finish

__ Furred ceilings and soffits

__ Slopes of ceiling

__ Ceiling mounted fixtures

MEDIAN CONTENT

Includes three to five phases of drafting and checking such as 30% completion, 60% completion, 100% completion with 80% to 90% completion check for larger projects.

Includes previously listed content plus:

__ Recesses

__ Valances and detail keys

__ Overhead door and folding partition tracks and detail keys

__ Noise barriers at low partition lines

__ Ceiling-mounted casework and detail keys

__ Ceiling mounted alarms, TV cameras, other special electric fixtures

__ Access panels and detail keys

__ Chases, chutes, future shafts and detail keys

COMPREHENSIVE CONTENT

May include five or more phases of drafting and checking: 15% completion, 30% completion, 60% completion, 90% completion, 100% completion plus final pre-bid or post-bid check. Quality control checking may be provided separate from progress phase checking and may be provided by outside consultants.

Includes previously listed content plus:

__ Dash line indications of framing above ceiling

__ Dash line indications of hatches, wells, shafts, etc. above ceiling

__ Fire sprinkler plan

__ Ceiling heating systems

__ Exposed ductwork

SCHEDULES: FINISH, DOOR, WINDOW, ETC.

MINIMUM CONTENT

Includes one to two-phase drafting and checking: 50% completion and/or final completion.

__ Notes on plan identifyng finishes, door types and sizes, and window types and sizes

MEDIAN CONTENT

Includes three to five phases of drafting and checking such as 30% completion, 60% completion, 100% completion with 80% to 90% completion check for larger projects.

Includes previously listed content plus:

__ Schedules as lists identifying generic substructure and finishes with references to specifications

__ Door and window symbols referenced to door and window frame schedules which are referenced in turn to detail keys showing sill, jamb, and head conditions

COMPREHENSIVE CONTENT

May include five or more phases of drafting and checking: 15% completion, 30% completion, 60% completion, 90% completion, 100% completion plus final pre-bid or post-bid check. Quality control checking may be provided separate from progress phase checking and may be provided by outside consultants.

Includes previously listed content plus:

__ General note explaining schedule system

__ Photo details of textures

__ Photocopy illustrations of manufactured doors and windows

__ Hardware schedule with door schedule

__ Short-hand door and window schedules included on floor plan sheets for convenient contractor reference

Large, full sheet schedules are not necessarily more comprehensive, especially if they have highly repetitive elements. Simpler schedule formats that fully convey all materials combinations in each room, keyed to a simple symbol, are preferred.

CONTENT CONSTRUCTION DETAILS

MINIMUM CONTENT

Includes one to two-phase drafting and checking: 50% completion and/or final completion.

__ Details provided primarily only as required by building department

__ Special condition details as required for building safety, waterproofing, and solidity of connections

__ Standard details as suited to illustrate typical construction

__ Detail drawings in profile outline with materials indications and materials identification notes

MEDIAN CONTENT

Includes three to five phases of drafting and checking such as 30% completion, 60% completion, 100% completion with 80% to 90% completion check for larger projects.

Includes previously listed content plus:

__ Plan and/or vertical cross sections of:
    __ Openings and penetrations
    __ Connections between different types of construction
    __ Joint connections between the same materials
    __ Profiles of ready-made products or fabrications connected to construction
__ Isometric or multi-view details where needed to clarify three dimensional detail situations

__ Details with assembly notes as needed for clarification and where not in conflict or duplication of specifications

__ General reference notes to related construction, specifications, and to drawings of other disciplines

COMPREHENSIVE CONTENT

May include five or more phases of drafting and checking: 15% completion, 30% completion, 60% completion, 90% completion, 100% completion plus final pre-bid or post-bid check. Quality control checking may be provided separate from progress phase checking and may be provided by outside consultants.

Includes previously listed content plus:

__ Specific reference notes to related drawings and specification section reference notes

__ Photodetails of existing conditions or of actual similar or related construction

__ Reference to detail construction mock-ups

PRODUCTION MANAGEMENT
DISCIPLINES COORDINATION AND DOCUMENT CHECKING

MINIMUM SERVICES

__ Identify any new consultants required for this phase, and negotiate contracts

__ Before finalizing new consultant contracts, review consultant service and contract terms with the client and obtain written client approval

__ Transmit updated information on building occupancies and design changes to consultants; make sure the architectural design team has identical updated information

__ Submit progress prints of architectural and consultants' work to all consultants

MEDIAN SERVICES

Includes the above listed services plus:

__ Review previous decisions on structural, construction, mechanical, and other systems for possible economies and improvements

__ Confirm that the various selected engineering and construction systems are compatible with one another

__ Obtain updated estimates of spatial requirements for appurtenances and engineered systems

__ Confirm that consultants, client, or others are handling the acquisition of approvals and permits for all utility services:
    __ Gas
    __ Electric
    __ Water
    __ Sewer
    __ Telephone
    __ Cable TV
    __ Computer link
    __ Utility-supplied steam or other heating medium
    __ Utility-supplied cooling medium

__ Obtain or update lists of special building equipment and fixtures required by the client that may affect consultants' work, and distribute to appropriate consultants

COMPREHENSIVE SERVICES

Includes the above listed services plus:

__ Conduct group meetings to allow consultants to compare their drawings with one another

__ Obtain update of the consultants' estimates of building operating costs

__ Review with the client the consultants' building operating cost estimates; obtain from the client written approval of the proposed mechanical and electrical systems

PRODUCTION MANAGEMENT
ARCHITECTURAL DESIGN AND DOCUMENTATION

MINIMUM SERVICES

__ Confirm and update the program's functional, occupancy, and spatial requirements with the client

__ Review changes in the program and note possible impact on the project design and documentation

__ Review the Design Development documents, updates of the design, and changes in the program for possible violations of codes and regulations

__ Continue and update the code search

__ Inform the client of necessary applications for approvals and permits

MEDIAN SERVICES

Includes the above listed services plus:

__ Review the Design Development documents, updates of the design, and changes in the program for possible conflicts with the original design intent or with fundamental engineering decisions

__ Compare the developed design with the client's construction budget

__ Confirm the type of construction contract to be used, such as single or separate contracts, and evaluate the effect of the contract type on drawing, specifications content and format

__ Prepare and coordinate final specifications (SEE THE SECTION ON SPECIFICATIONS)

__ Confirm the construction budget and review any contradictions between stated program needs and available funding

__ Enforce coordination check points to confirm that the architectural production team is fully informed of the most up-to-date consultants' information and vice versa

__ Determine and note reasons for changes in the design

__ Confirm client preferences or requirements for types of construction bidding and contracting that might affect the format of construction drawings and specifications

__ Review preferred construction methods for impact on design and documentation

__ Confirm the dates for submittal of all construction documents (drawings, calculations, contracts, specifications, and updates on construction cost estimates) to the client

COMPREHENSIVE SERVICES

Includes the above listed services plus:

__ Confirm with the client whether a detailed construction cost estimate, such as a quantity survey, is desired with the final working drawings

__ Compare architectural working drawings with the structural, mechanical, electrical, transportation, and other consultants' drawings by means of transparency overlays

__ Identify possible or definite bid alternates and plan content and organization of documents accordingly

__ Prepare data on costs and availability of special equipment and furnishings

STRUCTURAL DESIGN AND DOCUMENTATION

MINIMUM SERVICES

__ Coordinate progress checking of structural drawings and comparison with all other disciplines

__ Identify changes in the scope of structural work that occurred during the Design Development Phase

MEDIAN SERVICES

Includes the above listed services plus:

__ Review and reach agreement with the structural engineer on the number and content of structural Construction Documents:
    __ Design criteria
    __ Structural grid or system
    __ Structural framing plan(s) and sections(s)
    __ Foundation plan
    __ Calculations
    __ Required clearances for other work
    __ Structural details
    __ Materials schedules
    __ Specifications

__ Conduct structural, mechanical, civil, and architectural drawing cross-checking meetings

COMPREHENSIVE SERVICES

Includes the above listed services plus:

__ Create structural framing study models as necessary to clarify structure and construction methods

__ Observe structural materials testing to confirm design intent

__ Coordinate specialized consultants for non-standard structural design

MECHANICAL DESIGN AND DOCUMENTATION

MINIMUM SERVICES

__ Identify changes in the scope of mechanical work that occurred during the Design Development Phase

__ Coordinate progress checking of mechanical drawings and comparison with all other disciplines

MEDIAN SERVICES

Includes the above listed services plus:

__ Review and reach agreement with the mechanical engineer on the number and content of final mechanical construction documents:

    __ Building plans, sections, and other drawings to show:
        __ Noise and vibration control
        __ HVAC system type(s) and standard(s)
        __ Fire protection system(s)
        __ Plumbing supply and drain types and standards
        __ Equipment sizes and locations
        __ Chase sizes and locations
        __ Duct sizes and locations
        __ Mechanical equipment spatial requirements in plan
        __ Mechanical equipment spatial requirements in section
        __ Mechanical fixture and equipment schedules
        __ Mechanical construction details

    __ HVAC heat load and cooling calculations
        __ Energy use and conservation calculations
        __ Equipment and materials schedules
        __ Specifications
        __ Mechanical systems operations and maintenance instructions

__ Confirm with the mechanical consultant the acquisition of necessary approvals and permits for utilities:
    __ Gas
    __ Water
    __ Sewer
    __ Utility supplied steam or other heating medium
    __ Utility supplied cooling medium

__ Conduct mechanical, structural, and architectural drawing cross-checking meetings

__ Confirm with the mechanical consultant the compliance of the building mechanical and plumbing system design with codes and utility company requirements

__ Determine the impact on project time and cost as well as construction cost of revisions in mechanical work

__ Acquire estimates for probable construction costs of the building's mechanical systems

COMPREHENSIVE SERVICES

Includes the above listed services plus:

__ Acquire estimates for probable operating costs of the building's mechanical systems

__ Coordinate specialized consulting regarding indoor air pollution and related environmental hazards

__ Coordinate detailed energy-saving analysis and/or solar energy design

ELECTRICAL DESIGN AND DOCUMENTATION

MINIMUM SERVICES

__ Progress phase checking of electrical drawings

__ Coordination check between electrical drawings with architectural, structural, and mechanical work

MEDIAN SERVICES

Includes the above listed services plus:

__ Conduct multidiscipline and architectural drawing cross-checking meetings

__ Identify changes in the scope of electrical work that occurred during the Design Development Phase

__ Determine the impact on cost of revisions in electrical work

__ Confirm that changes in the electrical design comply with legal requirements

__ Review and reach agreement with the electrical engineer on the number and content of Electrical Construction Documents:

    __ Building plans and sections to show:
       __ Reflected ceiling lighting plans
       __ Power and switching
       __ Fire detection and alarm systems
       __ Security system
       __ Communications equipment, chases, and outlets
       __ Electrical equipment sizes, locations, and capacities
       __ Electrical vaults, transformer rooms
       __ Chase sizes and locations
       __ Duct sizes and locations
       __ Fixture schedules
       __ Electrical construction details

    __ Electrical, communications, security, fire, and related systems and equipment maintenance instructions

    __ Specifications

__ Arrange the assistance of the electrical engineer in obtaining approvals and permits for electrical and communications services

__ Obtain updated final estimates for probable electrical systems construction costs

COMPREHENSIVE SERVICES

Includes the above listed services plus:

__ Coordinate specialized communications systems such as for computers, environmental sensors, intelligent building appurtenances, etc.

__ Select, purchase, and check delivery of specialized or designer fixtures

__ Coordinate specialized lighting consultants such as for theatrical and exhibit lighting

__ Custom design and mockup assembly of electrical fixtures

__ Field testing and observation of specialized lighting effects

CIVIL DESIGN AND DOCUMENTATION

MINIMUM SERVICES

__ Provide all data necessary for Civil Engineer to complete grading and excavation plans

__ Confirm that site soils tests conform to requirements of local building authorities

__ Review engineer's site observation and soil test report with structural engineer to coordinate foundation plan

__ Recommend additional tests if original observations and tests are inconclusive and/or unacceptable to the prime design firm or structural engineer

MEDIAN SERVICES

Includes the above listed services plus:

__ Confirm that results of all previously requested site tests have been received and transmitted to the client, consultants, and the design team

__ Identify additional tests that may be required

__ Update the Test Log and file

__ Identify changes in the scope of civil engineering construction that have occurred through the Design Development Phase

__ Determine the impact on cost of revisions in civil work

__ Confirm that changes in the civil engineering design comply with legal requirements

__ Review and reach agreement with the civil engineer on the number and content of civil engineering Construction Documents:

    __ Site plans and sections to show:
        __ Cut and fill
        __ Excavations
        __ Irrigation
        __ Drainage
        __ Site-related construction
        __ Civil engineering construction details
        __ Specifications

__ Schedule completion dates for interim and final civil working drawings and specifications

__ Check and confirm compliance of sitework and civil engineering design with codes and regulations

__ Acquire updated estimates for probable civil engineering-related construction costs

COMPREHENSIVE SERVICES

Includes the above listed services plus:

__ Conduct civil, structural, landscaping, and architectural drawing cross-checking meetings

__ Design and coordination of special-purpose excavation

__ Underground construction design

LANDSCAPE DESIGN AND DOCUMENTATION

MINIMUM SERVICES

__ Provide schematic landscape plan

__ Provide site plan for landscape consultant with designer's recommendations

__ Review landscape plan proposed by consultant or nursery company and provide recommendations

MEDIAN SERVICES

Includes the above listed services plus:

__ Review and reach agreement with the landscape architect on the number and content of landscape Construction Documents:

    __ Landscape plans
    __ Sitework construction details
    __ Site-related plumbing work
    __ Site-related electrical work
    __ Specifications
    __ Landscape care instructions

__ Identify special-order planting that must be ordered early, to assure delivery and installation before the completion date

__ Schedule completion dates for interim and final landscape working drawings and specifications

__ Update estimates for probable landscape development costs

COMPREHENSIVE SERVICES

Includes the above listed services plus:

__ Conduct landscape architectural, civil engineering, mechanical, and electrical coordination meetings

__ Survey, mark, and diagram existing landscape requiring storage and protection during construction

__ Provide field search for specialized landscape elements

__ Custom design of specialized landscape features such as fountains, artificial streams, grottos, etc.

__ Plan and proposal for full landscape planting and landscape service through separate contract

INTERIOR DESIGN AND DOCUMENTATION

MINIMUM SERVICES

___ Provide floor plans for use by interior design consultants

MEDIAN SERVICES

Includes the above listed services plus:

___ Review and reach agreement with the interior designer on the number and content of interior design Construction Documents:

    ___ Interior partition landscaping
    ___ Furniture selection and planning
    ___ Fixtures selection and finishes palette
    ___ Materials and finishes palette
    ___ Color schedule
    ___ Interior design detailing
    ___ Specifications
    ___ Furnishings and finish material maintenance and cleaning instructions

___ Schedule completion dates for the final interior drawings and specifications

___ Review architectural, electrical, mechanical features for impact on Interior Design drawings

___ Update estimates for probable costs of interior design furnishings and fixtures

___ List and schedule special-order furnishings (such as carpet) that must be ordered early, to assure delivery and installation before move-in date

COMPREHENSIVE SERVICES

Includes the above listed services plus:

___ Complete color coordination and finish schedule

___ Art and decorative component purchase consultation:
    ___ Selection
    ___ Purchase
    ___ Delivery scheduling
    ___ Final approval on delivery

___ Signage design and construction details

___ Appurtenance and furnishings design and construction details

PROJECT DEVELOPMENT SCHEDULING

MINIMUM SERVICES

__ Provide estimated probable time and cost for completion of design services

__ Review with client the Scope of Work for possible revisions and additions to design services

MEDIAN SERVICES

Includes the above listed services plus:

__ Create or update the job calendar of estimated phase starts and completions:
   __ Construction Documents
   __ Bidding/Negotiation
   __ Contract Administration
   __ Post Construction

__ Distribute copies of the new or updated job calendar to all job participants

__ Create a schedule for budget and progress reviews

COMPREHENSIVE SERVICES

Includes the above listed services plus:

__ Provide CPM or Pert diagram charts and computer service for project scheduling and monitoring

__ Provide collaboration on scheduling for Construction Management contract and/or Fast Track construction

ESTIMATING PROBABLE CONSTRUCTION COSTS

MINIMUM SERVICES

__ Provide statement of possible construction cost based on comparative square footage or similar building type bids

MEDIAN SERVICES

Includes the above listed services plus:

__ Have all consultants prepare construction cost estimates for their phases of work

__ Provide "Statement of Probable Construction Cost" by square footage cost comparison with other similar projects

COMPREHENSIVE SERVICES

Includes the above listed services plus:

__ Obtain all consultants' construction cost estimates

__ Provide "Statement of Probable Construction Cost" through more precise itemized estimation method such as quantity survey

__ Obtain detailed bid estimates for negotiated contract

__ Provide cost estimates for contracting/design-build service as separate contract

PRESENTATIONS

All presentation and meeting time that is unpredictable in frequency, duration, extensions, or delays is presumed to be charged hourly as agreed by client and design firm.

MINIMUM SERVICES

___ Participation in public agency meetings or loan application meetings

___ Preparation and presentation of Design Development graphics as required for permits and financing

___ Review and negotiation meetings as required for permits and financing

MEDIAN SERVICES

Includes the above listed services plus:

___ Conduct all Construction Document presentations:
    ___ Interim presentations to client
    ___ Presentations to financing agencies
    ___ Presentations to regulatory agencies
    ___ Presentations to advisory boards/committees

___ Revisions in working drawings as required for permits and financing

___ Identify any contradictions between requested design changes and the original design program or prior client/designer decisions

___ Note any extended repercussions from design changes, and review with the client any extensions of the Scope of Work and any required changes in design service time and cost

___ Update building floor area calculations, building volume, usable area ratios, and other numerical comparisons with program requirements

___ Present the Design Development documents and cost data

___ Obtain approvals required for next phase: PRE-BIDDING and BIDDING

COMPREHENSIVE SERVICES

Includes the above listed services plus:

___ Prepare presentation materials in a form which will maximize possible reuse in working drawing production as well as later marketing:

    ___ Base/overlay drawing
    ___ Rendering entourage on overlays
    ___ Variations and options on overlays

___ Review possible future client uses of working drawing material, such as base sheet floor plans for promotional graphics

__ Update CPM or Pert charts for working drawings

__ Assist in charting and planning construction schedule

__ Apply a formal Value Engineering process to find potential construction improvements and economies

__ Review the next stage planning with consultants for methods of expediting and economizing on the bidding and construction administration processes

BID ADMINISTRATION

MINIMUM SERVICES

__ If the client is also the contractor, no specification may be required

__ For small projects, specifications may be handled as general notes in the drawings

__ For larger jobs, the design firm may have to review the following decisions and options with the client:

    __ Decide with client the type of construction contract:
        __ Competitive Bidding -- Open
        __ Competitive Bidding -- Selected Contractors
        __ Negotiated Contract
        __ Single Prime Contract
        __ Multiple Separate Contracts
        __ Stipulated Lump Sum
        __ Cost Plus Fee

    __ Decide with client the contractor fee types:
        __ Fixed Fee
        __ Fixed Fee with Guaranteed Maximum
        __ Percentage of Construction

    __ Decide among related options:
        __ Phased Construction
        __ Fast Track
        __ Construction Management
        __ Design-Build
        __ Contractor prepared construction documents

    __ Confirm the type of specification:
        __ Open/Contractor's Option
        __ Closed/Proprietary
        __ Product Approval
        __ Substitute Bid
        __ Approved Equal
        __ Product Description
        __ Performance
        __ Work Procedure

__ If documents require any more than standardized or outline specification, the quality and quantity of services will most likely shift to Median or Comprehensive

MEDIAN SERVICES

Includes the above listed services plus:

__ Conduct review meetings and/or drawing checks to coordinate decisions and alternatives on:
  - __ Room functions and relationships
  - __ Room finish schedule
  - __ Construction system
  - __ Structural system
  - __ Mechanical system
  - __ Lighting
  - __ Vertical transportation

  - __ Exterior materials:
    - __ Roofing
    - __ Walls
    - __ Fenestration

  - __ Interior partitioning system
  - __ Cabinetry
  - __ Site appurtenances

  - __ Materials, finishes, and fixture quality:
    - __ Superior
    - __ Middle Grade
    - __ Economy Grade
    - __ Mixed Grades

__ Identify and write specification sections that can be completed early in the working drawing process

__ Make checklist of latest special trade association standards required for the project

__ Confirm use of latest applicable product literature

__ Confirm use of latest applicable testing agency standards

__ Confirm use of all latest applicable codes and regulations

__ Acquire from the client any previous relevant specifications

__ Create a Project Manual binder for preliminary organization of specification information (use index tabs following the CSI Masterformat)

COMPREHENSIVE SERVICES

Includes the above listed services plus:

__ Provide completely detailed specification divisions and sections incorporating the following data:

- __ Materials:
  - __ Generic name
  - __ Proprietary name with manufacturer
  - __ Description by use
  - __ Description by performance criteria
  - __ Description by reference standard

- __ Required characteristics of materials:
  - __ Gauge or weight
  - __ Sizes, nominal or finished
  - __ Type of finish
  - __ Allowable moisture content

- __ Components or proportions of components of materials:
  - __ Mixes
  - __ Care in handling
  - __ Temperature protection
  - __ Moisture protection

- __ Installed location on the job if not fully indicated in the drawings

- __ Preparation for installation:
  - __ Pre-job inspection
  - __ Coordination with other subcontractor(s)
  - __ Cleaning
  - __ Preparation of surfaces

- __ Installation:
  - __ On-site fabrication
  - __ Connection to other work
  - __ Adjusting and fitting
  - __ Finishing

- __ Coordination:
  - __ Broadscope working drawing sheet reference
  - __ Detail drawing sheet reference
  - __ Consultant's drawing sheet reference
  - __ Related and/or connecting work by other trades or subcontractors
  - __ Related other specifications sections

- __ Workmanship standards and tolerances:
  - __ Quantified measurements
  - __ Referenced to published standards
  - __ Approval by inspection

343

___ Inspections and tests (may be combined with workmanship standards and tolerances)
___ Repair and patching
___ Clean-up, preparation for other work
___ Warranties, bonds, or guarantee requirements
___ Postconstruction adjustments or service

___ Review "Scope of Work" and "Work Not Included" articles in each section

___ Verify all references to work in other sections

___ Review the Special Conditions

___ Distribute copies of specifications for content review by department heads and/or job captains, and the designated project site representative(s)

BID ADMINISTRATION

MINIMUM SERVICES

__ For smaller projects, bids may be taken and negotiated without formalities

__ Clients may act as their own contractors and handle all subcontractor bids themselves

__ Larger projects in open bidding may require the following even as MINIMUM SERVICES:

ADMINISTRATION -- PRE-BID

__ Prepare and assemble bid documents:
    __ Bid Notice
    __ Bid Advertisement (if separate from the Invitation to Bid)
    __ Invitation to Bid
    __ Instructions to Bidders
    __ Contractor's Qualification Statement
    __ Bid Form
    __ Owner-Contractor Agreement
    __ Bid Documents Deposit
    __ Bid Security/Bid Bond
    __ Performance Bond/Labor and Material Payment Bond
    __ General and Supplementary Conditions
    __ Construction Documents: Drawings, Specifications, and Addenda

__ Confirm that all necessary permits and approvals from regulatory agencies have been obtained

__ Confirm that all necessary permits and approvals from public utilities have been obtained

FOR NON-OPEN BIDDING, NEGOTIATED CONTRACTS

__ Establish the criteria for contractor qualifications and acceptable contract terms

__ Review selection(s) of preferred contractor(s) and discuss choices with the client

__Establish terms for bargaining and acceptable alternatives in scheduling and budgeting construction

BIDDING MATERIALS -- FOR INVITED BIDDING

__ Establish criteria for qualifications of acceptable contractors

__ Select the preferred contractor(s) for negotiation, and review the selection with the client for approval

__ Create invited bidders list

__ Notify invited bidders

BIDDING MATERIALS -- FOR OPEN BIDDING

__ Provide Instructions to Bidders and/or advertising for bidders

__ Begin and maintain a Register of Bid Documents

__ Identify favored prospective prime contract bidders

__ Identify media for bid advertising: construction periodical(s), newspapers, plan rooms, etc.

__ Write the criteria for acceptable bidder qualifications, and confirm criteria with the client

__ Write the Invitation to Bid, and obtain client's approval

__ Identify the surety or bid bond to be required of bidders

__ Publish and distribute the Invitation to Bid

__ Obtain statements of qualification from prospective bidders

__ Notify selected bidders

__ Identify and list all bid documents to be distributed to bidders, and identify the amount of bid document deposit

__ Review the complete bid package with the client

__ Identify the quantity of bid documents to distribute to each bidder and the total number of bidders

__ Identify those other than bidders who will receive bid documents, such as the client, consultants etc.

__ Distribute bid documents to bidders, plan rooms, client, and all other concerned parties

__ Schedule a pre-bid conference to review documents with prospective bidders

__ Maintain a log of distributed documents, including bidders' deposit/security payments and refunds

__ Review proposed substitutions according to formal procedures established in the Instructions to Bidders

ADDENDA

__ Prepare an addendum log in the Register of Bid Documents

__ Distribute addenda to all bidders according to procedures established in the Bid Documents

__ When responding to any bidder's request for clarification or additional data, send copies of the clarification data as an addendum to all other bidders

BIDDING AND NEGOTIATIONS

__ Hold pre-bid meeting(s) with prospective bidders and client

__ Prepare a report on the pre-bid meeting(s) and send copies to all concerned parties

__ Prepare a Bid Tabulation Form

__ Receive bids according to procedures in the Instructions to Bidders

__ Check all bids to confirm the validity of the contractors' and major subcontractors' licenses

__ Confirm that the rules regarding bid security are enforced

ANALYSIS OF ALTERNATES AND SUBSTITUTIONS

__ Create a "Confirmation Form" memo to set down in writing all verbal interpretations, instructions, and confirmations (establish a time limit in which copies of such memos must be distributed)

__ Establish a record of consultations with the client on changes and alternates, with space for notes confirming client approvals of alternates

__ Establish a record of notifications to the Contractor(s) of approved and not approved alternates

__ Notify all bidders of accepted substitutions

MEDIAN SERVICES

Includes the above listed services plus:

__ Identify alternates that concern the work of consultants

__ Ask consultants to identify favored prospective subcontractors

__ If the selected contract form is Cost Plus Fee, establish the accounting and record keeping procedures to be used to monitor the contractor's performance

__ If the project is out-of-state, consult with the client's legal counsel on the existence of any special laws regarding the bidding process, construction documents, and forms of agreement

__ Identify and confirm the design firm and client's separate responsibilities in advertising for bids, receiving bids, negotiation, and acceptance

__ Identify insurance coverage the client should have prior to the execution of the contract

__ Identify insurance and bonds the client will require the contractor to have

__ List and confirm what materials, equipment, and furnishings are supplied by the client and installed by the contractor

__ List and confirm what materials, equipment, and furnishings are supplied by the client and installed by anyone other than the contractor

__ Confirm that the client has supplied an accurate site survey, site legal description, and a soil and subsurface condition report, all to be included with the construction documents

BID EVALUATION

__ Analyze the bids; check for errors or omissions

__ Write a comparison of bid tabulations with the latest design firm construction cost estimates

__ Review significant discrepancies between the bid tabulations and the last previous construction cost estimate

__ Review bids with the client and advise on bid acceptance and rejection

__ Obtain the client's acceptance of a bid or rejection of all bids

__ Record reasons for acceptance/rejections in the project manual

__ Advise the client how to draft a notice of acceptance that states an intent to execute the contract without specifically awarding the contract

__ Notify all bidders of acceptance or rejection

__ Return received documents and refund the bid deposits/security to unsuccessful bidders

CONSTRUCTION CONTRACT AGREEMENTS

__ Advise the client on construction contract format and content

__ Have consultants assist on preparation of separate prime contracts

__ Provide the client with a checklist of separate designer/client/contractor responsibilities as stated in the contract

__ Advise client and contractor of their insurance responsibilities

__ Schedule times for confirmation of required insurance coverage

__ Obtain performance and labor and material payment bonds from the contractor; review and forward copies of bonds to client

__ Obtain the contractor's certificate of insurance; review and forward copies of certificate to client

__ Obtain the client's property insurance policy; review and forward copies to the contractor

__ Identify and review the establishment of any non-typical insurance arrangements between the client and the contractor, include descriptions of such arrangements in the contract

__ Obtain the post-bid information from the accepted contractor as required in the Instructions to Bidders

__ Review the construction plan and time schedule with the client and contractor for inclusion in the contract

__ Consult with and assist the client in negotiating and executing the final contract

POST-BIDDING ADMINISTRATION

__ Create a log for recording all change orders and modifications to the contract

__ Provide all necessary contract documents, specified equipment brochures, and related project data to the contractor

__ Identify bid tabulation data, special agreements addenda, and memos, reports, minutes, and correspondence that should be included in the final Project Manual as part of construction contracts or construction documents

COMPREHENSIVE SERVICES

Includes the above listed services plus:

__ If all bids are rejected, confer with the client to establish the next step of bidding invitations or negotiations

__ If the client wants to proceed with limited interim construction prior to awarding the final contract, advise the client on the form and content of this type of letter of intent

__ Investigate whether other major projects are coming up for bid at the same time, and, if necessary, modify the bid date

__ Establish whether the design firm or the client's legal representative is to identify special governing laws for out-of-state bidding, contracts, and construction

__ Evaluate building construction systems proposed by contractors for phased construction

__ Assist the client in establishing criteria and schedules for phased construction or multiple contracts

MORE ACKNOWLEDGMENTS

We are indebted to many dozens of companies, architects, engineers, drafters, and designers for information, drawings, and charts used throughout this book.
Many are quoted or acknowledged in the text.
Others include:

Larry Jenks, The Klipp Partnership
Frank Mascia, Architect
Ned Abrams, FAIA, Architect
Bill Smith -- Jarvis Putty Jarvis
Edgar Powers, Jr., Gresham, Smith and Partners
AIA Task Force on CAD Layer Guidelines
AutoDesk
Dennis Neeley, ASG
Banks Upshaw, Architect
AIA Northern California Chapter committee on Production Office Proceedures
John Rattenberry, Architect, Taliesin Architects
Hansen Lind Meyer, Architects
Gorder/South Group, AIA
Wilscam Mullins Birge, inc.
Saunders-Thalden Associates, inc.
AECK Associates Architecture
The Ikoy Partnership, Architects
Moreland Christopher Myles Architects
Lellyett & Rogers Co.

Special thanks to the OCE Corporation for the use of their advanced reprographic equipment illustrated in Chapter 17.